Biochemistry and Molecular Biology of Vitamin B_6 and PQQ-dependent Proteins

Edited by
Ana Iriarte
Herbert M. Kagan
Marino Martinez-Carrion

Springer Basel AG

Editors

Prof. Dr. Ana Iriarte
Prof. Dr. Marino Martinez-Carrion
Division of Molecular Biology
and Biochemistry
School of Biological Sciences
University of Missouri
Kansas City, MO 64110
USA

Prof. Dr. Herbert M. Kagan
Department of Biochemistry
Boston University School of Medicine
80E Concord Street
Boston, MA 02118-2394
USA

Library of Congress Cataloging-in-Publication Data
Biochemistry and molecular biology of vitamin B_6 and PQQ-dependent proteins / edited by Ana Iriarte, Herbert M. Kagan, Marino Martinez-Carrion.
p. cm.
Includes bibliographical references and index.
ISBN 978-3-0348-9549-1 ISBN 978-3-0348-8397-9
DOI 10.1007/978-3-0348-8397-9
1. Vitamin B_6--Congresses. 2. PQQ (Biochemistry)--Congresses. 3. Enzymes--Regulation--Congresses. I. Iriarte, Ana. II. Kagan, Herbert M. III. Martinez-Carrion, Marino.
QP772.P9 B535 2000
612.3'99--dc21

Deutsche Bibliothek Cataloging-in-Publication Data
Biochemistry and molecular biology of vitamin B_6 and PQQ-dependent proteins / ed. by Ana Iriarte ... - Basel ; Boston ; Berlin : Birkhäuser, 2000
ISBN 978-3-0348-9549-1

Originally published by Birkhäuser Verlag in 2000
Softcover reprint of the hardcover 1st edition 2000
Printed on acid-free paper produced from chlorine-free pulp. TFC ∞

ISBN 978-3-0348-9549-1

9 8 7 6 5 4 3 2 1

Since the meeting in Capri in 1994, we have lost the vitality and input of the following members of the International Advisory Council. This volume is dedicated to their memory:

Jorge Churchich
Paolo Fasella
Alton Meister

CONTENTS

PREFACE

Since the first international meeting on Vitamin B_6 involvement in catalysis took place in 1962, there have been periodic meetings every three or four years. In 1990, scientists studying another cofactor, PQQ, which had already attracted the scientific community's interest for its possible involvement in amino acid decarboxylation and reactions involving amino groups, joined forces with those investigating pyridoxal phosphate-dependent enzymes. Since then, the international PQQ/quinoproteins meetings have been held jointly. In the years following the original meeting 37 years ago in Rome, Italy, the scientific gatherings have taken place in Moscow, Russia (1966); Nagoya, Japan (1967); Leningrad (St. Petersburg), Russia (1974); Toronto, Canada (1979); Athens, Greece (1983); Turku, Finland (1987); Osaka, Japan (1990); and Capri, Italy (1996). For the first time in the history of these symposia, the international meeting was held in the United States, from October 31 through November 5, 1999, in Santa Fe, New Mexico.

The scientific program focus shifted significantly beyond the original emphasis on catalysis to aspects such as cellular and genetic regulation of events involving proteins that require pyridoxal phosphate or quinoproteins. The growing awareness of the involvement of these proteins in biotechnology processes and fundamental physiological events, as well as their implication in diseases, was also represented, with emphasis on the molecular basis of these events.

The meeting was symposium S278, sponsored by the International Union of Biochemistry and Molecular Biology (IUBMB).

This volume incorporates highlights of the presentations made during the meeting, which have been speedily processed in order to share the information with the scientific community in a timely fashion.

The Editors

Ana Iriarte
Herbert M. Kagan
Marino Martinez-Carrion

ACKNOWLEDGEMENTS

Honorary President
ESMOND E. SNELL (USA)

International Advisory Board
Osao Adachi (Japan)
Robert Barouki (France)
Temir Berezov (Russia)
Francesco Bossa (Italy)
Philipp Christen (Switzerland)
Jorge Churchich (Hong Kong)
Krishnamurti Dakshinamurti (Canada)
Shawn Doonan (United Kingdom)
Johannis Duine (The Netherlands)
Athanasios Evangelopoulos (Greece)
Toshio Fukui (Japan)
Robert John (United Kingdom)
Hiroyuki Kagamiyama (Japan)
Herbert M. Kagan (USA)
Jack F. Kirsch (USA)
Timo Korpela (Finland)
Johan Jansonius (Switzerland)
Gennaro Marino (Italy)
Marino Martinez-Carrion (USA)
Donald McCormick (USA)
David Metzler (USA)
Yoshimasa Morino (Japan)
Verne Schirch (USA)
Klaus Schnackerz (Germany)
Kenji Soda (Japan)
Yuri Torchinsky (Israel)
Edith Wilson-Miles (USA)

U.S. Organizing Committee
Marino Martinez-Carrion, President
Laura Batenic, Secretary/Treasurer

Jorge Churchich
James Manning
Paul Cook
Scott Matthews
Leodis Davis
Donald McCormick
David Dooley
Edith Wilson-Miles
Michael Dunn
David Metzler
Herbert Kagan
Verne Schirch
Jack Kirsch

Sponsors
International Union of Biochemistry and Molecular Biology (IUBMB); Bayer Corporation, Pharmaceutical Division; BioMolecular Instruments, Division of Thermo Instrument Systems; Chiron Corporation; Merck, Sharp & Donne Research Laboratories; MitoKor; Parke-Davis Pharmaceutical Division of Warner-Lambert Corp.; School of Biological Sciences, University of Missouri-Kansas City

MOLECULAR REGULATION OF ENZYMES CONTROLLING LEVELS OF VITAMIN B_6

Biochemistry and Molecular Biology of Vitamin B_6
and PQQ-dependent Proteins
ed. by A. Iriarte, H.M. Kagan and M. Martinez-Carrion

Genetic and Genomic Approaches for Delineating the Pathway of Pyridoxal 5'-Phosphate Coenzyme Biosynthesis in *Escherichia coli*

Malcolm E. Winkler*
Department of Microbiology and Molecular Genetics, University of Texas Houston Medical School, Houston, TX 77030, U.S.A.
*Current address: Lilly Research Laboratories, Drop Code 1543, Indianapolis, IN 46285, U.S.A.

Pyridoxal 5'-phosphate (PLP) biosynthesis has been studied extensively in *Escherichia coli* K-12 by a combination of biochemical and genetic approaches, and the *E. coli* pathway serves as a benchmark for understanding PLP biosynthesis in other organisms. The goal of this brief review is to highlight some of the genomic and genetic approaches that have contributed to delineating the *de novo* and salvage pathways of PLP biosynthesis in *E. coli*. Classical genetic approaches provided the initial *pdx* mutants, and the combination of genomics and genetics has helped to elucidate the functions and redundancy of enzymes involved in PLP biosynthesis. Genetic approaches have also verified intermediates in the PLP biosynthetic pathway in bacterial cells.

Pyridoxal 5'-phosphate (PLP) is the active form of vitamin B_6 and acts as an essential, ubiquitous coenzyme in many aspects of amino acid and cellular metabolism (see [1,2]). PLP is synthesized *de novo* in *E. coli* by converging branched pathways that lead to 4-phosphohydroxy-L-threonine (4PHT) and D-1-deoxyxylulose phosphate (DXP), which are condensed to form pyridoxine 5'-phosphate (PNP) (Fig. 1) [2-9]. PNP is then oxidized by PdxH oxidase to form PLP, the active coenzyme (Fig. 1) [1,2,10,11]. PLP is converted to pyridoxamine 5'-phosphate (PMP) by the half-reaction of transaminases (Fig. 1) (see [1,2,6]). PMP is recycled back to PLP by the second half-reaction of transaminases and by PdxH PNP/PMP oxidase in the PLP salvage pathway (see below). Important new findings suggest that part or all of the *de novo* pathway for PLP biosynthesis is different in fungi, plants, and some bacteria from that in *E. coli* [12,13].

PLP can also be synthesized in *E. coli* by a salvage pathway that utilizes pyridoxal (PL), pyridoxine (PN), and pyridoxamine (PM) taken up from the growth medium (Fig. 2)

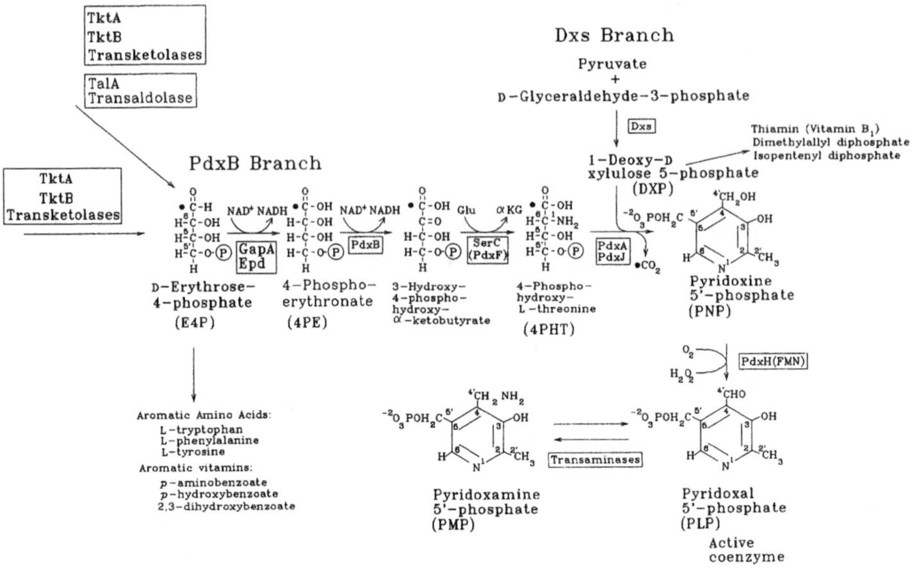

Figure 1. *de novo* PLP biosynthetic pathway in *E. coli* K-12. See text for details.

[10,11,14,15]. In the salvage pathway, PL, PN, and PM are first phosphorylated by the PdxK PN/PL/PM kinase or the PdxY PL kinase to form PLP, PNP, and PMP, respectively (Fig. 2) [14,15]. PNP and PMP are oxidized by the PdxH oxidase, which functions in both the salvage and *de novo* pathways [10,11]. Similar salvage pathways are present in other microbes and in mammalian cells that lack a *de novo* PLP biosynthetic pathway (see [10,11,14,15]). In mammalian cells, PLP homeostasis is further maintained by the offsetting activities of PL kinases and a PLP-specific phosphatase (P'ase) [16]. A cytoplasmic PLP P'ase activity has been detected in *E. coli* K-12, but it has not yet been determined whether this P'ase is specific for PLP [Y. Yang and M. Winkler, unpublished result].

Extensive early hunts for mutants of *E. coli* B or *E. coli* K-12 that require PN or PL for growth were carried out using classical methods, such as antibiotic enrichment, by Walter Dempsey and coworkers (see [1,2]). The *pdx* biosynthetic genes were grouped into five different linkage groups, which later complementation studies showed each consist of a single distinct gene [17]. *pdxB* mutants of *E. coli* K-12 require PN, PL, or glycolaldehyde (CHO-CH_2OH) (GA)

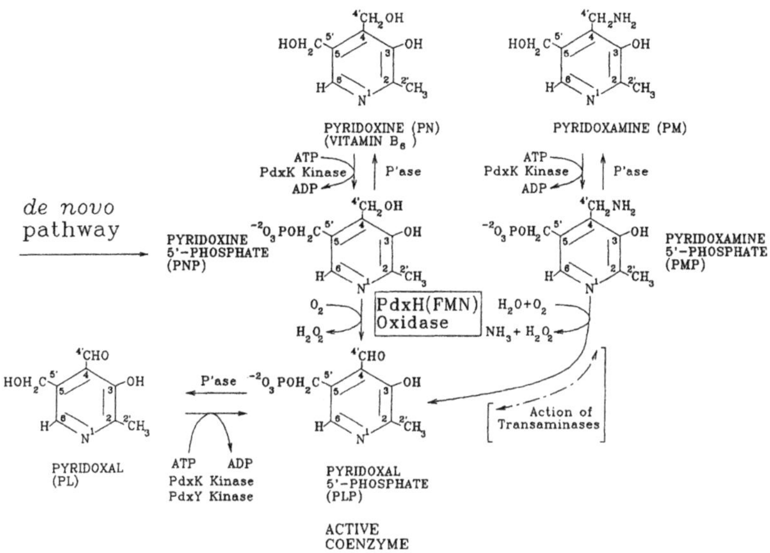

Figure 2. PLP salvage pathway in *E. coli* K-12. See text for details.

for growth (see below and [3]). Point mutations in the *serC(pdxF)* gene, which encodes the transaminase used for both PLP and serine biosynthesis (Fig. 1), cause a dual requirement for PN, PL, or GA and serine [3]. In contrast, *pdxJ* and *pdxA* mutants grow when supplemented with PN or PL, but not GA [2,3]. This distinct auxotrophy pattern led to the hypothesis that the PdxA and PdxJ gene products participate in closure of the PLP ring (Fig. 1) using an intermediate formed by the PdxB and SerC(PdxF) enzymes or synthesized by an alternative pathway from GA in *pdxB* or *serC(pdxF)* mutants [3]. Involvement of PdxA and PdxJ in ring closure was recently confirmed biochemically [7,8]. *pdxH* mutants, which are defective in the function of PNP/ PMP oxidase, were distinguished by growth on media containing PL, but not on PN (Figs. 1 and 2) [1,2,18].

Most of the genes that encode PLP biosynthetic enzymes in *E. coli* K-12 are located in complex superoperons with genes that mediate other functions (see [2,3,15,17,18]). In some cases, this arrangement may allow integration of PLP biosynthesis with other metabolic pathways and processes. The presence of PLP biosynthetic genes in superoperons limited the isolation of polar insertion mutations in some *pdx* genes. Polar insertion mutations were readily isolated in *pdxB*, *pdxA*, and *pdxH*, because these genes are located upstream of internal promoters in superoperons [see (2,17)]. In contrast, insertion mutations could not be readily isolated in *serC(pdxF)* or *pdxJ* for different reasons [3,17]. *serC(pdxF)* is the first gene of a two-gene superoperon that contains the *aroA* gene involved in the common aromatic compound biosynthetic pathway; consequently, *serC(pdxF)* insertion mutants require supplementation by eight different compounds for growth (PL, PN, or GA, serine, the three aromatic amino acids, and the three aromatic vitamins) [3]. *pdxJ* is in a very complicated gene arrangement [17]. *pdxJ* is expressed from an internal promoter that also drives the expression of the downstream *acpS* gene, which encodes acyl carrier protein synthase essential for fatty acid biosynthesis.

The breakthrough in ordering the PdxB branch of the *de novo* PLP biosynthetic pathway in *E. coli* K-12 came when genomic information was combined with results from isotope tracer and precursor feeding experiments (see [2,3]). DNA sequence data showed that *pdxB* encodes a dehydrogenase that is an evolutionary paralogue of the SerA 3-phosphoglycerate dehydrogenase involved in serine biosynthesis (see [3]). This finding led to the hypothesis that the PLP and serine biosynthetic pathways are parallel and utilize analogous phosphorylated four and three carbon substrates, respectively (see [3]). This reasoning placed the oxidation catalyzed by PdxB before the transamination catalyzed by SerC(PdxF) (Fig. 1) and was consistent with isotope tracer and feeding data implicating 4PHT or 4-hydroxythreonine (4HT) as the product of this branch of the PLP biosynthetic pathway [2-5]. The proposed substrate of PdxB could be generated by oxidation of erythrose 4-phosphate (E4P), which is the important precursor of the aromatic vitamins and amino acids (Fig. 1) (see [2,3,6,19]).

Several different genetic approaches were used to verify the PdxB branch of the PLP biosynthetic pathway. If E4P is the precursor of 4PHT, then mutants deficient in transketolase activity should require B_6 vitamers for growth. This hypothesis was verified in *tktA tktB* double mutants that lack function of the two redundant transketolase genes in *E. coli* [9]. The assignment of the GapA and Epd(GapB) genes to the first committed step of the PdxB branch of PLP biosynthesis was achieved by a combination of genomics, biochemistry, and genetics [6,19]. The structural similarity between glyceraldehyde 3-phosphate (GA3P) and E4P suggested that the missing enzyme was the GapA GA3P dehydrogenase or a paralogue of GapA [19].

Examination of genomic sequences revealed that *E. coli* contained a paralogue of GapA designated GapB. Protein purification demonstrated that GapB was indeed an E4P dehydrogenase, and *gapB* was renamed *epd* [19]. Construction and characterization of *gapA*, *epd,* and *gapA epd* mutants showed that both GapA and Epd are involved in PLP biosynthesis in *E. coli* [6]. The redundancy of GapA and Epd(GapB) function accounts for why single mutations were never identified that block the first step of PLP biosynthesis (Fig. 1).

A genetic approach was also used to demonstrate that 4PHT, not 4HT, is the obligatory intermediate in PLP biosynthesis *in vivo* [4]. In *E. coli* K-12, glycolaldehyde (GA) is not an obligatory intermediate in PLP biosynthesis; however, GA can be condensed with glycine to form 4HT [4]. If 4PHT is the obligatory intermediate, then 4HT formed from GA would need to be phosphorylated by a kinase to allow PLP biosynthesis in a *pdxB* mutant (Fig. 1). Purified ThrB homoserine kinase was known to use 4HT as a substrate, and it was postulated that ThrB converts 4HT to 4PHT *in vivo* [4]. This hypothesis was verified by the demonstration that *thrB pdxB* mutants fail to grow on minimal media supplemented with threonine and GA or 4HT, but did grow on media containing threonine and PN [4]. Ring closure with 4PHT would lead directly to PNP as the first B_6 vitamer synthesized *de novo*. This conjecture suggested that the PL/PM/PN kinases are not required in the *de novo* pathway of PLP biosynthesis, but instead are confined to a salvage pathway [4,5]. These ideas were recently confirmed when *pdxK* and *pdxY* kinase mutants were identified and characterized (see below; [14,15]) and when ring closure by purified PdxA and PdxJ was shown to produce PNP directly [7,8].

The involvement of DXP in PLP biosynthesis was discovered by isotope tracer experiments by the laboratories of Hill and Spenser (see [2]). DXP is formed by the condensation of pyruvate and GA3P [2,20,21]. Searches for mutants defective in DXP synthesis were unsuccessful, because it turned out that DXP is a precursor of PLP, thiamin (vitamin B_1), and isopentenyl pyrophosphate (IPP) and dimethylallyl diphosphate (DMADP), which are precursors of isoprenoid compounds essential for cell growth [20,21]. DXP as a precursor of IPP and DMADP was surprising, because many organisms synthesize IPP and DMAPP by the classical acetate/mevalonate pathway [20,21]. The *dxs* gene that encodes DXP synthase was located in *E. coli* by applying genomics. Because of the condensation catalyzed, DXP synthase was hypothesized to contain motifs found in transketolases [20,21]. One candidate ORF was in a cluster of genes encoding enzymes known to catalyze reactions that use IPP as a substrate [20,21]. Cloning, overexpression, and purification of the protein encoded by this candidate ORF showed that it indeed encoded DXP synthase [20,21]. With knowledge of its function, it has been possible to construct genetic systems in which *dxs* is conditionally expressed under the control of a tightly regulated promoter [Q. Sheng and M. Winkler, unpublished result]. Feeding studies of these conditional mutants demonstrates that DXP is required only for PLP, thiamin, and IPP biosynthesis in *E. coli* cells [Q. Sheng and M. Winkler, unpublished result].

Recently genetic approaches have been instrumental in identifying the enzymes involved in the PLP salvage pathway (Fig. 2). The PdxH PNP/PMP oxidase plays roles in both pathways (Figs. 1 and 2). *pdxH* mutants do not grow without oxygen [18]. Suppression that allows *pdxH* null mutants to grow without oxygen is caused by a single amino acid change in the PdxJ enzyme involved in ring closure [22]. This mutant PdxJ bypasses the normal biosynthetic pathway [22]. This bypass phenotype was the first indication of an interaction between PdxJ and DXP in ring closure [22]. The electron acceptor used by the PdxH FMN-dependent oxidase in cells in the absence of oxygen is an important unanswered question. Other uncharacterized phenotypes of *pdxH* mutants are their tendency to form long filamentous cells containing unsegregated nucleoids and to excrete significant amounts of L-glutamate and α-ketoisovalerate [18].

Genetic approaches led to the first identification of a gene encoding a PL/PM/PN PdxK kinase in any organism [14]. This identification was accomplished by using the fact that GA is converted to phosphorylated 4PHT, which provides the phosphate ester group of PNP (Fig. 1). If this idea is correct, then *pdxB* mutants lacking the regular *de novo* PLP biosynthetic pathway and missing the gene encoding PN kinase should grow when supplemented with GA, but not with PN [14]. Following transposon mutagenesis and screening, an insertion mutation was found in *pdxK*, which encodes PL/PN/PM kinase [14]. Isolation of a *pdxK* mutant by this scheme not only corroborated that PNP is the first B_6 vitamer synthesized *de novo* by the PLP biosynthetic pathway, but indicated that the PL/PM/PN PdxK kinase functions in the salvage pathway instead of the *de novo* biosynthetic pathway [14]. The PdxK sequence led to the rapid identification of its mammalian orthologue (see [15]) and showed that PdxK has signature motifs of the PfkB superfamily of carbohydrate kinases, which includes phosphofructokinases and ribokinases [14,15].

The identification of the *pdxY* gene, which encodes a previously unsuspected PL kinase, provides another example of the power of genomic and genetic approaches to elucidate the PLP salvage pathway [15]. *pdxK* kinase mutants still contained an unidentified PL kinase activity [14]. This observation led to the hypothesis that *E. coli* must contain two B_6 vitamer kinases [14,15]. This hypothesis was confirmed when amino acid alignments of PdxK with all other *E. coli* proteins deduced from the complete genome sequence led to a candidate PL kinase gene, which was named *pdxY* [15]. Construction and characterization of the phenotypes of *pdxY* null mutants, *pdxK pdxY* double mutants, and *pdxK pdxY pdxB* triple mutants led to the strong conclusion that the PdxK PL/PM/PN kinase and the PdxY PL kinase are the only physiologically important B_6 vitamer kinases in *E. coli* and their function is confined to the PLP salvage pathway [15]. Moreover, *pdxY* is located downstream from *pdxH* (encoding PNP/PMP oxidase) and essential *tyrS* (encoding aminoacyl-tRNA^{Tyr} synthetase) in a multifunctional operon that may link together PLP cofactor amount and amino acid supply and charging to tRNA for translation [15]. The role of the PdxK and PdxY kinases and cytoplasmic PLP P'ases in setting cellular PLP and PMP levels (Fig. 2) is currently under investigation and will be greatly advanced by the mutants and genomic and genetic approaches described herein.

This work was supported by PHS grant GM37561 from the National Institute of General Medical Sciences of the U.S.A. and by the Lilly Research Laboratories. I gratefully acknowledge and thank P.J. Arps, P.V. Schoenlein, A. van der Zel, E. Tancula, B.B. Roa, D.M. Connolly, H.-M. Lam, A.J. Pease, G. Zhao, Y. Yang, W. Luo, Q. Sheng, and H.-C. T. Tsui for their important contributions to this research area while working in my laboratory.

References

1. **Dempsey, W. B.** 1987. Synthesis of pyridoxal phosphate, p. 539-543. *In* F. C. Neidhardt, *et al.* (ed.), *Escherichia coli* and *Salmonella typhimurium*: cellular and molecular biology, vol. 1. ASM Press, Washington, D. C.

2. **Hill, R. E., and I. D. Spenser. 1996**. Biosynthesis of vitamin B_6, p. 695-703. *In* F. C. Neidhardt, *et al.* (ed.), *Escherichia coli* and *Salmonella*: cellular and molecular biology, 2nd ed, vol. 1. ASM Press, Washington, D. C.

3. **Lam, H. M., and M. E. Winkler.** 1990. Metabolic relationships between pyridoxine (vitamin B_6) and serine biosynthesis in *Escherichia coli* K-12. J. Bacteriol. **172:**6518-6528.

4. **Zhao, G., and M. E. Winkler.** 1996. 4-Phospho-hydroxy-L-threonine is an obligatory intermediate in pyridoxal 5'-phosphate coenzyme biosynthesis in *Escherichia coli* K-12. FEMS Microbiol. Lett. **135:**275-280.

5. **Drewke, C., M. Klein, D. Clade, A. Arenz, R. Muller, and E. Leistner.** 1996. 4-*O*-Phosphoryl-L-threonine, a substrate of the *pdxC(serC)* gene product involved in vitamin B_6 biosynthesis. FEBS Lett. **390:**179-182

6. Yang, Y., G. Zhao, T.-K. Man, and M. E. Winkler. 1998. Involvement of the *gapA*- and *epd(gapB)*-encoded dehydrogenases in pyridoxal 5'-phosphate coenzyme biosynthesis in *Escherichia coli* K-12. J. Bacteriol. **180**: 4294-4299.

7. Cane, D. E., Y. Hsiung, J. A. Cornish, J. K. Robinson, and I. D. Spenser. 1998. Biosynthesis of vitamin B_6: the oxidation of 4-(phosphohydroxy)-L-threonine by PdxA. J. Am. Chem. Soc. **120:**1936-1937.

8. Laber, B., W. Maurer, S. Scharf, K. Stepusin, and F.S. Schmidt. 1999. Vitamin B_6 biosynthesis: formation of pyridoxine 5'-phosphate from 4-(phosphohydroxy)-L-threonine and 1-deoxy-D-xylulose-5-phosphate by PdxA and PdxJ protein. FEBS Lett. **449**: 45-48.

9. Zhao, G., and M. E. Winkler. 1994. An *Escherichia coli* K-12 *tktA tktB* mutant deficient in transketolase activity requires pyridoxine (vitamin B_6) as well as the aromatic amino acids and vitamins for growth. J. Bacteriol. **176:**6134-6138.

10. Zhao, G., and M. E. Winkler. 1995. Kinetic limitation and cellular amount of pyridoxine (pyridoxamine) 5'-phosphate oxidase of *Escherichia coli* K-12. J. Bacteriol. **177:**883-891

11. Di Salvo, M., E. Yang, G. Zhao, M. E. Winkler, and V. Schirch. 1998. Expression, purification, and characterization of recombinant *Escherichia coli* pyridoxine 5'-phosphate oxidase. Prot. Expr. Purif. **13**: 349-356.

12. Ehrenshaft, M., P. Bilski, M.Y. Li, C.F. Chignell, and M.E. Daub. 1999. A highly conserved sequence is a novel gene involved in *de novo* vitamin B_6 biosynthesis. Proc. Natl. Acad. Sci. USA **96**: 9374-9378.

13. Osmani, A.H., G.S. May, and S.A. Osmani. 1999. The extremely conserved *pyroA* of *Aspergillus nidulans* is required for pyridoxine synthesis and is required indirectly for resistance to photosensitizers. J. Biol. Chem. **274**: 23565-23569.

14. Yang, Y., G. Zhao, and M. E. Winkler. 1996. Identification of the *pdxK* gene that encodes pyridoxine (vitamin B_6) kinase in *Escherichia coli* K-12. FEMS Microbiol. Lett. **141:**89-95.

15. Yang, Y., H.-C. T. Tsui, T.-K. Man, and M. E. Winkler. 1998. Identification and function of the *pdxY* gene, which encodes a novel pyridoxal kinase involved in the salvage pathway of pyridoxal 5'-phosphate biosynthesis in *Escherichia coli* K-12. J. Bacteriol. **180:**1814-1821.

16. Fonda, M.L. 1992. Purification and characterization of vitamin B_6-phosphate phosphatase from human erythrocytes. J. Biol. Chem. **267**: 15978-15983.

17. Lam, H. M., E. Tancula, W. B. Dempsey, and M. E. Winkler. 1992. Suppression of insertions in the complex *pdxJ* operon of *E. coli* K-12 by *lon* and other mutations. J. Bacteriol. **174:**1554-1567.

18. Lam, H. M., and M. E. Winkler. 1992. Characterization of the complex *pdxH-tyrS* operon of *Escherichia coli* K-12 and pleiotropic phenotypes caused by *pdxH* insertion mutations. J. Bacteriol. **174:**6033-6045.

19. Zhao, G., A. J. Pease, N. Bharani, and M. E. Winkler. 1995. Biochemical characterization of *gapB*-encoded erythrose 4-phosphate dehydrogenase of *Escherichia coli* K-12 and its possible role in pyridoxal 5'-phosphate biosynthesis. J. Bacteriol. **177:**2804-2812.

20. Sprenger, G.A., U. Schorken, T. Wiegart, S. Grolle, A.A. deGraaf, S.V. Taylor, T.P. Begley, S. Bringer-Meyer, and H. Sahm. 1997. Identification of a thiamin-dependent synthase in *Escherichia coli* required for the formation of the 1-deoxy-D-xylulose 5-phosphate precursor to isoprenoids, thiamin, and pyridoxol. Proc. Natl. Acad. Sci. USA **94**: 12857-12862.

21. Lois, L.M., N. Campos, S.R. Putra, K. Danielsen, M. Rohmer, and A. Boronat. 1998. Cloning and characterization of a gene from *E. coli* encoding a transketolase-like enzyme that catalyzes the synthesis of isoprenoid, thiamin, and pyridoxol biosynthesis. Proc. Natl. Acad. Sci. USA **95**:2105-2110.

22. Man, T.-K., G. Zhao, and M.E. Winkler. 1996. Isolation of a *pdxJ* point mutation that bypasses the requirement for the PdxH oxidase in pyridoxal 5'-phosphate coenzyme biosynthesis in *Escherichia coli* K-12. J. Bacteriol. **178**:2445-2449.

Biochemistry and Molecular Biology of Vitamin B_6
and PQQ-dependent Proteins
ed. by A. Iriarte, H.M. Kagan and M. Martinez-Carrion

ENZYMES CATALYSING FORMATION OF PYRIDOXAL PHOSPHATE FROM VITAMIN B_6

Donald B. McCormick

Department of Biochemistry, Emory University, Atlanta, GA 30322, U.S.A.

Summary: The evolutionary retention of homologous portions of pyridoxal kinases, both broadly and narrowly specific, and the occurrence of a highly conserved region of pyridoxine (pyridoxamine) phosphate oxidase with catalytically involved residues, emphasize the importance of these enzymes that are potential agents for release of pharmalogic effectors.

The interactions by which the three, natural, free forms of vitamin B_6 (PN, PL, and PM) can be routed to the coenzyme pyridoxal 5'-phosphate (PLP) require the successive actions of pyridoxal kinase and pyridoxine (pyridoxamine) 5'-phosphate oxidase. Additionally phosphatases catalyze hydrolytic reversions of the vitaminic 5'-phosphates to free vitamers. These essential interconversions are summarized in Figure 1.

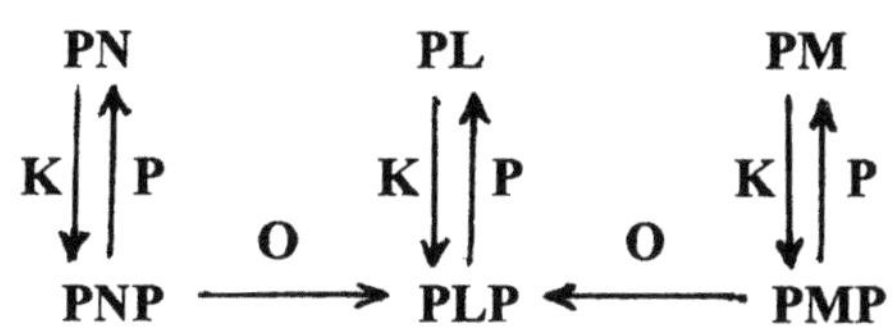

Fig. 1. Interconversions of B_6 vitamers with their 5'-phosphates noting steps catalyzed by kinase (K), oxidase (O), and phosphatases (P).

Pyridoxal Kinases

Our purifications and characterizations of the pyridoxal kinases responsible for catalyzing the phosphorylation of PN and PM as well as PL from both prokaryotic and eukaryotic sources, revealed that most higher organisms preferred Zn^{2+} rather than Mg^{2+} as the ATP-chelated cosubstrate and indicated additional activation by K^+ [1,2]. Extentions of these findings have been made by M. Fonda, J. Churchich, and others. Confirmation of the stimulatory effect of K^+ was recently reported for the kinase in human erythrocytes [3]. A pharmacologically interesting expansion of our earlier studies that demonstrated the inhibition of pyridoxal kinase by carbonyl reagents [1,4,5] is the recent finding that the mammalian kinase is also a benzodiazepine-binding protein [6]. Drugs that inhibit pyridoxal kinase have now been divided into three groups, *viz.* substrate competitive types, e.g. theophylline and progabide; those that form covalent complexes, e.g. cycloserine, dopamine, isoniazid, and thiamphenicol glycinate; and a third group comprised of levodopa, D-penicillamine, and muzolimine [7].

It was found by M. Winkler's group [8] that *Escherichia coli* has two kinases for B_6, *viz.* one (classic) using all three vitamers and another that uses only PL as substrate. Both function in the salvage pathway of PLP biosynthesis. In *E. coli*, unlike mammals, formation of PNP can be achieved via ring closure of 4-phospho-hydroxy-L-threonine [9]. PNP would then be oxidized to PLP by the oxidase, and the kinases would be used to recover PL and PN after hydrolytic cleavages of their phosphates. Expression of mammalian kinase cDNA is reported in this symposium by L. Nutter.

Pyridoxine (Pyridoxamine) Phosphate Oxidases

Early observations of an oxidase that catalyzes the second step in the formation of PLP from PNP and PMP in liver were reported by Pogell [10] and by Wada and Snell [11]. Our complete purification of this oxidase from rabbit liver, and its certain characterization as an FMN-dependent enzyme [12], was followed by a fairly extensive delineation of its properties [13], including ascertaining amino acid residues that may be involved in binding of the coenzyme and are likely participating in the catalytic mechanism [14]. The oxidase has since been isolated from other tissues and other organisms, but its general characteristics are similar even at the level of *E. coli* [9]. With the sequences of pyridoxine (pyridoxamine) phosphate oxidase from several sources, and using computer techniques applied in molecular biology, it was feasible to find out more about this enzyme. This was done in a few cases with microorganisms [15]; however, extending this comparison to more eukaryotic species would allow a greater certainty of the essential residues. A full-length cDNA clone from the *Schizophyllum commune* library was isolated by us using a DNA hybridization probe amplified by polymerase chain reaction with degenerate primers, the design of which was based on conserved regions in sequences of the oxidase from other organisms [16]. The cDNA clone possessed an open reading frame encoding a polypeptide of 229 amino acids. The amino acid sequence exhibits a striking degree of identity with the corresponding enzymes from *Saccharomyces cerevisiae* (76 %) [15], *Haemophilus influenzae* (51 %) [17], *Myxococcus xanthus* (48 %) [18], and *E. coli* (44 %) [19]. The sequence of the oxidase from rat tissues was reported by Ngo et al. [20] and will be expanded upon by L. Nutter in this symposium. Comparisons of aligned sequences around the highly conserved signature pattern are shown for these oxidases in Figure 2.

E. coli	**^{1}MSD-----^{185}LEQIEFWQGGEHRLHDRFLY^{204}-----LAP^{218}**
M. xanthus	**^{1}MAL-----^{237}PDRIEFWHAQESRLHDRHVY^{256}-----LYP^{270}**
H. influenzae	**^{1}MIL------^{196}PETVEFWQGRPSRLHDRIRY^{215}------LSP^{229}**
S. cerevisiae	**^{1}MTK-----^{193}PLEIEFWQGRPSRLHDRFVY^{212}------LAP^{228}**
S. commune	**^{1}MAT-----^{174}PLEIEFWQGRPSRLHDRFVY^{193}------LAP^{229}**
R.. norvegicus	**^{1}MTC-----^{213}PQVMEFWQGQTNRLHDRIVF^{232}----LAP^{261}**

Fig.2. Alignment of homologous sequences of PNP/PMP oxidases. Superscripts depict residue positions from the N-blocked M at 1 to the C-terminal P.

Yet another similar eukaryotic oxidase sequence, reported as Swiss-Prot: Q20939 in the EMBL/GenBank data files, is from *Caenorhabditis elegans*. ScanProsite sequence searches showed that all known sequences of the oxidase contain protein kinase C phosphorylation sites. There are also varyingly phosphorylation sites for casein kinase II and tyrosine kinase. There are N-myristoylation sites near the amino terminus of the oxidases from all known sources as well. We had shown earlier that the oxidase from rabbit liver was N-terminal blocked [21]. Clearly the oxidase is subject to regulation.

As regards amino acid residues involved in binding and catalysis, our spectral and modification studies using analogues of coenzyme, substrate, and product indicate the FMN is so situated within the catalytic cleft as to have the dimethylbenzenoid edge toward solvent [22], and substrate can bear a sizeable secondary amine function [12,23,24] instead of the natural hydroxymethyl or aminomethyl group at position 4 of the 5'-phosphopyridoxyl moiety. FMN is bound at the side chain by cationic interaction with the dianionic 5'-phosphate of the coenzyme and by stereospecific hydrogen bonding to secondary hydroxyls of the D-ribityl chain [12,22,25], ring association occurs with a W residue as reflected by mutual quenching of fluorescence [26]. Substrate and product are bound through their dianionic 5'-phosphate to a guanidium function of an R [27]. Catalysis seems to involve the base function of a H which serves to facilitate abstraction of a proton from the 4'-group of substrate [28], and a cationic group centered near N^1-O^2 of FMN [22], thereby enhancing the oxidative propensity of the coenzyme by stabilizing the dihydroflavin anion generated. Hence, the conservation of W and R residues for binding of coenzyme and substrate/product, and the absolute requirement for the H residue within the conserved sequence are all compatible with the mechanism we have proposed earlier [14]. The kinetics reflect a binary (ping-pong) mechanism with PNP from which hydrogen abstraction is rate limiting, whereas a ternary complex (sequential) mechanism proceeds with PMP [29]. Ultimate structural elucidation obtained by crystallographic means should clarify the exact positioning of residues within the ternary complex of protein, coenzyme, and substrate.

Kinase/Oxidase Facilitated Delivery of Bioactive Amines

The action of the broad specificity pyridoxal kinase from humans and most eukaryotes, followed by action of pyridoxine (pyridoxamine) phosphate oxidase, which we have found competent to catalyze formation of PLP from numerous N-(5'-phospho-4'-pyridoxyl)amines as well as PMP and PNP, provide a model for transporter-enhanced delivery of bioactive compounds [30]. Generally for cells there is facilitated entry with relative specificity for water-soluble vitamins including the B_6 group [31]. Our work on the uptake of B_6 by renal proximal tubular cells [32], their brush-border membrane vesicles [33], and binding proteins in the membranes [34] has shown that all three natural non-phosphorylated forms of B_6 gain facilitated entry. Cells from the kidney and liver also import N-(4'-pyridoxyl)amines that were synthesized by condensing amines with pyridoxal and reducing the Schiff base product [30,35,36]. The outline of such events, using pyridoxal and an effector amine, to generate a stable, transportable compound that gains facilitated entry into cells where successive actions of kinase and oxidase then liberate the effector amine and PLP is illustrated in Figure 3. Similar techniques are being applied in the ongoing search to effect better delivery of pharmacologic agents.

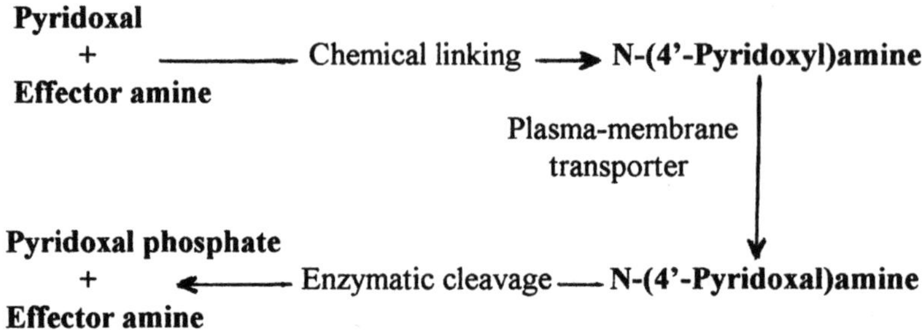

Fig. 3. Transport and metabolism of synthetic N-(4'-pyridoxyl)amines for delivery of potential effectors.

Acknowledgements

Most of this work was supported by N.I.H. grants awarded to the principal investigator over a forty-year span of research. Thanks are also due to the numerous coworkers, many cited in the references, who have helped garner much of the information presented.

References

1. McCormick, D.B., and Snell, E.E. (1959) Proc. Natl. Acad. Sci. USA **45**, 1371-1379.
2. McCormick, D.B., Gregory, M.E., and Snell, E.E. (1961) J. Biol. Chem. **236**, 2076-2084.
3. Laine-Cessac, P., and Allain, P. (1996) Enzyme & Protein **49**, 291-304.
4. McCormick, D.B., and Snell, E.E. (1961) J. Biol. Chem. **236,** 2085-2088.
5. McCormick, D.B., Guirard, B.M., and Snell, E.E. (1960) Proc. Soc. Exp. Biol. Med. **104**, 554-557.
6. Hanna, M.C., Turner, A.J., and Kirkness, E.F. (1997) J. Biol. Chem. **272**, 10756-10760.
7. Laine-Cessac, P., Cailleux, A., and Allain, P. (1997) Biochem. Pharmacol. **54**, 863-870.
8. Yang, Y., Tsui, H.C., and Winkler, M.E. (1998) J. Bacteriol. **180**, 1814-1821.
9. Zhao, G., and Winkler, M.E. (1995) J. Bacteriol. **177**, 883-891.
10. Pogell, B.M. (1958) J. Biol. Chem. **232**, 761-776.
11. Mada, H., and Snell, E.E. (1961) J. Biol. Chem. **236**, 2089-2095.
12. Kazarinoff, M.N., and McCormick, D.B. (1975) J. Biol. Chem. **250**, 3436-3442.
13. McCormick, D.B., and Merrill, A.H., Jr. (1980) In: Vitamin B_6. Metabolism and Role in Growth (H.P. Tryfiates, Ed.), Food and Nutrition Press, Westpoint, CT, pp. 1-26.
14. Bowers-Komro, D.M., and McCormick, D.B. (1984) In: Flavins and Flavoproteins (R.C. Bray, P.C. Engel, and S.G. Mayhew, Eds.), Walter de Gruyter, New York, NY, pp. 581-584.
15. Loubbardi, A., Marcireau, C., Karst, F., and Guillotin, M. (1995) J. Bacteriol. **177**, 1817-1823.
16. McCormick, D.B., and Chen, H. (1999) J. Nutr. **129**, 325-327.
17. Fleischman, R.D., Adams, M.D., White, O., et al. (1995) Science **269**, 496-512.
18. Hagen, T.J., and Shimkets, L.J. (1990) J. Bacteriol. **172**, 15-25.

19. Lam, H.M., and Winkler, M.E. (1992) **174**, 6033-6045.
20. Ngo, E.O., LePage, G.R., Thanassi, J.W., Meisler, N., and Nutter, L.M. (1998) Biochemistry **37**, 7741-7748.
21. McCormick, D.B., Kazarinoff, M.N., and Tsuge, H. (1976) In: Flavins and Flavoproteins (T.P. Singer, Ed.),Walter de Gruyter, New York, NY, pp. 708-719.
22. Merrill, A.H., Jr., Kasai, S., Matsui, K., Tsuge, T., and McCormick, D.B. (1979) Biochemistry **18**, 3635-3641.
23. Kazarinoff, M.N., and McCormick, D.B. (1973) Biochem. Biophys, Res. Commun. **52**, 440-446.
24. DePecol, M.E. , and McCormick, D.B. (1980) Anal. Biochem. **101**, 435-441.
25. Kazarinoff, M.N., and McCormick, D.B. (1974) Biochim. Biophys. Acta **359**, 282-287.
26. McCormick, D.B. (1977) Photochem. Photobiol. **26**, 169-182.
27. Choi, J.D., and McCormick, D.B. (1981) Biochemistry **20**, 5722-5728.
28. Horiike, K., Tsuge, H., and McCormick, D.B. (1979) J. Biol. Chem. **254**, 6638-6643.
29. Choi, J.D., Bowers-Komro, D.M., Davis, M.D., Edmondson, D.E., and McCormick, D.B. (1983) J. Biol. Chem. **258**, 840-845.
30. Zhang, Z., and McCormick, D.B. (1991) Proc. Natl. Acad. Sci. USA **88**, 10407-10410.
31. McCormick, D.B., and Zhang, Z. (1993) Proc. Soc. Exp. Biol. Med. **202**, 265-270.
32. Bowman, B.B., and McCormick, D.B. (1989) J. Nutr. **119**, 745-749.
33. Bowman. B.B., McCormick, D.B., and Smith, E.R. (1990) In: Vitamin B_6 (K. Dakshinamurti, Ed.), Ann. NY Acad. Sci. **585**, 106-109.
34. McCormick, D.B., Bowers-Komro, D.M., Bonkovsky, J., Larsen, C., and Zhang, Z. (1991) In: Enzymes Dependent on Pyridoxal Phosphate and Other Carbonyl Compounds as Cofactors (T. Fukui, H. Kagamiyama, K. Soda, and H. Wada, Eds.), Pergamon Press, Oxford, England, pp 609-611.
35. Zhang, Z., and McCormick, D.B. (1992) Arch. Biochem. Biophys. **294**, 394-397.
36. Zhang, Z., and McCormick, D.B. (1992) In: 1st Internat'l. Cong. On Vitamins and Biofactors in Life Sciences (T. Kobayashi, Ed.), Center for Academic Publications, Osaka, Japan, pp 208-211.

Biochemistry and Molecular Biology of Vitamin B_6
and PQQ-dependent Proteins
ed. by A. Iriarte, H.M. Kagan and M. Martinez-Carrion

A Divergence in the Biosynthetic Pathway and a New Role for Vitamin B6

M. Ehrenshaft[1], M.E. Daub[1], P. Bilski[2], M.Y. Li[1,2], C.F. Chignell[2], A.E. Jenns[1], and K.R. Chung[1]

North Carolina State University[1], Raleigh NC 27695 and National Institute for Environmental and Health Sciences[2], RTP, NC 27709.

Summary

Studies on resistance of the fungus *Cercospora nicotianae* to singlet oxygen generating photosensitizers revealed new insights into the biological role of vitamin B6. Heterologous expression of a gene originally isolated because it complemented photosensitizer-sensitive mutants revealed that this gene (*PDX1*) is an essential component in vitamin B6 synthesis. Data base analysis has revealed that organisms contain either *PDX1* homologues or homologues to *Escherichia coli* biosynthetic genes but not homologues to both, suggesting the existence of divergent pathways. Furthermore, vitamin B6 is an efficient quencher of singlet oxygen and quenching occurs through a chemical interaction during which vitamin B6 is consumed.

Introduction

Although the filamentous fungus *Cercospora nicotianae* produces high amounts of a toxic photosensitizer, cercosporin, it is completely resistant to it (5, 6, 14). The toxicity of cercosporin and other photosensitizers is due to their ability to absorb and transfer light energy to molecular oxygen to produce toxic oxygen species (19). Under reducing conditions, a number of radical species can be produced while direct energy transfer produces the extremely toxic, but non-radical, singlet oxygen ($^1O_2$). There are numerous biological resistance mechanisms against radical oxygen species, but very few defenses against $^1O_2$ (5, 6). *C. nicotianae* is also resistant to other photosensitizers, and thus exhibits an uncommon general photosensitizer resistance.

Identification of *SOR1/PDX1*

Functional complementation of a *C. nicotianae* strain sensitive to cercosporin and other photosensitizers (13, 14) led to the isolation of a gene, originally called *SOR1*, encoding a protein with extensive homology to numerous proteins in diverse organisms (9, 10). The extent of the homology and its cosmopolitan nature was puzzling because it was far more ubiquitous than photosensitizer resistance itself. Although cercosporin is almost universally toxic at low concentrations (5, 6), some of the *SOR1* homologues were from photosensitizer sensitive organisms while others were from anaerobes. The protein sequence itself lacks conserved functional domains, and so provided no clue to its biological role.

To determine if heterologous expression of *SOR1* could enhance photosensitizer resistance, the ORF was fused to a constitutive fungal promoter and transformed into a highly photosensitizer sensitive *Aspergillus flavus* strain. Intriguingly, we recovered numerous transformants with *SOR1*, but none with the vector control. Comparison of the selection conditions with the phenotype of the *A. flavus* strain indicated the strain was a pyridoxine auxotroph and that our medium lacked that essential ingredient. Our apparent complementation of a pyridoxine auxotroph using *SOR1* led us to test our *sor1* mutants for a comparable phenotype. While wild type *C. nicotianae* grows on a minimal medium lacking vitamins, *sor1* mutants cannot unless it is supplemented with pyridoxine (8). Because *sor1* mutants are clearly pyridoxine auxotrophs the name of this gene has been changed to *PDX1*. All organisms require vitamin B6, whether they synthesize it or obtain it nutritionally. Plants and microbes, with the exception of some obligate parasites, synthesize it. The synthesis profile of vitamin B6, therefore, solved the mystery of why *SOR1/PDX1* was so ubiquitous and well conserved.

PDX1 represents a divergence in the vitamin B6 synthesis pathway

The solution to one puzzle led to another. In *E. coli*, pyridoxine auxotrophs have been used to dissect the biosynthetic and genetic pathways (7, 11, 15, 17, 22). All the *E. coli* pyridoxine genes have been cloned and characterized, but none has any homology to *PDX1*, nor is there any gene in the entirely sequenced *E. coli* genome with homology to *PDX1*. *PDX1*, however, clearly plays a role pyridoxine synthesis in both *C. nicotianae* and *A. flavus* suggesting that *PDX1* represents a divergence from the previously characterized pathway.

Homology searches using *PDX1* and the *E. coli* genes *pdxA* and *pdxJ* were performed using all available databases. In organisms with completely sequenced genomes, *PDX1* is found in fungi, archaebacteria and a subset of eubacteria, while the *E. coli* genes are found only in eubacteria (Table 1). Without exception, no organism contains both a homologue to *PDX1* and homologues to the *E. coli* genes. Some completely sequenced genomes lack homologues to either (data not shown). These include *Caenorhabditis elegans*, the sole animal with a completely sequenced genome, and seven eubacterial obligate parasites with reduced coding capacity, presumably due to the fact that they obtain nutrients, such as pyridoxine, from their hosts. In organisms with incompletely sequenced genomes, *PDX1* homologues are found in nine plants, four fungi, *Dictyostelium discoidum*, two archaebacteria and eight eubacteria while *pdxA* and *pdxJ* are found in eighteen eubacteria. As with those with completely sequenced genomes, no organism has homologues to both *PDX1* and the *E. coli* genes.

Table 1. Occurrence of *PDX1* and the *Escherichia coli* pyridoxine biosynthetic genes *pdxA* and *pdxJ* in organisms with completely sequenced genomes.

ORGANISM	*PDX1*	*pdxA/J*	Taxon
Saccharomyces cerevisiae	+	-	Fungi
Methanococcus jannaschii	+	-	Archaebacteria
Pyrococcus horikoshii	+	-	Archaebacteria
Pyrococcus abyssi	+	-	Archaebacteria
Methanobacterium thermoautotrophicum	+	-	Archaebacteria
Archaeoglobus fulgidus	+	-	Archaebacteria
Aeropyrum pernix	+	-	Archaebacteria
Haemophilus influenzae	+	-	Eubacteria
Bacillus subtilis	+	-	Eubacteria
Mycobacterium tuberculosis	+	-	Eubacteria
Thermotoga maritima	+	-	Eubacteria
Deinococcus radiodurans	+	-	Eubacteria
Escherichia coli	-	+	Eubacteria
Synechocystis	-	+	Eubacteria
Helicobacter pylori	-	+	Eubacteria
Aquifex aolicus	-	+	Eubacteria

These data indicate that a divergence in pyridoxine synthesis genes that occurred during eubacterial evolution led to the majority of organisms (plants, fungi, archaebacteria and some eubacteria) using the divergent rather than the well characterized *E. coli* route. There is currently insufficient information to determine if the divergence is broader than a single step. ^{15}N-labeling data, however, had previously hinted that yeast pyridoxine synthesis utilized a different nitrogen donor than *E. coli* (20). Additionally, we have two other *C. nicotianae* pyridoxine auxotrophs that are not *PDX1* mutants. Because our fungus has no sexual stage, we cannot determine if these two mutants represent one or two genes in the *C. nicotianae* pyridoxine pathway. Work is now underway to isolate the gene(s) mutated in these two strains.

Pyridoxine and $^1O_2$

The relationship between auxotrophy and $^1O_2$-sensitivity was clarified by examining the response of one of our *pdx1* null strains to cercosporin and other photosensitizers in pyridoxine dose response experiments. While 10 ng/ml of pyridoxine was sufficient to restore wild-type growth to the pyridoxine auxotroph in the absence of photosensitizer, 100 μg/ml was necessary to support the same level of growth in photosensitizer-containing medium (data not shown). An exogenous source of pyridoxine, therefore, could complement both phenotypes, but photosensitizer resistance required 1000-fold higher concentrations for full restoration.

Our discovery that B6 vitamers quench $^1O_2$ clarified this disparity. Pyridine, the chemical backbone of vitamin B6, does not quench $^1O_2$, so even extensive surveys of $^1O_2$ quenchers did not examine vitamin B6 (2, 21). Our measurements, however, showed that vitamin B6 vitamers are surprisingly efficient quenchers of $^1O_2$ phosphoresence. We determined quenching constants for four forms of vitamin B6 (pyridoxine, pyridoxal, pyridoxal-5'-phosphate and pyridoxamine) and showed they quenched $^1O_2$ as effectively or better than some of the most effective biological antioxidants known (e.g. vitamins C and E) (8).

The above work, however, measured total quenching which is the sum of all physical and chemical quenching. Physical quenchers (e.g. carotenoids and azide) absorb energy form $^1O_2$ and dissipate it as heat, remaining unchanged. In contrast, chemical quenchers undergo a chemical reaction with $^1O_2$, and are altered and/or consumed. To determine if vitamin B6 quenches via a chemical mechanism, the fate of pyridoxine in an irradiated photosensitizer-containing solution was followed spectrophotometrically. Increased irradiation of a solution containing both photosensitizer and pyridoxine resulted in decreased pyridoxine concentration. This decrease could be blocked by adding increasing concentrations of sodium azide, a potent physical quencher (3). This indicates that pyridoxine loss in the first part of the experiment was due to $^1O_2$ exposure, blocked by addition of azide.

The identification of vitamin B6 as a chemical quencher of $^1O_2$ explains the disparity in pyridoxine levels needed to support growth of a pyridoxine auxotroph in the presence and absence of photosensitizer. Increasing interaction with $^1O_2$ results in decreasing amounts of pyridoxine, thus skewing the growth curve of the auxotroph. A similar observation was made in

Aspergillus nidulans (16), where light preincubation of pyridoxine and photosensitizer-containing medium produces medium unable to support growth of a pyridoxine auxotroph.

A facile explanation for the inability of *pdx1* mutants to grow in the presence of photosensitizer is their pyridoxine auxotrophy. Other evidence, however, suggests that vitamin B6 plays an active role against photosensitization. First, the efficient quenching of $^1O_2$ by vitamin B6 suggests its synthesis and accumulation has importance additional to its co-factor role. Secondly, studies in other organisms strongly suggest correlations with active oxygen defense. In yeast, a *PDX1* homologue shows greatly increased transcript and protein accumulation as cells enter stationary phase, a part of the cell cycle associated with increased production of active oxygen species (4, 12). In rubbertree, transcript amounts for *PDX1* homologues increase upon exposure to salicylic acid and ethylene (18), two compounds correlated with pathogen response, which in turn is correlated to increases in active oxygen species. Finally, in *Bacillus subtilis*, paraquat, a superoxide inducer, elicits an increase in the corresponding PDX1 protein (1).

Current efforts are focussed on further dissection of the *C. nicotianae* pyridoxine biosynthetic pathway with an emphasis on isolation of further genes in that pathway. We are also continuing to dissect the relationship between pyridoxine and photosensitizer resistance. Studies are underway to delineate regulatory signals that modulate *PDX1* expression and also to determine cellular concentrations of B6 vitamers. In addition to uncovering what appears to be the recruitment of an essential compound into a specialized, defensive role, we have shown that pyridoxine is not synthesized by a universal pathway in all organisms.

Acknowledgements

This work was supported by grants from the National Science Foundation and the USDA National Competitive Grants Program.

References

1. Antelmann H, Bernhardt J, Schmid R, Mach H, Volker U, Hecker M. 1997. First steps from a two dimensional protein index towards a response-regulation map for *Bacillus subtilis*. *Electrophoresis* 18:1451-63

2. Bellus D. 1979. Physical quenchers of singlet molecular oxygen. *Adv. Photochem.* 11:105-205

3. Bilski P, Li MY, Ehrenshaft M, Daub ME, Chignell CF. 2000. Vitamin B6 (pyridoxine) and its derivatives are efficient singlet oxygen quenchers and potential fungal antioxidants. *Photochem. Photobiol.* 71:In Press

4. Braun EL, Fuge EK, Padilla PA, Werner-Washburne M. 1996. A stationary-phase gene in *Sacchromyces cerevisiae* is a member of a novel, highly conserved gene family. *J. Bacteriol.* 178:6865-72
5. Daub ME, Ehrenshaft M. 2000. The photoactivated *Cercospora* toxin cercosporin: Contributions to plant disease and fundamental biology. *Annu. Rev. Phytopathol,* In press.
6. Daub ME, Ehrenshaft M, Jenns AE, Chung KR. 1998. Active oxygen in fungal pathogenesis of plants: the role of cercosporin in *Cercospora* diseases. In *Phytochemical Signals and Plant-Microbe Interactions, Recent Advances in Phytochemistry*, Vol. 32, ed. KR Downum, R Verpoorte, pp. 31-56. New York: Plenum Press
7. Drewke C, Klein M, Clade D, Arenz A, Muller R, Leistner E. 1996. 4-O-phosphoryl-L-threonine, a substrate of the pdxC(serC) gene product involved in vitamin B6 biosynthesis. *FEBS Lett.* 390:179-82
8. Ehrenshaft M, Bilski P, Li M, Chignell CF, Daub ME. 1999. A highly conserved sequence is a novel gene involved in *de novo* vitamin B6 biosynthesis. *Proc. Natl. Acad. Sci. (USA)* 96:9374-8
9. Ehrenshaft M, Chung KR, Jenns AE, Daub ME. 1999. Functional characterization of *SOR1*, a gene required for resistance to photosensitizing toxins in the fungus *Cercospora nicotianae*. *Curr. Genet.* 34:478-85
10. Ehrenshaft M, Jenns AE, Chung KR, Daub ME. 1998. *SOR1*, a gene required for photosensitizer and singlet oxygen resistance in *Cercospora* fungi is highly conserved in divergent organisms. *Molec. Cell* 1:603-9
11. Hill RE, Spenser ID. 1996. Biosynthesis of vitamin B6. In *Escherichia coli and Salmonella typhimurium: cellular and molecular biology*, Vol. 1, ed. FC Neidhardt, R Curtiss, JL Ingraham, ECC Lin, KB Low, B Magasanik, WS Reznikoff, M Riley, M Schaechter, HE Umbarger, pp. 695-703. Washington, D.C.: ASM Press
12. Jamieson DJ. 1995. The effect of oxidative stress on *Saccharomyces cerevisiae*. *Redox Rept.* 1:89-95
13. Jenns AE, Daub ME. 1995. Characterization of mutants of *Cercospora nicotianae* sensitive to the toxin cercosporin. *Phytopathology* 85:906-12
14. Jenns AE, Scott DL, Bowden EF, Daub ME. 1995. Isolation of mutants of the fungus *Cercospora nicotianae* altered in their response to singlet-oxygen-generating photosensitizers. *Photochem. Photobiol.* 61:488-93
15. Lam HM, Winkler ME. 1990. Metabolic relationships between pyridoxine (vitamin B6) and serine biosynthesis in *Escherichia coli* K-12. *J. Bacteriol.* 172:6518-28
16. Osmani AH, May GS, Osmani SA. 1999. The extremely conserved *pyroA* gene of *Aspergillus nidulans* is required for pyridoxine synthesis and is required indirectly for resistance to photosensitizers. *J. Biol. Chem.* 274:23565-9
17. Schoenlein PV, Roa BB, Winkler ME. 1989. Divergent transcription of *pdxBi*and homology between *pdxB* and *serA* gene products in *Escherichia coli* K-12. *J. Bacteriol.* 171:6084-92
18. Sivasubramaniam S, Vanniashingham VM, Tan CT, Chua NH. 1995. Characterization of HEVER, a novel stress-induced gene from *Hevea brasiliensis*. *Plant Mol. Biol.* 29:173-8
19. Spikes JD. 1989. Photosensitization. In *The Science of Photobiology*, ed. KC Smith, pp. 79-110. NY: Plenum Press
20. Tazuya K, Adachi Y, Masuda K, Yamada K, Kumaoka H. 1995. Origin of the nitrogen atom of pyridoxine in *Saccharomyces cerevisiae*. *Biochim. Biophys. Acta.* 1244:113-6
21. Wilkinson F, Helman WP, Ross AB. 1995. Rate constants for the decay and reactions of the lowest electronically excited singlet state of molecular oxygen in solution. An expanded and revised compilation. *J. Phys. Chem. Ref. Data* 24:663-1021
22. Zhao G, Winkler ME. 1996. 4-Phospho-hydroxy-L-threonine is an obligatory intermediate in pyridoxal 5'-phosphate coenzyme biosynthesis in *Escherichia coli* K-12. *FEMS Microbiol. Lett.* 135:275-80

Biochemistry and Molecular Biology of Vitamin B_6
and PQQ-dependent Proteins
ed. by A. Iriarte, H.M. Kagan and M. Martinez-Carrion

Molecular Cloning and Catalytic Properties of Human Brain Pyridoxal Kinase

H.S. Lee[1], B.J. Moon[1], S.Y. Choi[2] and O.S. Kwon[1]

[1]Department of Biochemistry, College of Natural Sciences, Kyungpook National University, Taegu 702-701, Korea.
[2]Department of Genetic Engineering, Division of Life Sciences, Hallym University, Chunchon 200-702, Korea.

Summary

cDNA fragments of ovine liver pyridoxal kinase were amplified by PCR using degenerate oligonucleotide primers based on partial amino acid sequence data. Using the PCR product as probes, we have isolated a full-length cDNA encoding the human brain pyridoxal kinase. The recombinant enzyme was overexpressed in *E. coli*, and the recombinant enzyme has maximum catalytic activity in the narrow pH range of 5.5-6.0. The Km values for two substrates pyridoxal and ATP are 97 μM and 12 μM, respectively. Zn^{2+} is the most effective divalent cation in the phosphorylation of pyridoxal. Results from intrinsic fluorescence quenching measurements showed that the tryptophanyl residues in the recombinant protein are more accessible to the neutral acrylamide than to the negatively charged iodide.

Introduction

Pyridoxal 5'-phosphate (PLP) is a cofactor required by numerous enzymes that catalyze transamination, decarboxylation and racemization reactions (Snell, 1990). The formation of PLP from pyridoxal and ATP is catalyzed by pyridoxal kinase (EC 2.7.1.35) (McCormick *et al.*, 1961). Pyridoxal kinase has been detected in all mammalian species as well as in many microorganisms. The enzyme isolated from sheep brain has been identified to be a dimmer with a molecular weight of 80 kDa (Kwok and Churchich, 1979; Kwok *et al.*, 1987). The physiological significance of pyridoxal kinase is not fully understood, but this enzyme is of particular interest because of the intimate relationship of vitamin B_6 metabolism to brain disorders such as convulsive seizures and Down syndrome. Further progress in the physiological and functional studies depends upon detailed information on the structure and action mechanism of the enzyme.

cDNAs encoding pyridoxal kinase have been recently isolated from various species (Hanna *et al.*, 1997; Gao *et al.*, 1998; Yang *et al.*, 1998), but the tertiary structure of the enzyme is still not available. In order to further our understanding of pyridoxal kinase, we have isolated a full-length cDNA encoding the human brain enzyme. Furthermore we expressed a recombinant human pyridoxal kinase in *E.coli*, and characterized its catalytic properties. A successful over-expression of soluble and catalytically active human enzyme should facilitate future studies involving the utilization of mutagenesis and X-ray crystallography.

Materials and Methods

Degenerate oligonucleotide primers, 5'-CA(TC)GT(AGTC)AA(CT)CA(AG)TA(CT)GA (CT)TA-3' (forward) and 5'-GC(AG)TC(AGTC)AC(CT)TT(AG)TGCAT(CT)TCCAT-3' (reverse) corresponding to the sequence of two peptides from ovine pyridoxal kinase, were used in PCR amplification. The amplified ovine liver cDNA fragment of approximately 0.5kb was labeled by nick translation, and used to screen the human brain cDNA library constructed in λZAP II vector (Stratagene).

To facilitate an expression vector construction, two restriction sites of NdeI and BamHI were introduced by PCR at both ends of the open reading frame. The amplified gene was subcloned into the pGEM vector, and NdeI fragment was then inserted into pT7-7 vector (pT7-7/HPK). To construct pMAL-c2/HPK, a BamHI fragment of the pT7-7/HPK-r containing the reverse oriented gene was inserted into pMAL-c2 expression vector. The bacterial cells JM109, harboring pMAL-c2/HPK, were cultured in LB media with ampicillin at 37°C until an OD_{600} of the culture reached 0.5. Then IPTG was added to a final concentration of 0.3mM, and the cells incubated overnight at 25°C. For purification of the expressed protein, the cell lysate was applied to an amylose column for affinity chromatography, and eluted with the column buffer containing 10mM maltose. The purified fusion protein was treated with factor Xa, and the sample was then applied to hydroxyapatite column to remove the maltose and the protease. Finally the second amylose affinity chromatography yielded highly purified human pyridoxal kinase.

For the enzyme assay, the initial rate of formation of pyridoxal-5'-P was measured by following the increase in absorbance at 388nm at least 3 min. The standard assay mixture contains pyridoxal (0.5mM), ATP (0.5mM) and $ZnCl_2$ (0.1mM) in 70mM potassium phosphate buffer (pH 6.0). A unit of enzymatic activity is equivalent to the formation of 1nmol of PLP/min catalyzed by the enzyme at 37°C.

Results and Discussion

cDNA Cloning and Expression of human brin pyridoxal kinase

Using two degenerate oligonucleotide primers corresponding to the amino acids sequence of the ovine brain pyridoxal kinase, PCR amplifications were performed to screen an ovine liver λgt11 cDNA library. The amplified cDNA fragment was used for screening of a human brain cDNA library. One of the isolated clones consists of 1115bp encoding a human brain pyridoxal kinase. As shown in Fig.1, the deduced amino acid sequence from an open reading frame corresponds to a polypeptide of 312 amino acid residues with a molecular weight of 35102, which is identical to the results from the testis cDNA reported by Hanna *et al* (1997). Comparison of the deduced amino acids sequence with other mammalian enzymes suggests that the pyridoxal kinase is highly homologous in mammalian species (Fig. 2).

To investigate enzymatic properties of the human pyridoxal kinase, the recombinant enzyme was overexpressed in *E.coli* as a fusion protein with maltose binding protein. Most of the expressed protein was found in the soluble extract after lysis of the cells. Based on the specific activity and SDS-PAGE analysis, it was estimated that the expressed protein presents more than 10% of the cellular proteins. After two steps of affinity chromatography, we usually obtain about 10 mg of pure pyridoxal kinase from 1 liter of cultured cells.

Properties of recombinant human brain pyridoxal kinase

Maximum catalytic activity was observed over the pH range 5.5-6.0 when pyridoxal was used as

```
cccgagcgagcgcgagaccgtgcccccgcctcggccccgcgcgcccgcgccaggcccggcATGGAGGAGGAGTGCCGGGTGCTCTCCATA   90
                                                             M  E  E  E  C  R  V  L  S  I     10
CAGAGCCACGTCATCCGCGGCTACGTGGGCAACCGGGCGGCCACGTTCCCGCTGCAGGTTTTGGGATTTGAGATTGACGCGGTGAACTCT  180
 Q  S  H  V  I  R  G  Y  V  G  N  R  A  A  T  F  P  L  Q  V  L  G  F  E  I  D  A  V  N  S     40
GTCCAGTTTTCAAACCACACAGGCTATGCCCACTGGAAGGGCCAAGTGCTGAATTCAGATGAGCTCCAGGAGTTGTACGAAGGCCTGAGG  270
 V  Q  F  S  N  H  T  G  Y  A  H  W  K  G  Q  V  L  N  S  D  E  L  Q  E  L  Y  E  G  L  R     70
CTGAACAACATGAATAAATATGACTACGTGCTCACAGGTTATACGAGGGACAAGTCGTTCCTGGCCATGGTGGTGGACATTGTGCAGGAG  360
 L  N  N  M  N  K  Y  D  Y  V  L  T  G  Y  T  R  D  K  S  F  L  A  M  V  V  D  I  V  Q  E    100
CTGAAGCAGCAGAACCCCAGGCTGGTGTACGTGTGTGATCCAGTCTTGGGTGACAAGTGGGACGGCGAAGGCTCGATGTACGTCCCGGAG  450
 L  K  Q  Q  N  P  R  L  V  Y  V  C  D  P  V  L  G  D  K  W  D  G  E  G  S  M  Y  V  P  E    130
GACCTCCTTCCCGTCTACAAAGAAAAAGTGGTGCCGCTTGCAGACATTATCACGCCCAACCAGTTTGAGGCCGAGTTACTGAGTGGCCGG  540
 D  L  L  P  V  Y  K  E  K  V  V  P  L  A  D  I  I  T  P  N  Q  F  E  A  E  L  L  S  G  R    160
AAGATCCACAGCCAGGAGGAAGCCTTGCGGGTGATGGACATGCTGCACTCTATGGGCCCCGACACCGTGGTCATCACCAGCTCCGACCTG  630
 K  I  H  S  Q  E  E  A  L  R  V  M  D  M  L  H  S  M  G  P  D  T  V  V  I  T  S  S  D  L    190
CCCTCCCCGCAGGGCAGCAACTACCTGATTGTGCTGGGGAGTCAGAGGAGGAGGAATCCCGCTGGCTCCGTGGTGATGGAACGCATCCGG  720
 P  S  P  Q  G  S  N  Y  L  I  V  L  G  S  Q  R  R  R  N  P  A  G  S  V  V  M  E  R  I  R    220
ATGGACATTCGCAAAGTGGACGCCGTCTTTGTGGGCACTGGGGACCTGTTTGCTGCCATGCTCCTGGCGTGGACACACAAGCACCCCAAT  810
 M  D  I  R  K  V  D  A  V  F  V  G  T  G  D  L  F  A  A  M  L  L  A  W  T  H  K  H  P  N    250
AACCTCAAGGTGGCCTGTGAGAAGACCGTGTCTACCTTGCACCACGTTCTGCAGAGGACCATCCAGTGTGCAAAAGCCCAGGCCGGGGAA  900
 N  L  K  V  A  C  E  K  T  V  S  T  L  H  H  V  L  Q  R  T  I  Q  C  A  K  A  Q  A  G  E    280
GGAGTGAGGCCCAGCCCCATGCAGCTGGAGCTGCGGATGGTGCAGAGCAAAAGGGACATCGAGGACCCAGAGATCGTCGTCCAGGCCACG  990
 G  V  R  P  S  P  M  Q  L  E  L  R  M  V  Q  S  K  R  D  I  E  D  P  E  I  V  V  Q  A  T    310
GTGCTGTGAgggccccgccgcttgcccgtgacgaccagcgcgttggtgtctccgtgtttgtccctgtgaaaacatgtaacgtctgcctta 1080
 V  L  *                                                                                     312
gagccatgaccgaaacttgatattttttctttca                                                           1115
```

Fig. 1. ***Nucleotide and deduced amino acid sequence of human brain pyridoxal kinase.*** The 5' and 3' flanking regions are indicated as lowercase whereas the open reading frame as uppercase letters. The predicted amino acid sequence is shown below the nucleotide sequence.

```
                       |         |         |         |         |         |         |        |
HBPK             MEEECRVLSIQSHVIRGYVGNRAATFPLQVLGFEIDAVNSVQFSNHTGYAHWKGQVLNSDELQELYEGLR
RLPK             MEGECRVLSIQSHVVRGYVGNRAAMFPLQVLGFEVDAVNSVQFSNHTGYAHWKGQVLTSQELHALYEGLK
PBPK   MQAGSWVVGGGDSDSRVLSIQSHVVRGYVGNRAATFPLQVLGFEVDAVNSVQFSNHTGYAHWKGQVLNSDELHALYEGLK
                 ....*********.********* *********.***********************.*.**..*****.

HBPK   LNNMNKYDYVLTGYTRDKSFLAMVVDIVQELKQQNPRLVYVCDPVLGDKWDGEGSMYVPEDLLPVYKEKVVPLADIITPN
RLPK   ANNVNKYDYVLTGYTRDKSFLGMVVDIVQELKQQNSRLVYVCDPVMGDKWNGEGSMYVPQDLLPVYREKVVPMADIITPN
PBPK   LNNVNQYDYVLTGYTRDKSFLAMVVDIVRELKQQNPRLVYVCDPVMGDKWDGEGSMYVPEDLLPVYREKVVPVADIITPN
       .**.*.***************.******.******.*********.****.********.******.*****.*******

HBPK   QFEAELLSGRKIHSQEEALRVMDMLHSMGPDTVVITSSDLPSPQGSNYLIVLGSQRRRNPAGSVVMERIRMDIRKVDAVF
RLPK   QFEAELLSGRKIHSQEEAFAVMDVLHRMGPDTVVITSSDLPSPKGSDYLMALGSQRMRKPDGSTVTQRIRMEMRKVDPVF
PBPK   QFEAELLTGRRIHSEEEALAVMDMLHAMGPDTVVITSSDLPSPRGKDYLIALGSQRTRSPDGSVATQRIRMEICKVDAVF
       *******.**.***.***. ***.** ****************.*..**..*****.*.*.**.. .****.. ***.**

HBPK   VGTGDLFAAMLLAWTHKHPNNLKVACEKTVSTLHHVLQRTIQCAKAQAGEGVRPSPMQLELRMVQSKRDIEDPEIVVQATVL
RLPK   VGTGDLFAAMLLAWTHKHPDNLKVACEKTVSAMQHVLQRTIRCAKAEAGEGQKPSPAQLELRMVQSRKDIEDPEIVVQATVL
PBPK   VGTGDLFAAMLLAWTHKHPNNLKVACEKTVSAMHHVLRRTIQCAKAKAGEGLKPSPAQLELRMVQSKRDIEDPEVVVQATVL
       *******************.***********...***.***.****.**** .*** *********..******.*******
```

Fig. 2. ***Comparison of amino acid sequences of mammalian pyridoxal kinases.*** The sequence for human brain pyridoxal kinase (HBPK) is from present study (or GenBank accession No. U89606), rat liver (RLPK) from AF020346 and porcine brain enzyme (PBPK) from AF041255. Asterisks represent identical amino acids (78%), single dots are high similarity, and blank spaces are no homology.

the substrate. The activity decreases gradually above pH 6, and more dramatically below pH 5.5. The stability of the enzyme at various pH was taken into consideration in the analysis of the kinetic parameters. At pH values below 5.5, a gradual irreversible inactivation was observed. On the other hand, exposure of the enzyme to alkaline buffers prior to the enzymatic assays has no effect on protein stability. At optimum pH value, the specific activity of recombinant human pyridoxal kinase is 2470 units/mg and it exhibits Km values of 97 μM and 12 μM for pyridoxal and ATP, respectively (Table I). In addition, pyridoxal kinase requires a divalent ion for its catalytic activity. Of the divalent ions examined, Zn^{2+} is the most effective, and the order of activation is $Zn^{2+} > Co^{2+} > Mn^{2+} > Mg^{2+} > Ca^{2+}$ (Table II).

Table I, ***Catalytic parameters of recombinant human pyridoxal kinase***

Specific activity (units/mg)	2470
Km (uM)	
Pyridoxal	97
ATP	12

Table II, ***Effect of divalent cations on the kinase activity***

Compound	Ralative activity (%)
Zn^{2+}	100
Co^{2+}	85
Mn^{2+}	45
Mg^{2+}	15
Ca^{2+}	5

The degree of exposure of the tryptophanyl residues to collisional encounters with small molecules was derived from intrinsic fluorescence measurements under conditions of iodide (I^-) and acrylamide quenching. Upon excitation at 295nm and emission at 350nm, we obtained a typical Stern-Volmer plot: straight lines for both quenchers with Ksv values for acrylamide and KI of 9.7 and 4.5M^{-1}, respectively. Judging from these results, it appears that the three tryptophanyl residues in the recombinant human pyridoxal kinase are more accessible to neutral acrylamide molecules than to the negatively charged iodide ions.

Acknowledgements

This article is dedicated to the deceased professor J.E.Churchich. This work was supported by the Genetic Engineering Research Grant from the Ministry of Education, Korea to O.S. Kwon and in part by a Research Grant from Hallym Academy of Sciences to S.Y. Choi.

References

Gao, Z.G., Lau, C.K., Lo, S.C.L., Choi, S.Y., Churchich, J.E., Kwok, F. (1998) Porcine pyridoxal kinase: cDNA cloning, expression and primary sequence conformation. *Int. J. Biochem. Cell Biol. 30:* 1379-1388.

Hanna, M.C., Turner, A.J., and Kirkness, E.F. (1997) Human pyridoxal kinase. *J. Biol. Chem. 272:*10756-10760

Kwok, F., and Churchich, J.E. (1979) Brain pyridoxal kinase : purification and characterization. *J. Biol. Chem. 254:* 6489-6495

Kwok, F., Scholz, G., and Churchich, J.E. (1987) Brain pyridoxal kinase dissociation of the dimeric structure and catalytic activity of the monomeric species. *Eur. J. Biochem. 168:* 577-583

McCormick, D.B., Gregory, M.E., and Snell, E.E. (1961) Pyrodoxal phosphokinase I: assay, distribution, purification and properties. *J. Biol. Chem. 236:* 2076-2084

Snell, E.E. (1990) Vitamine B_6 and decarboxylation of histidine. *Ann. N.Y. Acad. Sci. 585:* 1-12

Yang, Y., Tsui, H.T., Man, T.K., and Winkler, M.E. (1998) Identification and function of the pdxY gene, which encodes a novel pyridoxal kinase involved in the salvage pathway of pyrioxal-5'-phosphate biosynthesis in *Escherichia coli* K-12. *J. Bacteriol. 180:* 1814-1821

REGULATION OF GENE EXPRESSION OF PLP-DEPENDENT PROTEINS

Biochemistry and Molecular Biology of Vitamin B_6
and PQQ-dependent Proteins
ed. by A. Iriarte, H.M. Kagan and M. Martinez-Carrion

Regulation of the aspartate and alanine aminotransferases in humans and rodents.

Martine Aggerbeck, Fadéla Beurton, Céline Tomkiewicz, Michèle Garlatti, Emmanuelle Plée-Gautier*, Bénédicte Antoine*, Claude Forest*, Françoise Muzeau and Robert Barouki. Unité INSERM 490, Centre Universitaire des Saints-Pères, 45 rue des Saints-Pères, 75006 Paris, France. * CEREMOD, CNRS, 9 rue Jules Hetzel, 92190 Meudon, France

Summary

We have studied the regulation of transaminase gene expression by hormones and drugs in humans and rodents. In rodents, most of the regulation is hormonal and is consistent with the role of these enzymes in gluconeogenesis. In humans, these genes are up-regulated by drugs such as fibrates, which may partially account for the increase in the activities in serum observed in some treated patients. Interestingly, fibrates down-regulate these genes in the rat, suggesting that there are important differences in the regulation of gene expression between rodents and humans.

Introduction

Alanine aminotransferase (E.C.2.6.1.2.) and aspartate aminotransferase (E.C.2.6.1.1.) are ubiquitous enzymes present as two isoenzymes, cytosolic and mitochondrial. They catalyze important reactions, mainly in amino acid metabolism and gluconeogenesis(1). Serum cytosolic aspartate aminotransferase (cAspAT) levels are important in the diagnosis of liver, heart and muscle diseases while serum alanine aminotransferase (AlaAT) levels are most useful in the diagnosis of liver diseases (2). It is assumed that an increase in the serum levels of these enzymes reflects leakage from the damaged cells. This is certainly the case when the increase in serum activities is great. In case of moderate and transient increases, an alternative possibility is that changes in serum levels could reflect changes in the expression of the genes of these enzymes due to physiological and pharmacological effects. These effects may be either positive, leading to increased serum activities in the absence of cytotoxicity or negative, leading to serum activities that may not reflect the intensity of the tissue damage. It is thus important to determine how these genes are regulated. In our laboratory, we have studied the regulation of the transcription of the genes encoding both the cytosolic aspartate and alanine aminotransferases in humans and in rodents by several hormones, nutrients and drugs.

I. Regulation of the rat cytosolic aspartate aminotransferase (cAspAT) gene promoter by glucocorticoids, insulin and glucose

In the rat liver and kidney and in the rat hepatoma cell line, Fao, glucocorticoids increase cAspAT activity and mRNA levels without changingthe activity of the mitochondrial form (3, 4). In Fao cells, this effect is potentiated by cAMP and inhibited by insulin (4). Streptozotocin-induced diabetes increased the activity and the mRNA level of cAspAT in the rat liver (5). These effects result from changes at the transcriptional level (6).

We have cloned and sequenced 2405 bp of the promoter of the rat cAspAT gene which displays properties of both a housekeeping gene and a tissue-specific regulated gene (6). The basal activity is restricted to 290 proximal basepairs (7) in which the cAMP responsive site is located. Glucocorticoids exert their action through two regions, a proximal region (-682/-26) containing both a glucocorticoid responsive element (GRE) and a nuclear factor I site (NFI) and a distal region (-1983/-1718) (6, 8, 9), which is responsive to insulin (Figure 1).

This distal region was studied in more detail by subcloning deleted fragments in front of heterologous promoters capable of expressing the chloramphenicol acetyltransferase (CAT) activity into Fao cells and the parent cell line, H4IIEC3. Measurement of CAT activity showed that a 282 (-1984/-1702) bp fragment contains the elements required for the induction of gene expression by glucocorticoids and the negative effect of insulin (data not shown).

Deletions of different sizes of the (-1984/-1702) fragment, cloned in front of the less complex thymidine kinase (Tk) promoter were transiently transfected into H4IIEC3 cells. A 142 bp fragment (1844/-1702) retained responsiveness to both hormones. In contrast, several other deletions led to dramatic changes in the action of glucocorticoids, insulin or both (10).

Several consensus motifs known to bind transcription factors were detected in the (-1844/-1702) sequence and footprint experiments using rat liver or H4II cell nuclear extracts revealed binding to three sequences, PI to PIII (Figure 1), (10).

We focused on the -1817/-1775 region which shows sequence similarities to insulin responsive elements (IREs) found in the promoters of other genes regulated by glucocorticoids and insulin in a similar manner (11). In this region, 5 different IREs are present which overlap with hepatocyte nuclear factor-3 (HNF-3) and nuclear factor I (NFI) consensus sequences (figure 1).

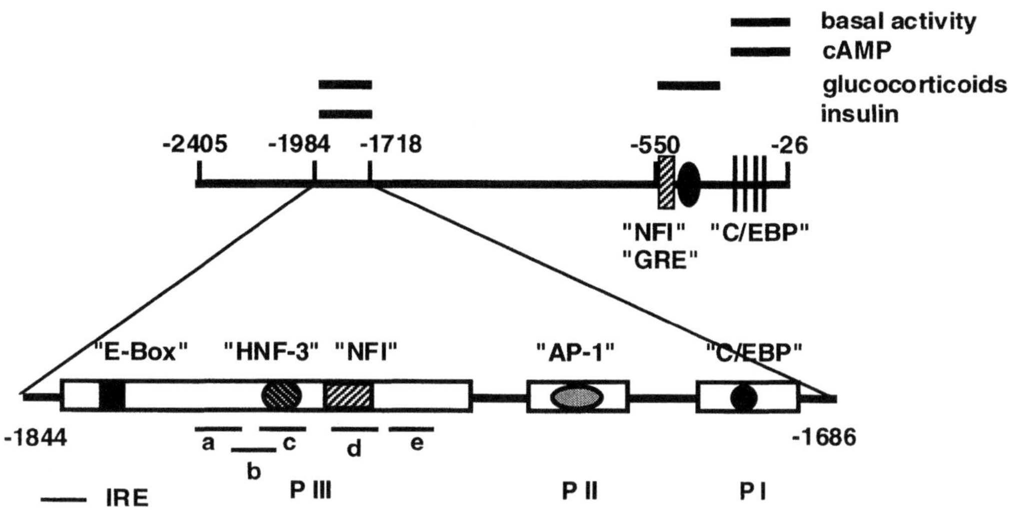

Figure 1. Location of consensus DNA binding sites, of the sites mediating hormonal effects in the cAspAT gene promoter and of footprints (PI to PIII).

Retardation assays revealed that members of the HNF-3 and NFI families bind to this region (10). We mutated these IREs and transfected the constructions into H4II cells. Only the combined mutation of 2 IREs (c and d) led to the loss of the insulin effect whereas a 3-fold stimulation by glucocorticoids was still seen. Additional studies are now under way to determine which proteins are necessary for the effects of these hormones.

cAspAT participates in gluconeogenesisin the liver and is expected to be a glyceroneogenic enzyme in the adipocyte (12). In the adipoblast cell line, 3T3-F442A, and in rat adipose tissue, glucose deprivation induces 3 to 4-fold cAspAT activity and mRNA level (13). This effect is cell-specific, since neither hepatoma nor muscle or kidney cells responded to glucose deprivation. Using permanent transfections, it was shown that the region (-1984/-1718), necessary for the effect of insulin on liver cells, is also required for the effect of glucose. cAspAT responded to stress in a manner similar to glucose-regulated proteins and is a new member of this family (14).

II. Regulation of rodent cytosolic aspartate aminotransferase (cAspAT) and alanine aminotransferase (cAlaAT) genes by fibrates

Fibrates were developed as drugs to decrease serum triglycerides and cholesterol in humans (15). They act through the peroxisome proliferator-activated receptor (PPAR) (16). The effect of fibrates was tested because treatment with these drugs trigger moderate elevations in serum transaminases in some patients. In rats and mice, fibrate treatment leads to an increase in the size and in the number of peroxisomes in the liver which, eventually, provokes tumors in the animals. In the mouse, treatment with fenofibrate decreased the level of the mRNAs encoding cAlaAT and cAspAT by 42 and 81% (17). PPARα deficient mice displayed a 2 to 3-fold increase of both cAspAT and cAlaAT compared to normal mice and fenofibrate treatment did not modify these levels (17). These results strongly suggest that the inhibition by fenofibrate is mediated by PPARα and that an endogenous ligand for PPARα may repress the transaminase mRNA levels in normal mice.

The activities of both transaminases were reduced by fenofibrate in the rat hepatoma cell line Fao, particularly following dexamethasone induction (50 and 70% decrease for cAlaAT and cAspAT, respectively). This was due to a decrease in the mRNA levels of both transaminases (unpublished data). A proximal 550 bp fragment of the cAspAT promoter subcloned in front of the CAT reporter gene was transfected into the rat hepatoma cell line, H4II. Treatment of the transfected cells by fenofibrate leads to a 40% inhibition of the promoter activity obtained in control cells (unpublished data), a result in agreement with the effect observed in mice.

III. Regulation of the human cytosolic aspartate aminotransferase (cAspAT) and alanine aminotransferase (cAlaAT) genes by fibrates

Several of the effects of fibrates in rodents, including increased peroxisome proliferation and tumorigenesis, have never been observed in man. However, as mentioned above, some patients treated with fibrates develop moderate and transient elevations of serum transaminases, which is interpreted as reflecting limited liver toxicity (15). In the human hepatoma cell line, HepG2, fenofibric acid increases the enzyme activity and the mRNA levels of both cAspAT and cAlaAT, an effect due an increase in the transcription of the genes (17).

We have recently cloned a 2660 bp fragment of the human cAspAT promoter and found a 65% homology between the rat and the human promoters in the 800 proximal basepairs. Fibrate treatment of HepG2 cells transfected with this 800 bp fragment cloned in front of the luciferase gene resulted in a 2-fold increase in luciferase activity (unpublished data), suggesting an increase in transaminase gene expression in human cells. This result could explain the moderate increases of the transaminases in the serum of some fibrate-treated patients, assuming that the normal serum levels reflect the general turnover of the cells and of their enzymatic contents. Experiments are now under way to analyze more precisely the region(s) of the promoter responsible for the stimulation by fibrates and to understand why these drugs act in opposite ways in man and in the rat.

Conclusions

For several years, our group has been studying the regulation of transaminase expression in humans and rodents. We have shown that, despite the fact that cAspAT is a housekeeping protein, the expression of its gene is regulated in a highly specific manner, at the level of the tissue and the cell. Glucocorticoids induce the expression of the enzyme only in the liver (5) and the kidney, (18). The inhibition by insulin is restricted to the liver whereas the stimulation by glucose deprivation is observed only in the adipose tissue (13). We are still studying the precise signalling pathways used by insulin and glucose to regulate cAspAT gene expression, as well as the nuclear targets of these compounds.

We compared the regulation of transaminase gene expression in humans and rodents. Several conclusions can be drawn : i) there is species specificity, since the regulation of the genes by fibrates in man is opposite to that in rodents, ii) there appears to be important differences in the promoter sequences, iii) expression of the human genes can be induced by drugs like fibrates, which could account, at least partially, for the increased activity in the serum of treated patients. Additional studies with different drugs are required to further support this hypothesis.

References

1. Cooper, A. J. L. and Meister, A. (1985) Cooper, A. J. L. and Meister, A. *Transaminases* (P. Christen and D. E. Metzler, eds), Wiley, New York, 534-557
2. Schmidt, E. and Schmidt, F. W. (1985) Schmidt, E. and Schmidt, F. W. *Transaminases* (P. Christen and D. E. Metzler, eds), Wiley, New York, 587-624
3. Pave-Preux, M., Ferry, N., Bouguet, J., Hanoune, J. and Barouki, R. (1988) Pave-Preux, M., Ferry, N., Bouguet, J., Hanoune, J. and Barouki, R. *J Biol Chem* 263 , 17459-17466

4. Barouki, R., Pave-Preux, M., Bousquet-Lemercier, B., Pol, S., Bouguet, J. and Hanoune, J. (1989) Barouki, R., Pave-Preux, M., Bousquet-Lemercier, B., Pol, S., Bouguet, J. and Hanoune, J. *Eur J Biochem* 186 , 79-85
5. Feilleux-Duche, S., Garlatti, M., Burcelin, R., Aggerbeck, M., Bouguet, J., Girard, J., Hanoune, J. and Barouki, R. (1994) Feilleux-Duche, S., Garlatti, M., Burcelin, R., Aggerbeck, M., Bouguet, J., Girard, J., Hanoune, J. and Barouki, R. *Am J Physiol* 266 , C911-918
6. Aggerbeck, M., Garlatti, M., Feilleux-Duché, S., Veyssier, C., Daheshia, M., Hanoune, J. and Barouki, R. (1993) Aggerbeck, M., Garlatti, M., Feilleux-Duché, S., Veyssier, C., Daheshia, M., Hanoune, J. and Barouki, R. *Biochemistry* 32 , 9065-9072
7. Garlatti, M., Tchesnokov, V., Daheshia, M., Feilleux-Duche, S., Hanoune, J., Aggerbeck, M. and Barouki, R. (1993) Garlatti, M., Tchesnokov, V., Daheshia, M., Feilleux-Duche, S., Hanoune, J., Aggerbeck, M. and Barouki, R. *J Biol Chem* 268 , 6567-6574
8. Garlatti, M., Daheshia, M., Slater, E., Bouguet, J., Hanoune, J., Beato, M. and Barouki, R. (1994) Garlatti, M., Daheshia, M., Slater, E., Bouguet, J., Hanoune, J., Beato, M. and Barouki, R. *Mol Cell Biol* 14 , 8007-8017
9. Garlatti, M., Aggerbeck, M., Bouguet, J. and Barouki, R. (1996) Garlatti, M., Aggerbeck, M., Bouguet, J. and Barouki, R. *J Biol Chem* 271 , 32629-32634
10. Beurton, F., Bandyopadhyay, U., Dieumegard, B., Barouki, R. and Aggerbeck, M. (1999) Beurton, F., Bandyopadhyay, U., Dieumegard, B., Barouki, R. and Aggerbeck, M. *Biochem J* 343 Pt 3 , 687-695
11. Streeper, R. S., Svitek, C. A., Chapman, S., Greenbaum, L. E., Taub, R. and O'Brien, R. M. (1997) Streeper, R. S., Svitek, C. A., Chapman, S., Greenbaum, L. E., Taub, R. and O'Brien, R. M. *J Biol Chem* 272 , 11698-11701
12. Reshef, L., Hanson, R. W. and Ballard, F. J. (1970) Reshef, L., Hanson, R. W. and Ballard, F. J. *J Biol Chem* 245 , 5979-5984
13. Plee-Gautier, E., Aggerbeck, M., Beurton, F., Antoine, B., Grimal, H., Barouki, R. and Forest, C. (1998) Plee-Gautier, E., Aggerbeck, M., Beurton, F., Antoine, B., Grimal, H., Barouki, R. and Forest, C. *Endocrinology* 139 , 4936-4944
14. Plee-Gautier, E., Grimal, H., Aggerbeck, M., Barouki, R. and Forest, C. (1998) Plee-Gautier, E., Grimal, H., Aggerbeck, M., Barouki, R. and Forest, C. *Biochem J* 329 , 37-40
15. Balfour, J. A., McTavish, D. and Heel, R. C. (1990) Balfour, J. A., McTavish, D. and Heel, R. C. *Drugs* 40 , 260-290
16. Vu-Dac, N., Schoonjans, K., Kosykh, V., Dallongeville, J., Fruchart, J. C., Staels, B. and Auwerx, J. (1995) Vu-Dac, N., Schoonjans, K., Kosykh, V., Dallongeville, J., Fruchart, J. C., Staels, B. and Auwerx, J. *J Clin Invest* 96 , 741-750
17. Edgar, A. D., Tomkiewicz, C., Costet, P., Legendre, C., Aggerbeck, M., Bouguet, J., Staels, B., Guyomard, C., Pineau, T. and Barouki, R. (1998) Edgar, A. D., Tomkiewicz, C., Costet, P., Legendre, C., Aggerbeck, M., Bouguet, J., Staels, B., Guyomard, C., Pineau, T. and Barouki, R. *Toxicol Lett* 98 , 13-23
18. Feilleux-Duche, S., Garlatti, M., Aggerbeck, M., Poyard, M., Bouguet, J., Hanoune, J. and Barouki, R. (1993) Feilleux-Duche, S., Garlatti, M., Aggerbeck, M., Poyard, M., Bouguet, J., Hanoune, J. and Barouki, R. *Am J Physiol* 265 , C1298-1305

Mimosine's Mechanism is Pyridoxal-Phosphate Independent

Emia Oppenheim and Patrick J. Stover
Cornell University, USA 14853

Summary: Mimosine is a toxic plant amino acid and a structural analog of both pyridoxal-phosphate (PLP) and tyrosine (figure 1) (1). Mimosine has been described as an effective inhibitor of DNA replication in mammalian cells (2). Several mechanisms have been proposed to explain mimosine's inhibitory effects; but none have been reliably confirmed. Our research disproves one of the current theories that mimosine acts as an inhibitor of serine hydroxymethyltransferase (SHMT) and thereby inhibits purine nucleotide biosynthesis in mammalian cells. Mimosine has been shown to decrease purine nucleotide pools in Chinese Hamster Ovary (CHO) cells, and this has been suggested to be the primary inhibitory effect of mimosine on mammalian cells. It has been suggested that mimosine limits SHMT activity, a PLP-dependent enzyme that is critical for the synthesis of DNA precursors, and this in turn results in the depletion of cellular deoxyribonucleotide purine pools. The results of this study indicate that while mimosine does inhibit human MCF-7 cell proliferation, it does not deplete purine or pyrimidine deoxyribonucleotide pools in MCF-7 cells. Mimosine's effects appear PLP independent with respect to SHMT inhibition.

Introduction: The plant species *Leucaena* contains the toxic non-protein amino acid mimosine (figure 1). Since the early 1990's, mimosine has been actively utilized to arrest the cell-cycle of mammalian cell cultures.

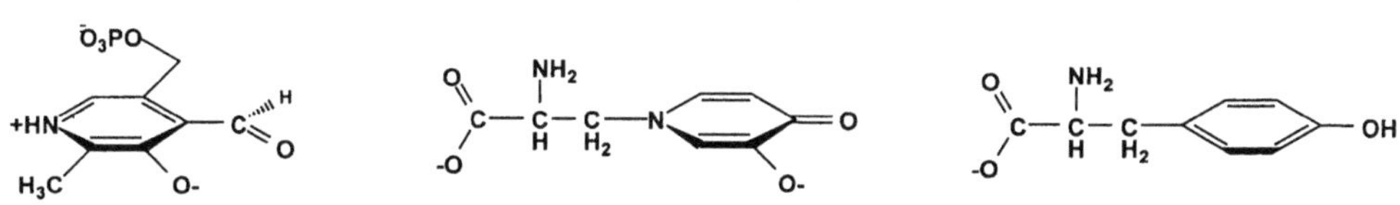

Pyridoxal-Phosphate **Mimosine** **Tyrosine**

Figure 1: Chemical Structure of Mimosine.

Recent studies have demonstrated that mimosine specifically crosslinks to mitochondrial SHMT in crude CHO cell extracts, suggesting that mimosine may target SHMT (3). SHMT is a PLP-dependent enzyme that catalyzes the reversible interconversion of serine and tetrahydrofolate to glycine and methylenetetrahydrofolate. This reaction generates single carbon units in the form of methylenetetrahydrofolate that are required for the synthesis of purines and thymidine, and is also required to remethylate homocysteine to methionine.

Several mechanisms have been proposed describing the role of mimosine in inhibiting DNA replication. Mimosine has been suggested to deplete dNTP pools by chelating intracellular iron. Chelation of intracellular iron would be expected to inhibit the iron-dependent enzyme, ribonucleotide reductase, thereby depleting dNTP pools (4). Alternatively, mimosine, a PLP antagonist, has been proposed to inhibit SHMT. This mechanism assumes that inhibition of SHMT would depress the folate-dependent synthesis of purines and deplete purine deoxyribonucleotide and dTTP pools. This would then inhibit cell-cycle progression. Finally, another mechanism suggests that mimosine depletes dNTP pools resulting in the induction of p21 expression, a cell-cycle dependent kinase inhibitor. It has been proposed that the induction of p21, and not the depletion of purine deoxyribonucleotide pools, directly causes cell-cycle inhibition (5). All of these mechanisms assume that mimosine depletes purine dNTP pools.

There are two isoforms of SHMT, one in the cytosol (cSHMT) and one in the mitochondria (mSHMT). The mSHMT isozyme, in conjunction with the mitochondria, is essential for the production of one-carbon units for intracellular reactions (6). The role cSHMT has yet to be defined but it may play a role in the regulation of folate metabolism. The cSHMT isozyme is poised in the folate pathway to regulate the flux of one-carbon units available for purine nucleotide synthesis, thymidine synthesis and homocysteine remethylation (7). A recent study found mimosine bound a 50 kDa protein later identified as mSHMT (3). Therefore, mimosine's effect on purine deoxyribonucleotide pools in CHO cells may result from an inhibition of ribonucleotide reductase and/or SHMT. In this study, we investigated the effects of mimosine on dNTP pools and purine ring biosynthesis in human MCF-7 cells. The results of this study suggest that although mimosine inhibits MCF-7 cell proliferation, it does not influence dNTP pools in these cells. These studies suggest that mimosine does not inhibit cell proliferation by depleting purine deoxyribonucleotide biosynthesis, as previous studies have suggested.

Materials and Methods:

Growth Studies/Hypoxanthine Rescue Studies: The efffect of mimosine on MCF-7 cell proliferation was determined by measuring [^{3}H]-thymidine incorporation into DNA. MCF-7 cells were cultured in α-MEM with 10% fetal bovine serum, at 37°C in a 5% CO_2 incubator. The cells were plated at 20% confluence and treated with mimosine (final concentration of 350 μM). At 24, 48, 72 and 96 h intervals, [^{3}H]-thymidine was added to the culture media (1 μCi/ml) and the cells were cultured for an additional 12 h. After labeling, the cells were washed 2x with 1 ml of phosphate-buffered saline and fixed with 500 μl of 10% trichloroacetic acid. The fixed cells were washed with 0.5 ml of 95% ethanol and resuspended with 0.3 ml of 0.2M NaOH. The amount of [^{3}H] incorporation into DNA was quantified using a liquid scintillation counter. The amount of label incorporated into DNA was correlated to the rate of cell growth. For the hypoxanthine rescue studies, the growth studies were repeated in the presence of mimosine and 1 mM hypoxanthine.

Determination of intracellular dNTP and NTP pools by HPLC : Human MCF-7 cells were cultured with α-MEM media lacking purines, pyrimidines, nucleotides and deoxyribonucleotides for 24 h. The following day, the media were replaced with media containing 350 μM mimosine and the cells were cultured for an additional 24 h. The cells were harvested by trypinization and the cell pellets were washed with 3 ml of phosphate-buffered saline. The cells were lysed with the addition of 0.5 ml 0.6M TCA, at 4°C. The lysed cells were incubated on ice for 20 min and the solution was clarified by centrifugation. Cell pellets were set aside for protein quantification by the method of Lowry. The supernatants were diluted with an equal volume of ice-cold freon containing 0.5M trioctylamine. The solution was vortexed and centrifuged. The solution formed two layers. The upper phase contained the nucleotides and was removed, lyophilized and stored at –70°C. When needed the lyophilized pellets were resuspended in 250 μl of 0.2M ammonium phosphate, pH 6.0. The samples were filtered through 0.2 μm acrodisc filter prior to HPLC analysis. NTPs and dNTPs were separated and quantified using a Shimadzu HPLC (LC-10AS) equipped with a C_{18} column as described previously (8). Absorbance was followed at 260nm with 4 nm bandwidth. Buffer A consisted of 0.2M $NH_4H_2PO_4$ in 1M KCl (pH 5.35) and buffer B was 0.2M $NH_4H_2PO_4$ with 1.25 M KCl and 10% methanol (pH 5). Samples were applied to the column pre-equilibrated with 100% buffer A at a flow rate of 0.8 ml/min. Fifteen minutes following the injection, a linear gradient to 100% buffer B was initiated. After 35 min, the system returned to 100% buffer A and the column was re-equilibrated for ten minutes in preparation for the next injection. The identification and concentration of nucleotides in each sample were determined from standard curves generated from nucleotide standards. All values were normalized to protein concentration.

Results / Discussion: The effects of mimosine on purine nucleotide biosynthesis was investigated in the terminally differentiated MCF-7 breast adenocarcinoma cell line. Initially, the effects of mimosine on MCF-7 cell proliferation was characterized (figure 2). Figure 2 shows that mimosine inhibits MCF-7 cell proliferation, consistent with previous studies demonstrating that mimosine inhibits CHO cell proliferation (9). MCF-7 cells are also inhibited by deferroxamine, a well-characterized iron chelator. Figure 2 also shows that hypoxanthine supplementation in the culture media does not rescue the MCF-7 cells. Purine nuclotide biosynthesis can occur through the SHMT independent salvage pathway in these cells, and supplementation of the media with hypoxanthine did not overcome the inhibitory effects of mimosine. These results indicate that inhibition of purine ribonucleotide biosynthesis is not the mechanism through which mimosine inhibits cell growth, and therefore inhibition of SHMT is not likely the mechanism through which mimosine inhibits cell growth.

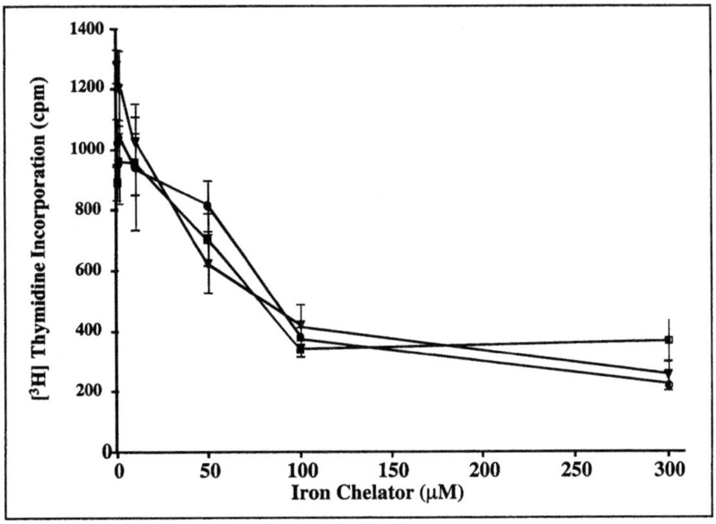

Figure 2. Effect mimosine or deferrioxamine on MCF-7 cell proliferation. The effect of mimosine (squares), deferrioxamine (triangles) and mimosine with 1 mM hypoxanthine (circles) exposure on the incorporation of $[^3H]$-thymidine into MCF-7 DNA was determined as described in *Materials and Methods*.

These studies were then extended to determine if mimosine inhibited the synthesis of deoxyribonucleotides in MCF-7 cells as previously demonstrated for CHO cells (9). We determined the levels of ribonucleotides and deoxyribonucleotides in MCF-7 cells as a function of mimosine exposure by HPLC (table I).

Table 1. Effect of 24 h mimosine exposure (350 μM) on nucleotide pools in MCF-7 cells. All values are expressed as pmoles nucleotide per μg protein.

	ATP + ADP	dATP	GTP + GDP	CTP	dTTP
Without mimosine	78 ± 1.5	1.8 ± 0.1	45 ± 17	42 ± 3	2.9 ± 0.2
With mimosine	65 ± 0.2	2.2 ± 0.1	68 ± 20	45 ± 3	1.7 ± 0.7

These results clearly demonstrate that a 24 h mimosine exposure does not dramatically alter purine nucleotide pools in MCF-7 cells, as has been demonstrated in CHO cells. The lack of change in adenine and guanine ribonucleotide pools confirms that inhibition of purine biosynthesis is not the mechanism through which mimosine inhibits cell proliferation. Additionally, the lack of change in dATP concentrations suggest that mimosine does not inhibit ribonucleotide reductase in MCF-7 cells. We were unable to accurately measure dGTP levels due to the sensitivity of our assay. Measurement of nucleotide pools following 72 h of mimosine exposure yielded similar results, although the experimental values contained more variability due to the increase in the dead cell population. These results suggest that mimosine does not influence nucleotide pools in MCF-7 cells as previously observed for CHO cells. These results also suggest that the effect of mimosine on CHO nucleotide pools is independent of its effects on cell proliferation.

In summary, our findings suggest that mimosine does inhibit MCF-7 cell proliferation, but the mechanism of inhibition is independent of nucleotide biosynthesis. The current literature suggests that purine pool depletion is an essential step in the overall mechanism for mimosine's effect on cell proliferation. Mimosine is a PLP antagonist, which could potentially inhibit the PLP-dependent enzyme, SHMT, and thereby deplete purine nucleotide pools. However, mimosine did not alter purine pools or requirements in MCF-7 cells. These results suggest that mimosine's mode of action is as of yet undefined.

References:

1. Tsai, W. C. and Ling, K. H. Toxic Action of Mimosine. *Toxicon* 9, 241. 1971.
2. Mosca, P. Dijkwel P. Hamlin J. The Plant Amino Acid Mimosine May Inhibit Initiation at Origins of Replication in Chinese Hamster Cells. *Molecular and Cellular Biology* 12(10), 4375-4383. 1992.

3. Ji, C., Marnett, L. J., and Pietenpol, J. Cell Cycle re-entry following chemically-induced cell cycle synchronization leads to elevated p53 and p21 levels. *Oncogene* 15, 2749-2753. 1997.
4. Kalejta, R. and Hamlin, J. The dual effect of mimosine on DNA replication. *Experimental Cell Research* 231, 173-183. 1997.
5. Girgis, S., Suh, J. R., Jolivet, J., and Stover, P. J. 5-formyltetrahydrofolate regulates homocysteine remethylation in human neuroblastoma. *The Journal of Biological Chemistry* 272, 4729-4734. 1997.
6. Girgis, S., Nasrallah, I., Suh, J. R., Oppenheim, E., Zanetti, K., Mastri, M., and Stover, P. J. Molecular cloning, characterization and alternative splicing of the human cytoplasmic serine hydroxymethyltransferase gene. *Gene* 210, 315-324. 1998.
7. Lin, H., Falchetto, R., Mosca, P., Shabanowitz, J., Hunt, D., and Hamlin, J. Mimosine targets Serine Hydroxymethyltransferase. *The Journal of Biological Chemistry* 271, 2548-2556. 1996.
8. James, S., Cross, D., and Miller, B. Alterations in nucleotide pools in rats fed diets deficient in choline, methionine and/or folic acid. *Carcinogenesis* 13(12), 2471-2473. 1992.
9. Gilbert, D., Neilson, A., Miyazawa, H., DePamphilis, M., and Burhans, W. Mimosine arrests DNA synthesis at replication forks by inhibiting deoxynucleotide metabolism. *The Journal of Biological Chemistry* 270, 9597-9606. 1995.

Biochemistry and Molecular Biology of Vitamin B_6
and PQQ-dependent Proteins
ed. by A. Iriarte, H.M. Kagan and M. Martinez-Carrion

Environmental Stimuli and Regulatory Factors Affecting the Expression of the Glutamic Acid Decarboxylase System in *Escherichia coli*

Angela Tramonti, Paolo Visca*, Francesco Bossa and Daniela De Biase

Dipartimento di Scienze Biochimiche "A. Rossi Fanelli" and Centro di Biologia Molecolare del Consiglio Nazionale delle Ricerche, Università di Roma "La Sapienza", 00185 Roma, and *Dipartimento di Biologia, Università di Roma Tre-IRCCS "Lazzaro Spallanzani", 00146 Roma, Italy.

Summary

In enterobacteria the *gad* system, which includes genes encoding PLP-dependent glutamic acid decarboxylases, is required to overcome the acid stress. Herein, we provide evidence that the *gad* system is strongly induced also during anaerobic growth and suggest the possible involvement in *gad* transcriptional control of two AraC-like proteins. Similarly to other decarboxylation/antiport systems, the *gad* system would confer a unique growth advantage under anaerobic conditions by generating an electrogenic potential which can be used to fuel ATP synthesis.

Introduction

In their natural habitats, pathogenic bacteria are faced with a wide array of environmental stresses. Understanding how they survive is not only a fundamental issue of biology, but is also crucial to gain insight into the mechanisms of pathogenicity, since bacteria with impaired stress response are also less virulent (1). Acidity is one of the most frequently encountered hostile condition for enteric bacteria. In humans, ingested bacteria are faced with transient acid conditions since: i) typically the residence time in stomach (pH <3) is 2 hours before reaching the gut and, ii) these bacteria must also endure an acidic stress within the lower intestine for the presence of volatile fatty acids (butyrate, propionate, acetate) produced from sugar fermentation. Thus, the ability to sense and respond to potentially lethal effects of low pH is essential for persistence of the intestinal microflora. It has recently been demonstrated that the *gad* system confers to the enteric bacteria *Escherichia coli*, *Shigella flexneri* and *Lactococcus lactis*, the ability to overcome the acid stress (2-4). Genes of the *gad* system include *gadA* and *gadB*, encoding two biochemically identical isoforms of the PLP-dependent glutamic acid decarboxylase (GadA and GadB), and *gadC*, which has been proposed to encode a putative antiporter implicated in the export of γ-aminobutyrate, the glutamic acid decarboxylation product. We investigated the regulation of *gad* during the *E. coli* growth cycle under different environmental conditions, and demonstrated that transcription increases at the entry into the stationary growth phase under aerobic conditions (2). At this stage, *gad* mRNAs are more abundant when bacteria are grown in acidic, hypo- and hyper-osmolar environments. By combining RNA blot analyses and 5'-mapping of *gad* transcripts, we found that *gadA*, *gadB* and *gadC* can be independently transcribed and that a polycistronic *gadBC* mRNA can also originate *in vivo* from the *gadB* promoter. We also found that H-NS, a histone-like protein known to act in the *E. coli* nucleoid as a global *trans*-acting transcriptional silencer, regulates *gad* expression (2). A correlation exists between the response of enterobacteria to the acidic stress and the onset of the pathogenic process, and there are clear evidences that the stationary phase factor σ^S affects both acid resistance and the expression of certain virulence genes (5). We suggest that H-NS indirectly regulates *gad* expression, through the post-transcriptional control of σ^S expression (6). We observed that *gad* transcription is confined to the stationary phase in the *hns* proficient background, and an insertional mutation in the *rpoS* gene abrogates *gad* expression during the stationary phase. In addition, the induction of *gad* occurring during the exponential growth in a *hns* mutant is completely reversed in a *hns rpoS* double mutant, independently of the growth stage. Based on our results, either RpoS is directly involved in *gad* transcription or it can be required for the expression of AppY-like (7) or CadC-like (8) regulatory protein(s) responsible for *gad* activation. Moreover, generation of individual *gadA*, *gadB* and *gadC E. coli* knockouts (2,3) demonstrated that expression of both decarboxylating enzyme (GadA or GadB) and cognate antiporter (GadC) is required to overcome the acid stress.

In the present report we provide evidence that *gad* genes expression is strongly induced during anaerobic growth. The possible involvement in *gad* transcriptional control of two ORFs, homologous to the AraC family of transcriptional regulators, located in the same orientation downstream of *gadA*, is also discussed.

Materials and Methods

Construction of gad::lacZ fusions and β-galactosidase assay. Translational fusions were generated for both *gad*A and *gadB* genes by cloning into the vector pMC1403 (9) PCR-generated fragments spanning from position +50 (codon 17) to positions -15, -96, and -203, relative to the ATG translation start codon of each gene. Primers used for PCR amplification were designed to allow directional cloning at the *Eco*RI-*Bam*HI sites of pMC1403, in frame with the *lacZ* gene (2). Plasmids pBs*gadA* and pBs*gadB* (10) were used as templates for PCR amplification of the *gad* promoter regions. Designation of plasmids carrying the *gad::lacZ* fusions are given by the gene name, followed by positions of first/last nucleotide of the cloned fragment, relative to the ATG start codon of *gadA* or *gadB*. Since transcription begins at position -27, relative to the ATG (2), the *gad*-15/+50::*lacZ* fusions were used to detect basal levels of β-galactosidase expression. The β-galactosidase activity assays were carried out according to Miller (11) and expressed as follows: 1,000 × OD_{420}/(OD_{600} culture× reaction time × volume).

Results and Discussion

As a premise for a better understanding of the *gad* regulatory network, the *gad::lacZ* fusions were used to determine a hierarchy of *gad* inducing environmental stimuli. The conditions investigated included growth under aerobic conditions in standard LB medium, pH 7.4, at different temperatures (30°C, 37°C, or 42°C), at different pHs (50 mM phosphate buffer, pH 4, 7.1 or 9), as well as in the absence of NaCl or in the presence of twice the standard NaCl concentration. Moreover, *gad* expression in standard LB medium supplemented with glutamate, aspartate or glutamine (10 mM) and under anaerobiosis at 37°C, was also investigated. β-Galactosidase assays on exponential cultures grown under the above conditions indicate that reporter activity increases upon growth in the absence of NaCl, under anaerobiosis and at acidic pH (Fig. 1). The other conditions tested did not alter the reporter gene activity compared with the standard growth condition (data not shown).

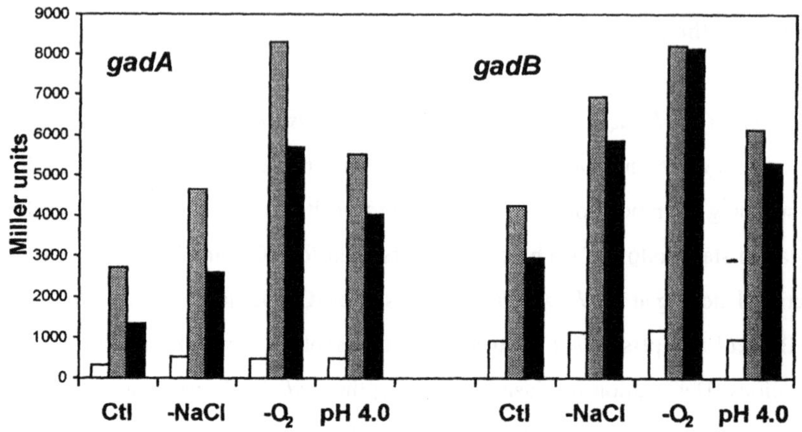

Figure 1. β-Galactosidase (LacZ) activity directed by different *gad::lacZ* translational fusions in *E. coli*. White bars: *gad*-15/+50::*lacZ*; grey bars: *gad*-96/+50::*lacZ*; black bars: *gad*-203/ +50::*lacZ*. Different growth conditions are indicated for both *gadA* and *gadB*. Ctl: control.

As in the case of *gad*, both arginine (*adi*) and the lysine (*cad*) decarboxylase systems are strongly induced at low pH under anaerobic conditions (12). Upstream of the *cadBA* operon and downstrem of *adiA*, genes have been identified coding for cognate transcriptional activators. CadC is the response regulator which acts on *cadBA* (8), while AdiY is an AraC-like protein which stimulates AdiA expression (13). Interestingly, sequence analysis of two ORFs located downstream of *gadA* predicts the presence of two AraC-like transcriptional regulators (*yhiX* and *yhiW*; 14). We have aligned their predicted protein products, YhiX and YhiW, with H-NS-regulated members of the AraC family of transcriptional activators, including AdiY, AppY and VirF (Fig. 2).The consensus sequence for the helix-turn-helix motif (15), involved in DNA binding, is also conserved in YhiX and YhiW. The identification of these AraC-like genes closely downstream of *gadA* confers an additional level of complexity to the *gad* regulatory network. Our findings raise the possibility that H-NS repression on the *gad* system could occur indirectly, likely by binding of this global regulatory protein to the control regions of *yhiX* and *yhiW*, as previously demonstrated for *appY* (6) and *virF* (16).

```
YhiX            MQSLHGNCL IAYARHKYIL TMVNGEYRY. ..FNGGDLVF ADASQIRVDK
YhiW               MTHVCS VILIRRS..F DIYHEQQKIS LHNESILLLE KNLADDFAFC
AdiY            MRICSDQPC IVLLTEKDVW IRVNGKEPIS LKANHMALLN CENNIIDVSS
AppY               MDYVCS VVFICQS..F DLIINRRVIS FKKNSLFIVS DKIRRELPVC
VirF MMDMGHKNKI DIKVRLHN.Y IILYAKRCSM TVSSGNETLT IDEGQIAFIE RNIQINVSIK

YhiX CVENF.... VFVSRDTLSL FLPMLKEEAL NLHA.HKKVS SLLVHHCSRD IPVFQEVAQL
YhiW SPDTR....R LDIDELTVCH YLQNI.RQLP RNLGLHSKDR LLI....NQS PPMPLVTAIF
AdiY LNNTL....V AHISHDIIKD YLRFLNKDLS QIPVWQRSAT PILTLPCLTP .DVFRVAAQH
AppY PSKLR....I VDIDKKTCLS FFIDVNNELP GKFTL.DKNG YIA....EEE PPLSLVFSLF
VirF KSDSINPFEI ISLDRNLLLS IIRIMEPIYS FQHSYSEEKR GLNKKIFLLS EEEVSI.DLF

YhiX SQNKNLRYAE MLRKRALIFA LLSVFLEDEH FIPLLLNVLQ PNMRTRVCTV INNNIAHEWT
YhiW DSFNESGVNS PILSNMLYLS CLSMFSHKKE LIPLLFNSI. STVSGKVERL ISFDIAKRWY
AdiY SMMPAETESE KERTRALLFT VLSRFLDSKK FVSLMMYMLR NCVSDSVYQI IESDIHKDWN
AppY EGIKIADSHS LWLKERLCIS LLAMFKKRES VNSFILTNI. NTFTCKITGI ISFNIERQWH
VirF KSIKEMPFGK .RKIYSLAC  LLSAVSDEEA LYTSISIASS LSFSDQIRKI VEKNIEKRWR

YhiX LARIASELLM SPSLLKKKLR EEETSYSQLL TECRMQRALQ LIVIHGFSIK RVAVSCGYHS
YhiW LRDIAERMYT SESLIKKKLQ DENTCFSKIL LASRMSMARR LLELRQIPLH TIAEKCGYSS
AdiY LSMVASCLCL SPSLLKKKLK SENTSYSQII TTCRMRYAVN ELMMDGKNIS QVSQSCGYNS
AppY LKDIAELIYT SESLIKKRLR DEGTSFTEIL RDTRMRYAKK LITSNSYSIN VVAQKCGYNS
VirF LSDISNNLNL SEIAVRKRLE SEKLTFQQIL LDIRMHHAAK LLLNSQSYIN DVSRLIGISS
                           T....I. ...R...A.. LL......I. DIA...GF.S

YhiX VSYFIYVFRN YYGMTPTEYQ ERSAQRLSNR DSAASIVAQG NFYGTDRSAE GIRL
YhiW TSYFINTFRQ YYGVTPHQFA QHSPGTFS
AdiY TSYFISVFKD FYGMTPLHYV SQHRERTVA
AppY TSYFICAFKD YYGVTPSHYF EKIIGVTDGI NKTID
VirF PSYFIRKFNE YYGITPKKFY LYHKKF
     ..YF...F.. ..G.TPS..R
```

Figure 2: Sequence alignment of related transcriptional regulators performed using PILEUP (GCG package). Gaps are indicated by dots. Amino acid residues conserved in the five proteins are boldfaced. The consensus sequence found in the C-terminal region in at least 50% of AraC-like proteins (15) is given below the alignment.

Why is the *gad* system so important for enteric bacteria? The most obvious answer relies on the maintainance of cellular pH homeostasis, an issue which has been addressed in detail elsewhere (2). However, acid resistance is not the only functional role played by the *gad* system. We have shown that Gad expression in *E. coli* responds to a variety of environmental stimuli, including anaerobiosis. The strong increase in Gad (reported as LacZ activity) under anaerobic conditions represents a novel finding. Anaerobiosis is well known to generate acidification of the environment due to volatile fatty acids production through fermentative metabolism. Components of the intestinal microflora facultatively or obligately grow under anaerobic conditions with severe limitations in their ability to drive ATP synthesis. It has recently been observed that resting *Lactobacillus* cells can generate ATP from amino acid decarboxylation/antiport systems (17). Amino acid decarboxylation is coupled with proton extrusion, thereby generating an electrogenic potential required for the F_0F_I-(H^+)-ATPase activity. In an acid environment CO_2 levels are reduced, due to chemical equilibrium, and ATP synthesis under O_2-limiting conditions requires to be coupled to alternative electrogenic systems. The increase in Gad activity under anaerobic conditions and at low pH would satisfy the CO_2 growth demand and, combined to the antiporter (GadC) action, contribute to generate an electrogenic gradient useful to fuel ATP synthesis. If this is applied to the facultative anaerobes *E. coli* and *Shigella*, the *gad* system would certainly confer a unique advantage for anaerobic growth, as it would make possible the formation of an electrogenic potential under conditions in which respiratory chain enzymes would be inactive. Under this point of view, the presence of glutamate decarboxylases in phylogenetically distant bacteria comprising the normal gut microflora provides an attractive explanation for the multiple roles played by the *gad* system.

References

1) Mahan, M.J., Slauch, J.M. and Mekalanos, J.J. (1996). In *Escherichia coli* and *Salmonella*. Cellular and molecular biology, 2nd Ed (Neidhart F.C.ed.) pp. 2803-2815, ASM Press, Washington D.C.
2) De Biase, D., Tramonti, A., Bossa, F. and Visca, P. (1999) *Mol. Microbiol. 32*: 1198-1211.
3) Castanie-Cornet, M., Penfound, T.A., Smith, D., Elliott, J.F. and Foster, J.W. (1999). *J. Bacteriol. 181*:3525-3535.
4) Sanders, J.W., Leenhouts, K., Burghoorn, J., Brands, J.R., Venema, G. and Kok, J. (1998) *Mol. Microbiol. 27*: 299-310.
5) Bearson, S., Bearson, B. and Foster, J.W. (1997) *FEMS Microbiol. Lett. 147*: 173-180.
6) Barth, M., Marschall, C., Muffler, A., Fischer, D. and Hengge-Aronis, R. (1995) *J. Bacteriol. 177*: 3455-3464.
7) Atlung, T., Sund, S., Olesen, K. and Brønsted, L. (1996) *J. Bacteriol. 178*: 3418-3425.
8) Watson, N., Dunyak, D.S., Rosey, E.L., Slonczewski, J.L. and Olson, E.R. (1992) *J. Bacteriol. 174*: 530-540.
9) Casadaban, M.J., Chou, J. and Cohen, S.N. (1980) *J. Bacteriol. 143*: 971-980.
10) De Biase, D., Tramonti, A., John, R.A. and Bossa, F.(1996) *Protein. Expr. Purif. 8*: 430-438.
11) Miller, J.H.(1972) Experiments in Molecular Genetics. Cold Spring Harbor Laboratory Press, Cold Spring Harbor, New York.
12) Auger, E.A., Redding, K.E., Plumb, T., Childs, L.C., Meng, S.Y. and Bennett, G.N (1989) *Mol. Microbiol. 3*: 609-620.
13) Stim-Herndon, K.P., Flores, T.M. and Bennett, G.N. (1996) *Microbiology 142*: 1311-1320.
14) Sofia,H.J., Burland, V., Daniels, D.L., Plunkett III, G. and Blattner, F.R. (1994) *Nucleic Acids Res. 22*: 2576-2586.
15) Gallegos, M-T., Michan, C. and Ramos, J.L. (1993) *Nucleic Acids Res. 21*: 807-810.
16) Prosseda, G., Fradiani, P.A., Di Lorenzo, M., Falconi, M., Micheli, G., Casalino, M., Nicoletti, M. and Colonna, B. (1998) *Res. Microbiol. 149*: 15-25.
17) Higuchi, T., Hayashi, H. and Abe, K. (1997) *J. Bacteriol. 179*: 3362-3364.

PQQ AND QUINOPROTEINS: BIOLOGY AND REGULATION

Biochemistry and Molecular Biology of Vitamin B_6
and PQQ-dependent Proteins
ed. by A. Iriarte, H.M. Kagan and M. Martinez-Carrion

The membrane glucose dehydrogenase of *Escherichia coli*

C. Anthony, R.A. Salleh, P.L. James and G.E. Cozier

School of Biological Sciences, University of Southampton, Southampton, UK, SO16 7PX.

Summary

The glucose dehydrogenase in the membrane of *E. coli* is a PQQ-containing quinoprotein dehydrogenase which oxidises glucose to the gluconolactone and thence gluconic acid in the periplasm. It is produced as the apoenzyme requiring incubation with PQQ and Mg^{++} for production of active holoenzyme. Its structure has been previously modelled on that of methanol dehydrogenase. Results of studies of mutant forms of the enzyme have confirmed the validity of the model and tested ideas about the mechanism and structure of these quinoprotein dehydrogenases.

Introduction

The membrane-bound glucose dehydrogenase (GDH) of *Escherichia coli* is a quinoprotein having pyrroloquinoline quinone (PQQ) as its prosthetic group (1, 2, 3, 4). It catalyses the oxidation of the pyranose form of D-glucose (at the C-1 position), and other monosaccharides to the lactone. The reaction occurs in the periplasm and the electron acceptor is ubiquinone in the membrane. A divalent metal (usually Mg^{2+}) is required for reconstitution of active holoenzyme from apoenzyme plus PQQ. The N-terminal region forms a membrane anchor with five transmembrane segments. The periplasmic region shows 26% identity of sequence with the α subunit of methanol dehydrogenase (MDH), the X-ray structure of which (5) has been used to produce a model of GDH (6). This model has been the basis of a number of studies using mutagenesis for investigating the process of production of active enzyme from apoenzyme, the mechanism, and some aspects of its structure (7-12). Production of mutant MDH is difficult in methylotrophs and GDH provides an appropriate enzyme for study by mutagenesis of this type of quinoprotein dehydrogenase. This is a brief review of the structure and function of GDH and of the properties of some of these mutants that have been used to test proposals relating to this type of PQQ-containing quinoprotein dehydrogenases.

The propeller fold superbarrel structure of GDH and MDH

The periplasmic part of GDH is a superbarrel made up of eight topologically identical four-stranded, twisted antiparallel β sheets (W-shaped), stacked radially around a pseudo eight-fold symmetry axis. This arrangement is known as a propeller fold, each β sheet motif representing a propeller blade (Figure 1). A feature likely to have a special function in maintaining the structure of the superbarrel is the final β sheet motif (W8) which is made up of strands A, B and C from the C-terminus plus the D strand from the N-terminus. Mutation in this C8 strand has been shown to alter the stability of GDH (11).

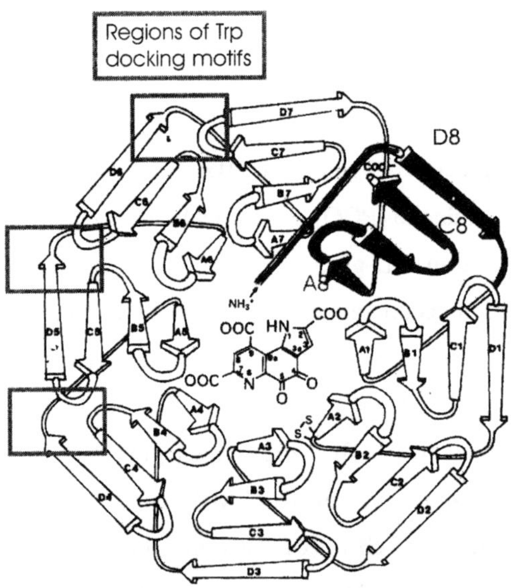

Figure 1. The topology of the periplasmic domain of GDH (6). The dark part is the W8 β-sheet which is made up of a D strand from the N-terminus, with A, B and C strands from the C-terminus. This figure also shows some of the regions of the 'tryptophan docking motifs'.

The 'propeller blades' are held together by a series of tryptophan-docking motifs involving eight 11-residue consensus sequences. The tryptophan at position 11 is stacked between alanine at position 1 of the same motif and the peptide bond between residue 6 and an invariant glycine (position 7) in the next motif. For example, motifs W6 and W7 are held together by Ala674 and Trp684 in W6 and Gly741 in W7 (Figure 2).

As predicted, mutation of Gly741 to serine leads to unstable and far less active enzyme (7). The motif W7 is in the most conserved region of sequence of these enzymes and other mutations in this region lead to production of inactive enzyme (eg mutation of Asp730) (7).

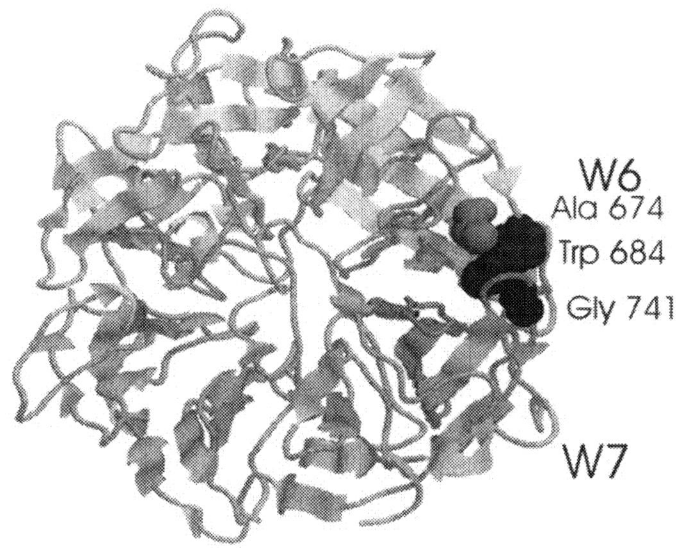

Figure 2. One of the tryptophan docking interactions in GDH. The components of one tryptophan docking motif are shown in spacefill mode, coloured dark. In this example, Trp684 (of W6) is held between Ala674 (W6) and Gly741 (W7).

The histidines in the active site of GDH

Figure 3 shows the way in which the PQQ is held between the coplanar Trp404 and His262 in the model GDH; when viewed from the outside the two histidine residues are seen partly to obscure the entrance to the active site. In the X-ray structure of MDH the PQQ is 'sandwiched' between tryptophan and a novel disulphide ring which had been proposed to be involved in electron transport from PQQ to the cytochrome electron acceptor. Mutation of His262 to tyrosine, however, led to no change in reaction with the ubiquinone electron acceptor in GDH (12). Similarly, there was little change in interaction with electron acceptor (phenazine ethosulphate) in the normal dye-linked assay system; the mutation did however lead to a large decrease in affinity for glucose and other substrates, consistent with its position in the entrance to the active site. The mutation led to no change in affinity for PQQ during reconstitution. By contrast, mutation of the second histidine at the entrance to the active site (His775) led to a large decrease in affinity for PQQ during reconstitution (H775R and H775A) and a large decrease in activity, although affinity for glucose was not markedly altered. remarkably, other mutations in His775 (H775N and H775D) led to an 'improved specificity' for glucose with a higher activity with this substrate (7).

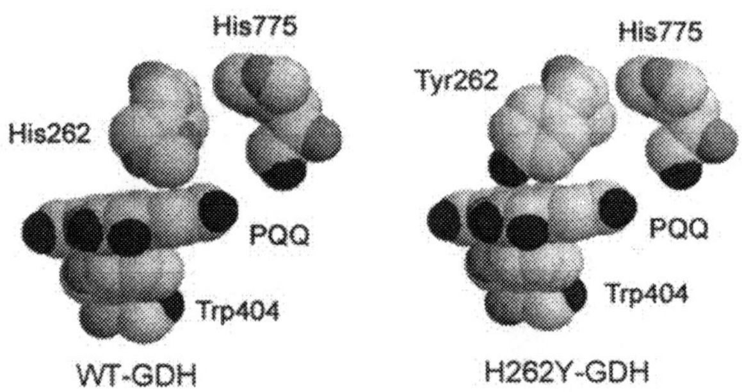

Figure 3. PQQ bonded between Trp404 and His262 in the active site. The H262Y mutant GDH is also shown. Both histidines are seen when the entrance to the active site is viewed from outside the molecule (6,12).

Mutations affecting the mechanism of GDH

Figure 4 shows the active site in the 'model' GDH (6). A Ca^{2+} ion is present in the active site of MDH (5) but it is possible that Mg^{2+} replaces Ca^{2+} in GDH. By analogy with the proposed MDH mechanism it is proposed that the reaction with substrate is initiated by proton abstraction (from the C1 anomeric hydroxyl) by Asp466; this is followed by either hydride or oxyanion transfer to the C5 carbon of PQQ. This is thought to be facilitated by the interaction of the Ca^{2+} with the oxygen of the C5 carbonyl group.

Mutation of Asp466 to asparagine leads to complete loss of activity, consistent with its proposed essential role in the initial proton abstraction process (8). Thr424 is proposed to play a role in binding the active site metal, and this is consistent with the large decrease in activity, and the altered metal ion specificity that is seen when this is mutated to asparagine (9).

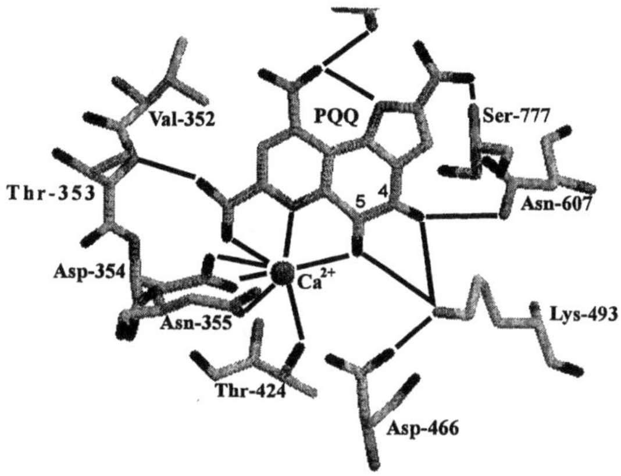

Figure 4. The active site of GDH. Although a Ca^{2+} ion is shown, it is possible that this is replaced by a Mg^{2+} ion.

Asn607 is shown to form a hydrogen bond with the C4 carbonyl oxygen of PQQ. Mutation of this residue to alanine, which would prevent any interaction, led to a small decrease in activity (to 33%), indicating, as expected, that this interaction is not crucial for activity (8).

As shown in Figure 4, Lys493 interacts with both carbonyl oxygen atoms of PQQ; this lysine replaces arginine in the MDH structure. When Lys494 was mutated to arginine in GDH its activity, and its affinity for PQQ during reconstitution, was decreased markedly (8). Spectroscopic measurements indicated that the reduction of PQQ by substrate is unaffected and that the mutation has its effect by preventing subsequent oxidation of the reduced quinol form of PQQ. This suggests that Lys494 may lie on a direct electron transfer pathway from PQQ to ubiquinone in GDH (8).

Acknowledgements

The work cited from the Authors' laboratory was supported by the BBSRC and The Wellcome Trust

References

1. Matsushita, K., and O. Adachi (1993a) Bacterial quinoproteins glucose dehdyrogenase and alcohol dehydrogenase. In "Principles and application of quinoproteins" (ed. V.L. Davidson) pp. 47-63, Marcel Dekker Inc, New York.

2. Anthony, C. and Ghosh, M. (1998) The structure and function of the PQQ-containing quinoprotein dehydrogenases. *Prog. in Biophys. Mol. Biol. 69:*, 1-21.

3. Goodwin, P.M. and Anthony, C. (1998) The biochemistry, physiology and genetics of PQQ and PQQ-containing enzymes. *Adv. Microb. Physiol. 40:* 1-80.

4. Anthony, C. (1996) Quinoprotein-catalysed reactions. *Biochem. J. 320*: 697-711.

5. Ghosh, M., Anthony, C., Harlos, K., Goodwin, M.G. and Blake, C.C.F. (1995) The refined structure of the quinoprotein methanol dehydrogenase from *Methylobacterium extorquens* at 1.94 A. *Structure 3*: 177-187.

6. Cozier, G.E. and C. Anthony (1995) Structure of the quinoprotein glucose dehydrogenase of *Escherichia coli* modelled on that of methanol dehydrogenase from *Methylobacterium extorquens*. *Biochem.J. 312:* 679-685.

7. Yamada, M., Inbe, H., Tanaka, M., Sumi, K., Matsushita, K. and Adachi, O. (1998) Mutant isolation of the *Escherichia coli* quinoprotein glucose dehydrogenase and analysis of crucial residues Asp-730 and His-775 for its function. *J.Biol.Chem. 273:* 22021-22027.

8. Elias, M.D., Tanaka, M., Izu, H., Matsushita, K., Adachi, O. and Yamada, M. (2000) Functions of amino acid residues in the active site of *Escherichia coli* PQQ-containing quinoprotein glucose dehydrogenase. *J. Biol. Chem. (In Press).*

9. Yoshida, Y. and Sode, K. (1997) Thr424 to Asn substitution alters bivalent metal specificity of pyrroloquinoline quinone glucose dehydrogenase. *J. Biochem. Mol Biol. & Biophys. 1:*, 89-93.

10. Yoshida, Y. and Sode, K. (1997) Improved substrate specificity and dynamic range for glucose measurement of *Escherichia coli* PQQ glucose dehydrogenase by site directed mutagenesis. *Biotechnol. Letts. 19:* 1073-1077.

11. Yoshida, Y., Kojima, K., Witarto, A. B. and Sode, K. (1999) Engineering a chimeric pyrroloquinoline quinone glucose dehydrogenase: improvement of EDTA tolerance, thermal stability and substrate specificity. *Protein Engineering 12:* 63-70.

12. Cozier, G.E., Salleh, R.A. and Anthony, C. (1999) Characterisation of the membrane glucose dehydrogenase from *Escherichia coli* and characterisation of a site directed mutant in which His262 has been changed to tyrosine. *Biochem. J. 340:* 639-647.

Biochemistry and Molecular Biology of Vitamin B_6
and PQQ-dependent Proteins
ed. by A. Iriarte, H.M. Kagan and M. Martinez-Carrion

Structural properties of homodimeric quinoprotein ethanol deyhdrogenase from *Pseudomonas aeruginosa*

H. Görisch[1], T. Keitel[2], A. Diehl[1], T. Knaute[2], Z. Dauter[3] and W. Höhne[2]

[1]Institut für Biotechnologie, Technische Universität Berlin, Berlin 13353, and [2]Institut für Biochemie, Humboldt-Universität zu Berlin, Berlin 10117, Germany, and [3]Brookhaven National Laboratory, Bldg. 725A-X9, Upton NY 11973, USA

Summary

The X-ray structure of the periplasmic quinoprotein ethanol dehydrogenase (QEDH) from *Pseudomonas aeruginosa* shows a homodimer, where each subunit forms a disk-shaped super-barrel composed of eight antiparallel twisted ß-sheets, arranged in a propeller-like fashion. The ß-propeller fold is stabilized by a repetitive Ala/Trp/Gly docking motif, which is modified in four of the eight docking contacts. The overall folding of the polypeptide is very similar to that known for the large α-subunit of the quinoprotein methanol dehydrogenase (QMDH) from methylotrophs. In QEDH phe-408 and leu-409 of the active site cavity replace the bulky trp-531 in QMDH, which results in a less well defined binding site for methanol with regard to ethanol as substrate.

Introduction

Most microbial alcohol dehydrogenases are cytoplasmic NAD(P)-dependent enzymes. Some Gram-negative bacteria in addition are able to produce periplasmic alcohol dehydrogenases which contain the prosthetic group pyrroloquinoline quinone (PQQ) [1]. The best studied example of the PQQ-dependent enzymes is the quinoprotein methanol dehydrogenase (QMDH) from methylotrophic bacteria. The enzyme is a heterotetramer of two large 60 kDa and two small 10 kDa subunits. QMDH shows a high pH-optimum above pH 9 and requires an alkylamine as activator. The enzyme oxidizes methanol and other primary alcohols, but secondary alcohols are not attacked [2]. The X-ray structures of two QMDHs from methylotrophs have been determined [3,4].

A different PQQ-dependent alcohol dehydrogenase has been isolated from ethanol grown cells of *Pseudomonas aeruginosa*. This quinoprotein ethanol dehydrogenase (QEDH) shows similar enzymatic properties as QMDH. But in contrast to the enzymes from methylotrophs it strongly discriminates between methanol and ethanol preferring ethanol as substrate [5]. Native QEDH is a homodimer of a 60 kDa subunit, which is coded for by the *exa*A gene of the ethanol oxidation gene cluster in *P. aeruginosa* [6,7]. The X-ray structure of QEDH reveals that its overall folding topology is very similar to that reported for the large subunit of QMDH.

Materials and Methods

Crystallization and data collection: QEDH from *P. aeruginosa* ATCC 17933 was purified and crystallized as described previously [5,8]. X-ray diffraction data were collected at the X31 synchroton radiation beamline of the EMBL outstation (DESY, Hamburg, Germany). Reflections were integrated, scaled and reduced by using MOSFLM [9].

Structure determination: The structure was solved by the method of molecular replacement choosing a large subunit of the heterotetrameric QMDH from *Methylophilus* W3A1 [10].

Docking calculations: The software package ICM [11] was used for docking calculations with methanol and ethanol to the crystal structures of QEDH from *P. aeruginosa* and QMDH from *Methylophilus* W3A1 [10] (PDB identification code 4aah). The alcohol molecules were placed in the active site with their hydroxyl group pointing toward the interior of the enzyme, but in a distance of about 7 Å from the putative reaction center. Monte-Carlo simulations were performed to obtain the optimal alcohol positions. The calculations were monitored by recording a stack of 50 low-energy conformations [12].

Results

The overall structure of QEDH from *P. aeruginosa* showed an 8-bladed ß-propeller superbarrel, where each propeller W_1 to W_8 was formed by four antiparallel ß-strands A, B, C, and D. Blade W_8 was a composite structure with strands A, B, and C beeing donated by the C-terminus while strand D stemed from the N-terminus of the polypeptide. The individual blades were docked together by close van-der-Waals contacts of an Ala/Trp/Gly triade, with the Ala and Trp side chains of blade W_n and a Gly back-bone of blade W_{n+1}. Modified docking contacts were found in four out of eight equivalent positions, Table 1 and Fig. 1a-d. Besides the calcium ion forming a complex with PQQ at the active site, a second calcium binding site was found at the

N-terminus of QEDH from *P. aeruginosa*. The two subunits of the homodimeric enzyme showed an extended interaction area with mainly hydrophobic interactions and hydrogen bonds.

Table 1. Docking of propeller blades in QEDH

Docking contact	Residues of propeller blade	
	W_n	W_{n+1}
D_1 ($W_1 + W_2$)	Ala-82 / Trp-92	Gly-137
D_2 ($W_2 + W_3$)	Ala-131 / Trp-141	Gly-194
D_3 ($W_3 + W_4$)	Ala-188 / Trp-198	Gly-301
D_4 ($W_4 + W_5$) modified	Gly-295 + Thr-259/ Trp-305	Gly-356
D_5 ($W_5 + W_6$) modified	Phe-348 / Phe-362	Gly-476
D_6 ($W_6 + W_7$)	Ala-470 / Trp-480	Gly-517
D_7 ($W_7 + W_8$) modified	Ala-511 / Trp-521	Leu-44
D_8 ($W_8 + W_1$) modified	Val-576 / Trp-48	Gly-88

At the active site the prosthetic group PQQ was fixed by a hydrogen bond network with several amino acid side chains and complex formation with a calcium ion. In addition, there was hydrophobic interaction with Trp-248 underneath in parallel orientation and van-der-Waals contact with the sulfur atoms of the disulfide bridge of the two adjacent residues of Cys-105 and Cys-106. The peptide bond of the 8-membered disulfide ring showed *trans* configuration. The walls of the active site cavity were formed by the hydrophobic residues Trp-270, Phe-408, Leu-409 and Ala-553. The residue Trp-557 formed kind of a lid closing the active site.

Docking calculations were performed with several alcohols at the active site of QEDH for *P. aeruginosa*. With methanol and ethanol, the OH-group of the alcohol in the lowest energy complex pointed to the Asp-316 carboxyl group via a hydrogen bridge. In order to fit higher alcohols like butanol in the active site a flip of the side chain orientation of Trp-557 was necessary. The binding of isopropanol is unfavorable in comparison to n-propanol. Ethanol in the active site showed close van-der-Waals contacts to Cys-106, Asp-316, Phe-408 and PQQ. Methanol however is bound loosely by direct contact only to Asp-316 and PQQ, Fig. 2a, b.

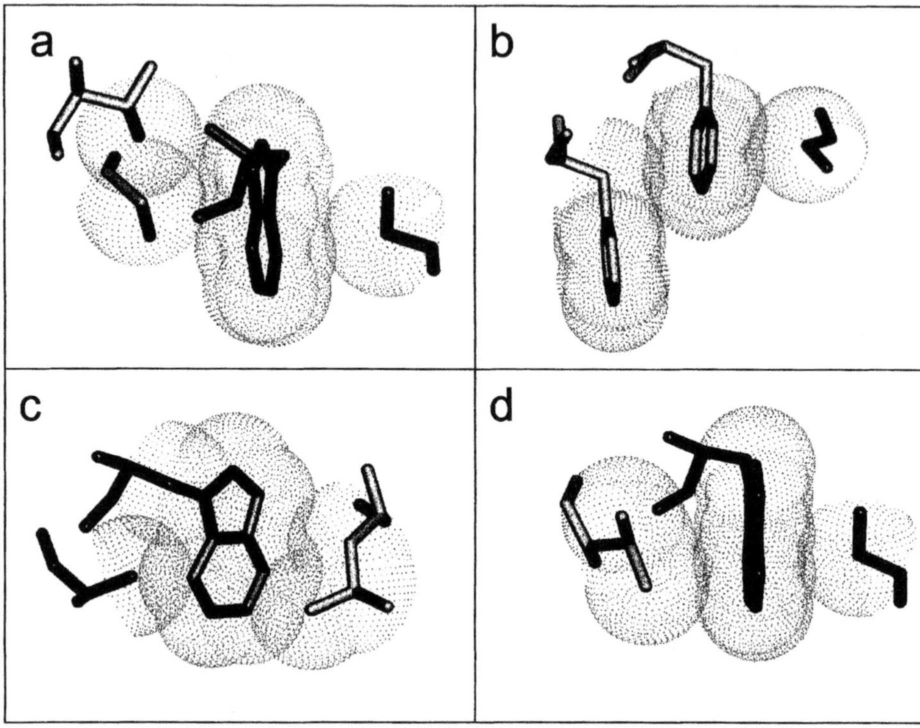

Fig. 1. Docking contacts between propeller blades. The four deviating motifs are shown: a) W_4/W_5: Gly-295/Thr-259/Trp-305/Gly-356; b) W_5/W_6: Phe-348/Phe-362/Gly-476; c)W_7/W_8: Ala-511/Trp-521/Leu-44; d) Val-576/Trp-48/Gly-88. Regular residues in dark gray, deviating residues in light gray.

Discussion

QEDH from *P. aeruginosa* shows very similar enzymatic properties in comparison to the quinoprotein methanol dehydrogenase (QMDH) from methylotrophs [5]. Despite a low sequence identity of 35% [6] the overall folding topology of these two enzymes is very similar. The 60 kDa subunits of both QEDH and QMDH form a 8-propeller superbarrel. QMDH however is a heterotetramer of two small 10 kDa and two large subunits [1], while QEDH is a homodimer. No electron density of a small subunit could be found. QEDH contains at the N-terminus an additional calcium ion binding site, which contributes to the thermal stability of the enzyme [13] and is not present in QMDH. This calcium site probably compensates for the lack of a stabilizing small subunit.

The main differences between the two enzymes are found in the active site cavity. A bulky Trp-531 in QMDH is replaced by Phe-408 and Leu-409 in QEDH, and Leu-547 in QMDH is replaced by Trp-557 in QEDH, where the side chain of this amino acid covers the active site like a lid.

Docking calculations with different alcohols in the active site of QEDH show, that ethanol makes more close contacts to amino acid residues lining the active site than does methanol, which suggests, that ethanol might be a better substrate than methanol. This assumption is in agreement with enzyme kinetic studies, which demonstrate that the K_m-value for methanol is increased by three orders of magnitude with respect to that for ethanol [5]. QMDH shows similar K_m values with methanol and ethanol as substrates [2]. Similar docking calculations with QMDH indeed indicate that the number of close contacts of both substrates in the active site are not very different.

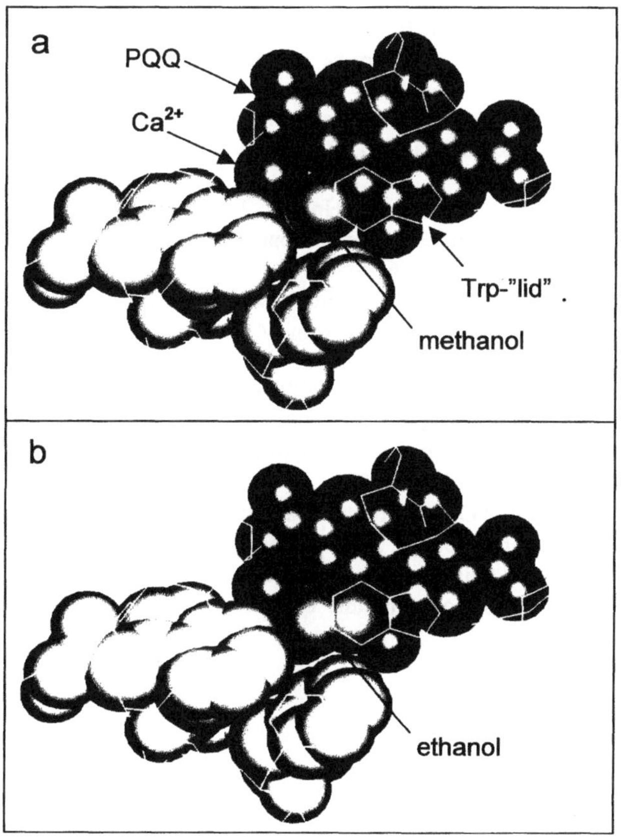

Fig. 2. Docking of alcohols in the active site of QEDH: a) ethanol; b) methanol. Calcium ion and PQQ in dark gray; active site residues in white

References

[1] Goodwin, P.M. and Anthony, C. (1998) The biochemistry, physiology and genetics of PQQ and PQQ containing enzymes. Adv. Microb. Physiol. 40, 1-80

[2] Anthony, C. (1982) The biochemistry of methylotrophs. Academic Press, London

[3] Xia, Z., Dai, W., Zhang, Y., White, S.A., Boyd, G.D. and Mathews, F.S. (1996) Determination of the gene sequence and the three-dimensional structure at 2.4 angstroms resolution of methanol dehydrogenase from *Methylophilus* W3A1. J. Mol. Biol. 259, 480-501

[4] Ghosh, M., Anthony, C., Harlos, K., Goodwin, M.G. and Blake, C. (1995) The refind structure of the quinoprotein methanol dehydrogenase from *Methylobacterium extorquens*. Structure 3, 177-187

[5] Rupp, M. and Görisch, H. (1988) Purification, crystallization and characterization of quinoprotein ethanol dehydrogenase from *Pseudomonas aeruginosa*. Biol. Chem. Hoppe-Seyler 369, 431-439

[6] Diehl, A., v. Wintzingerode, F. and Görisch, H. (1998) Quinoprotein ethanol dehydrogenase of *Pseudomonas aeruginosa* is a homodimer. Sequence of the gene and deduced structural properties of the enzyme. Eur. J. Biochem. 257, 409-419

[7] Schobert, M. and Görisch, H. (1999) Cytochrome c_{550} is an essential component of the quinoprotein ethanol oxidation system in *Pseudomonas aeruginosa*: cloning and sequencing of the genes encoding cytochrome c_{550} and an adjacent acetaldehyde dehydrogenase. Microbiology 145, 471-481

[8] Stezowski, J.J., Görisch, H., Dauter, Z., Rupp, M., Hoh, A., Englmeier, R. and Wilson, K. (1989) Preliminary X-ray crystallographic study of quinoprotein ethanol dehydrogenase from *Pseudomonas aeruginosa*. J. Mol. Biol. 205, 617-618

[9] Leslie, R. (1990) Crystallographic computing, Oxford University Press, Oxford

[10] Xia, Z.X., Dai, W.W., Xiong, J.P., Hao, Z.P., Davidson, V.L., White, S. and Mathews, F.S. (1992) The three-dimensional structures of methanol dehydrogenase from two methylotrophic bacteria at 2.6-Å resolution. J. Biol. Chem. 267, 22289-22297

[11] Abagyan, R.A., Totror, M.M. and Kuznetsor, D.N. (1994) ICM - a new method for protein modeling and design. Applications to docking and structure prediction from the distorted native conformation. J. Comp. Chem. 15, 488-506

[12] Abagyan, R.A. and Argos, P. (1992) Optimal protocol and trajectory visualization for conformational searches of peptides and proteins. J. Mol. Biol. 225, 519-532

[13] Mutzel, A. and Görisch, H. (1991) Quinoprotein ethanol dehydrogenase: preparation of the apo-form and reconstitution with pyrroloquinoline quinone and Ca^{2+} or Sr^{2+} ions. Agric. Biol. Chem. 55, 1721-1726

Biochemistry and Molecular Biology of Vitamin B_6
and PQQ-dependent Proteins
ed. by A. Iriarte, H.M. Kagan and M. Martinez-Carrion

PHYSIOLOGICAL IMPORTANCE OF PYRROLOQUINOLINE QUINONE

Robert Rucker‡, Tracy Stites‡, Francene Steinberg‡, and Alyson Mitchell§

Departments of Nutrition‡ and Food Science & Technology§, One Shields Avenue, University of California, Davis, CA, 95616

Summary

O-quinone cofactors derived from tyrosine and tryptophan are involved in biological reactions that range from oxidative deaminations to free-radical-related redox reactions. Among these cofactors, pyrroloquinoline quinone (PQQ) has excited interest because of its presence in foods, unusual chemical properties and role as a growth-promoting factor in animals. Oral supplementation of PQQ in the nmol/g-diet range has been shown to improve B- and T-cell responsiveness to mitogens, mitochondrial function and reproductive outcome in mice. PQQ is easily absorbed and efficiently taken up by cultured cells. Consequently, a case can be made that PQQ may have physiological importance in animals.

Introduction

The nutritional importance of pyrroloquinoline quinone (PQQ) is suggested by studies of improved neonatal growth and immune function in mice upon its addition to chemically defined diets devoid of PQQ (1). PQQ administration can counter the effects of various anti-inflammatory agents (2). PQQ can also substitute for riboflavin in reductases, such as heme reductase (3). Further, PQQ is present in tissues and fluids (4-6). Kumazawa et al. (4) report that the levels of "free" PQQ, based on ^{13}C-PQQ recovery, range from 4 to 60 ng/g, i.e. from 0.01-0.2 μM, in a wide variety of plant and animal tissues. Since PQQ reacts readily with amino acids to form imidazole derivatives (imidazolopyrroloquinoline quinone, IPQ), PQQ may exist in even higher concentrations when the various adducts of PQQ are included as a part of estimations (5).

PQQ's chemical properties are analogous to combining the properties of riboflavin, ascorbic acid and pyridoxal cofactors into one molecule. In particular, PQQ can catalyze redox cycling reactions such that pmol amounts of PQQ generate μmole amounts of product (6,7).

PQQ and Neonatal Growth and Development

A chemically defined diet, based on synthetic amino acids, has been used to study the effects of PQQ on murine growth and development (see Ref. 1 and Table 1). The PQQ-deficient diet is first fed to weanling females until mating. The mated mice are then maintained on the PQQ-deficient diet or assigned to a diet containing a specified amount, e.g. 2 mg PQQ/kg-diet or less.

In typical experiments only about half of f_1-generation mice produce litters, when fed the deficient diet (1)

Table 1. Growth of BALB/C mice in Response to PQQ Supplementation

Experiment	PQQ (ng/g diet)	Weight, g ± S. E. M. (Number of mice)[b]	Comments
1a	0	6.0 ± 0.1 (10)*	The diet is described in Ref. 1. Succinyl sulfathiazole was added at 20 g/kg in the Experiment designated 1a-1f. F_o-generation dams were assigned to one of the designated diets and fed the diet throughout pregnancy. The F_1 offspring were then switched to the same diet at weaning. The weights are for the offspring obtained ~35 days after conception.
1b	100	6.1 ± 0.2 (10)*	
1c	200	10.5 ± 0.9 (10)*	
1d	300	13.0 ± 0.8 (10)	
1e	1000	13.4 ± 0.6 (10)	
1f	5000	14.5 ± 0.4 (10)	
2	0	10.7 ± 1.8 (24)*	For Experiments 2-4, the diet composition was similar to that reported in Ref. 1. However, no antibiotics were used and the vitamin E concentration was adjusted from 400 mg to 40 mg vitamin E per kg-diet.
	2000	13.5 ± 1.1 (20)	
3	0	9.7 ± 0.6 (18)**	See comments for Exp. 2
	2000	12.7 ± 0.6 (18)	
4	0	7.0 ± 0.8 (18)*	See comments for Exp. 2
	2000	9.2 ± 0.7 (4)	

[a] With respect to these studies, it has been shown that PQQ (as PQQ or IPQ) is efficiently absorbed by the intestine (> 60 %; Ref. 8) and does not appear to be synthesized by microorganisms that are common to the gut microflora of the mouse (9).
[b] * = $P < 0.05$; ** = $P < 0.1$.

PQQ Concentrations in Tissues and Fluids

Most analytical methods underestimate the levels of PQQ in tissues because the contributions from IPQ adducts are ignored. For example, tissue levels of PQQ are frequently based on enzyme or redox cycling assays that do not measure contributions from IPQ and IPQ-amino acid derivatives, (IPQ/R derivatives, where R represents a given amino acid side-chain. When free PQQ is added to biological samples, it disappears rapidly, which suggests that PQQ in tissues is present in or as other products. We have recently utilized diode-array reverse-phase HPLC methods coupled to electrospray ionization mass spectrometry (ESI/MS) for the identification and quantification of PQQ and associated derivatives, such as IPQ and IPQ/R (cf. Table 2 and Ref. 5 for experimental details).

TABLE 2. Incubation of PQQ with the Amino Acids (AA) analyzed by Reverse-Phase HPLC and ESI Mass Spectrometry

	IPQ Derivatives			IPQ/R Derivatives		
AA[a]	% Yield (relative to total PQQ)	HPLC Retention (min)	Molecular Mass[b]	% Yield (relative to total PQQ)	HPLC Retention (min)	Molecular Mass[b]
L-Alanine	7	13.6	341	19	13.9	355
L-Aspartate	3	13.6	341	31	12.5	398
L-Arginine	3	13.5	341	11	12.4	422
L-Cysteine	5	13.6	341	6	12.8	387
L-Tyrosine	54	13.6	341	42	14.2	447
L-tryptophan	62	13.6	341	37	15.9	470
Glycine	98	13.6	341			
L-Glutamate	4	13.6	341			
L-Serine	67	13.6	341			
L-Threonine	64	13.6	341			
L-Glutamine				13	13.8	412
L-Methionine				20	14.5	415
L–Phenylalanine				45	16.0	431
L –Serine				17	13.8	370
L –Val				5	14.1	384
L –Histidine				64	12.1	421
L–Isoleucine				9	14.9	397
L –Lysine[c]				9	14.9	397
				5	14.4	409

[a] Reactions of PQQ with amino acids (1:5 molar ratio) were carried out in 0.1 M phosphate buffer (pH 7.4), for 24 hours at 80° C in the dark. Reactions were stopped by freezing the samples.
[b] Each mass is the average of two separate determinations.
[c] Products corresponding to PQQ reacting with both the α- and ε-amine functions of lysine.

With regard to tissues and fluids, we have analyzed human milk samples for PQQ, IPQ and IPQ/R derivatives (5). Sample preparation involved: 1) extraction with hexane to remove lipid; 2) separation from proteins and other solutes by dialysis against nanopure water; 3) concentration and further purification by absorption on C_{18} columns (pretreated with 0.001 N HCl followed by acetonitrile in 0.1% acetic acid) and elution of PQQ derivatives with 5% pyridine; 4) quantitation by reverse phase HPLC (5).

PQQ was most reactive toward Gly, yielding 98% of the IPQ product with λ_{max} at 422 nm (Table 2). ESI mass analysis of fractions corresponding to this product indicated that the reaction product was IPQ with a [M+H+] ion at m/z of 341. Reverse-phase HPLC chromatograms of PQQ incubated with amino acids resolved 16 products (Table 2). Ala, Asp, Cys, Gly, Glu, Ser, Thr, and Tyr form IPQ conjugates with PQQ. IPQ and IPQ/R derivatives were formed upon incubation with Ala, Arg, Cys, Ser, Trp, and Tyr.

With regard to human milk, HPLC chromatograms of samples spiked with PQQ indicated that spiked PQQ was recoverable (>90 %) as IPQ. Chromatograms of milk (not spiked) demonstrated the presence of both PQQ and IPQ. PQQ was present in milk at 140-180 ng/ml. On a milk solid basis, this corresponds to 500-750 ng/g; sufficient to optimize growth and reproduction in mice.

SYSTEMIC EFFECTS OF PQQ DEFICIENCY

The novel feature of nutritional studies involving PQQ is that observed responses result from PQQ deprivation rather than pharmacological administration. In addition to impaired or abnormal growth and reproductive performance, PQQ deficiency has a negative impact on B- and T- cell responsiveness to mitogens (1). We have also observed that PQQ deficiency affects mitochondrial function in mice. Mitochondrial size and number in mouse liver are affected by PQQ status as well as changes in the respiratory control rate (RCR). The average RCR for PQQ-deficient mice is ~ 2, whereas the RCR is ~2.5 in PQQ-supplemented mice (unpublished data). Further, over 90% of the preparations of mitochondria had sufficient integrity to obtain meaningful measurements compared to only 20% of the preparations from PQQ-deficient mice. The size and number of mitochondria from PQQ-deficient mice were reduced to 70 % that of supplemented mice.

DOES THE RESPONSE TO PQQ COMPARE WITH THAT OBSERVED FOR OTHER AROMATIC POLYPHENOLIC COMPOUNDS

Although other aromatic polyphenolic compounds, e.g. quercetin, catechin, epicatechin, possess some of the same chemical and biological properties as PQQ (10,11), there are a number of important differences. The stability of some polyphenols, particularly in the flavane family, e.g. catechin and epicatechin, is poor (10). Flavanes polymerize easily or interact with protein (11). This is not the case for PQQ, whose polyanionic character precludes easy polymerization. That

PQQ is also superior in redox cycling assays to flavonoids and enediols, such as ascorbic acid, is another important consideration (6,7). PQQ also appears to be readily taken up by cells (Fig. 1). PQQ is highly soluble in water, which possibly contributes to PQQ's apparent high availability, i.e. ease of cellular uptake as well as intestinal absorption (8,9).

Figure 1. Uptake of ^{3}H-PQQ by Human Fibroblasts. ^{3}H-labeled PQQ was prepared commercially (NEN® Life Science Products, Inc., Boston MA 02118 by means of catalytic exchange with tritium gas). The product was next purified by chromatography on columns of A-25; absorption, extensive washing and elution from C-18 silica columns, and as a last step reverse phase HPLC as described previously (5). The specific activity of the final product was ~2 Ci/mmol. The fibroblasts were cultured in Eagle's medium containing 5% calf serum. At confluence, the cultures (5 x 10^6 cells) were switched to serum-free medium and ~2 µCi of 3H-PQQ was added. The cells were incubated for the times indicated. The cells were fractionated using standard procedures. Radioactivity in a given fraction was estimated following absorption and elution from C-18 prep columns and scintillation counting. Note that about 0.1-0.2% of the total radioactivity added to cultures was taken up by the cells.

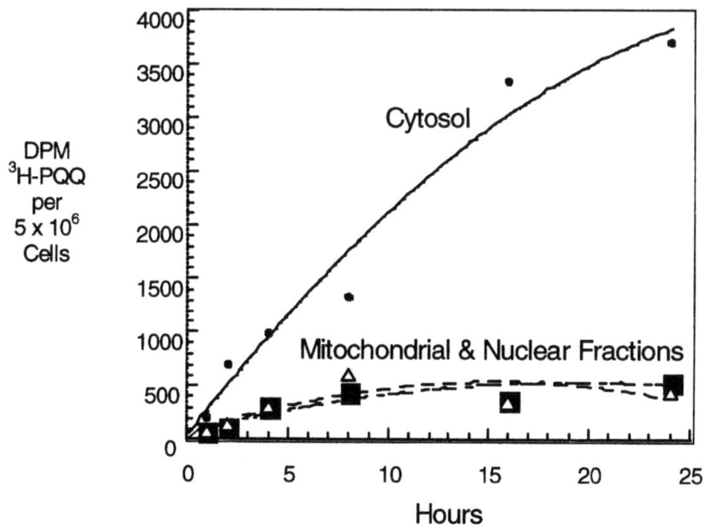

In contrast, radiocarbon from ^{14}C-quercetin is not detected in tissues 12 hours after oral administration. Consequently, it may be inferred that to maintain sustained tissue concentrations of most flavonoids at ~0.05 µM or greater, one or more grams of given flavonoids must be consumed daily. In comparison, a 70 Kg person should be able to maintain PQQ at ~0.05 µM (assuming uniform distribution) with an intake of 1-3 mg per day, i.e. in keeping with values in the literature for the PQQ tissues levels (4,5).

What remains lacking, however, is a clear definition of the biologic mechanism for PQQ's actions on growth, development, and various systemic functions. The most important goal of future work should be directed at defining such mechanism(s). Such an effort would not only clarify the question of PQQ's importance, but should aid in improving our understanding of the putative health benefits of other polyphenolic substances common in foodstuffs and the diet.

Acknowledgements
This work was supported by NIH grant DK56031 and a gift from M&M Mars.

References

1 Steinberg, F. M., Gershwin, E., & Rucker, R. B. (1994) Dietary pyrroloquinoline quinone: growth and immune response in BALB/c mice. *Journal of Nutrition 124:* 744-753.
2 Urakami, T., Yoshida, C., Akaike, T., Maeda, H., Nishigori, H., & Niki, E. (1997) Synthesis of monoesters of pyrroloquinoline quinone and imidazopyrroloquinoline, and radical scavenging activities using electron spin resonance in vitro and pharmacological activity in vivo. *J Nutr Sci Vitaminol (Tokyo) 43:* 19-33.
3 Xu, F., Mack, C. P., Quandt, K. S., Shlafer, M., Massey, V., & Hultquist, D. E. (1993) Pyrroloquinoline quinone acts with flavin reductase to reduce ferryl myoglobin in vitro and protects isolated heart from re-oxygenation injury. *Biochem Biophys Res Commun. 193:* 434-439.
4 Kumazawa, T., Sato, K., Seno, H., Ishii, A., & Suzuki, O. (1995) Levels of pyrroloquinoline quinone in various foods. *Biochem J. 307:* 331-3.
5 Mitchell, A. E., Jones, A. D., Mercer, R. S., & Rucker, R. B. (1999) Characterization of pyrroloquinoline quinone amino acid derivatives by electrospray ionization mass spectrometry and detection in human milk. *Analytical Biochemistry. 269:* 317-325.
6 Fluckiger, R., Paz, M. A., & Gallop, P. M. (1995) Redox-cycling detection of dialyzable pyrroloquinoline quinone and quinoproteins. *Methods Enzymol. 258:* 140-9.
7 Stites,T., Mitchell A. E., and Rucker, R. (2000) Physiological importance of quinoenzymes and the o-quinone family of cofactors. *J Nutrition* in press
8 Smidt, C. R., Bean-Knudsen, D., Kirsch, D. G., & Rucker, R. B. (1991) Does the intestinal microflora synthesize pyrroloquinoline quinone? *Biofactors. 3:* 53-9.
9 Smidt, C. R., Unkefer, C. J., Houck, D. R., & Rucker, R. B. (1991) Intestinal absorption and tissue distribution of [14C]pyrroloquinoline quinone in mice. *Proc Soc Exp Biol Med. 197:* 27-31.
10 Zhu Q.Y., Zhang, A., Tasang D., Huang Y., &Chen Z-Y (1997) Stability of Green tea catedchins. *J Ag food Chem.45*:4624-4628.
11 Formica, J.V. & Regelson, W. (1995) Review of the biology of quercetin and related bioflavonoids. *Fd Chem Toxic 33:* 1061-1080.

Biochemistry and Molecular Biology of Vitamin B_6
and PQQ-dependent Proteins
ed. by A. Iriarte, H.M. Kagan and M. Martinez-Carrion

Mechanism of Topa Quinone Biogenesis in Copper Amine Oxidase Studied by Site-Directed Mutagenesis and X-Ray Crystallography

Katsuyuki Tanizawa,[1] Hideyuki Matsunami,[1] and Hiroshi Yamaguchi[2]

[1]Institute of Scientific and Industrial Research, Osaka University, Ibaraki, Osaka 567-0047, Japan
[2]School of Science, Kwansei Gakuin University, Nishinomiya, Hyogo 662-8501, Japan

Summary

The copper-dependent, self-catalytic mechanism of the biogenesis of the topa quinone (TPQ) cofactor has been studied by site-specific mutagenesis of the three His residues (His431, His433, and His592) involved in binding of the Cu atom in the recombinant phenylethylamine oxidase from *Arthrobacter globiformis*. All three mutants, H431A, H433A, and H592A, mostly lacked the ability to form TPQ upon aerobic incubation with Cu^{2+} ion. However, free imidazole added exogenously to the H592A mutant rescued stable binding of the Cu^{2+} ion and TPQ formation, both of which were also confirmed by X-ray crystallography. In contrast, a tyrosinate-to-Cu^{2+} ion charge-transfer complex was formed transiently in the H433A mutant. These results demonstrate the functional non-equivalence of the three Cu-binding His residues and their importance for TPQ biogenesis.

Introduction

Copper-containing amine oxidases require a redox-active organic cofactor, topa quinone (TPQ), in catalyzing the oxidation of various primary amine substrates (1). TPQ is commonly encoded by a Tyr codon in all genes coding for the enzymes and is produced by the post-translational modification that proceeds in a Cu^{2+} ion-dependent, self-catalyzed reaction (2, 3). X-Ray crystallographic structures have been determined for tyramine oxidase from *Escherichia coli* (4-7), diamine oxidase from pea seedling (8), phenylethylamine oxidase from *Arthrobacter globiformis* (AGAO) (9), and methylamine oxidase from *Hansenula polymorpha* (10). These structural studies revealed their very high similarities in the overall polypeptide fold, active site structure, and coordination geometry of the Cu atom.

For AGAO, both structures of the inactive precursor form containing the unprocessed Tyr and the active mature form containing the Cu atom and TPQ have been solved (9). The two structures have been found almost superimposable, suggesting that the structural changes during the TPQ biogenesis are limited to the active site. In the precursor AGAO structure, the destined Cu site is comprised with the imidazole N atoms of three His residues (His431, His433, and His592) and the 4-hydroxyl O atom of Tyr382, the precursor to TPQ, positioned in an approximately tetrahedral geometry. In contrast, the TPQ ring in the active AGAO structure is tilted away from the Cu site and the Cu atom is coordinated with the same three His residues and two additional water molecules, arranged in a distorted square-pyramidal geometry. An important finding obtained from the structural comparison was that the imidazole ring of one of the three Cu-binding His residues, His592, could adopt two distinct conformations in the two forms, whereas those of His431 and His433 were restricted to only a single conformation. To elucidate the mechanism of TPQ biogenesis and the role(s) of the Cu-binding His residues, we have replaced each His of AGAO with Ala by site-specific mutagenesis and analyzed the mutant enzymes for their abilities to bind the Cu atom and generate the TPQ cofactor by UV-vis, electron paramagnetic resonance (EPR), and resonance Raman (RR) spectroscopies as well as by X-ray crystallography.

Experimental Procedures

Site-specific mutagenesis for His431, His433, and His592 of AGAO, each to be replaced by Ala, was performed by the two-step polymerase chain reaction using plasmid pPEAO2 encoding wild-type (WT) enzyme (11) as a template. WT and mutant enzymes were overproduced in *E. coli* BL21 (DE3) cells grown on an M9 medium supplemented with 0.4% tryptone, 0.4% glucose, and 50 μg/mL ampicillin. The enzymes were purified to homogeneity in the Cu/TPQ-free, precursor form according to the methods reported previously (12). Conversion of the precursor WT and mutant enzymes into the mature forms was generally performed by incubating each precursor enzyme (1 mM subunit) with 2 mM $CuSO_4$ in 25 mM Hepes (pH 6.8) for 24 h. The Cu contents in the proteins were analyzed with a Shimadzu AA-6400G atomic absorption spectrophotometer. UV-vis absorption spectra were measured with a Hewlett-Packard 8452A diode-array spectrophotometer. EPR spectra were measured on a home-built EPR spectrometer using a Varian X-band cavity equipped with an Oxford Instruments cryostat (ESR-900). RR scattering was excited at 514.5 nm with an Ar^+ ion laser (Spectra Physics, 2017) and detected with a triple polychromator (JASCO, NR-1800) equipped with a CCD detector (Princeton Instruments). Single crystals of WT and mutant enzymes were obtained by microdialysis against 0.2 M potassium sodium tartrate in 25 mM Hepes (pH 6.8). X-ray diffraction data of the crystals were collected using a CCD camera at the beam line 44B2 of the SPring-8, Japan Synchrotron Radiation Research Institute (Harima Science Garden City, Hyogo, Japan).

Results and Discussion

The H431A, H433A, and H592A mutant enzymes of AGAO have been constructed and purified in a Cu/TPQ-free, precursor form. When incubated aerobically with excess Cu^{2+} ions, TPQ is formed by self-processing in the WT enzyme, exhibiting an absorption maximum at 480 nm (2). In contrast, addition of Cu^{2+} ion to the H431A, H433A, and H592A mutants caused almost no spectral changes, even after 24 h of incubation (Fig. 1). This indicates that all of the three His residues are very important for TPQ formation, probably through binding of the Cu^{2+} ion. Indeed, atomic absorption analysis showed that the Cu content was decreased by mutation of either of the three His residues. However, when free imidazole was added exogenously, the formation of TPQ was significantly enhanced in the H592A mutant (Fig. 1); identities of TPQ in the WT and imidazole-rescued H592A mutant were confirmed by measurements of their RR spectra. Exogenous addition of imidazole had almost no effect on WT, Y382F (Phe mutant of the TPQ precursor), H431A, and H433A mutants. An optimum concentration of imidazole for TPQ formation in H592A was about 20 mM. Higher concentrations of imidazole were rather inhibitory for copper binding and TPQ formation, probably due to its chelating activity to the Cu^{2+} ion. Exogenous imidazole also rescued Cu binding to the H592A mutant, as shown by EPR spectroscopy which revealed typical features of the non-blue, type II Cu signals, although the copper coordination geometry appeared slightly different (more rhombic) from that in the WT enzyme. In accord with the X-ray structures of AGAO, in which the imidazole ring of His592 could have two distinct conformations (9), the mutation of His592 to Ala is assumed to have created enough of a cavity in the metal-binding site to accommodate a free imidazole molecule, which assists binding of the Cu atom. Although the imidazole-rescued H592A mutant contained both Cu^{2+} ion and TPQ, it showed very low catalytic activity, even in the presence of imidazole. When the substrate 2-phenylethylamine was added to the Cu/TPQ-containing H592A mutant, TPQ was reduced very rapidly, even under fully aerobic conditions. Therefore, the low activity of H592A is due to a decreased rate of the oxidative half reaction, which must be correlated with the position of the TPQ ring and the Cu coordination geometry, as described later.

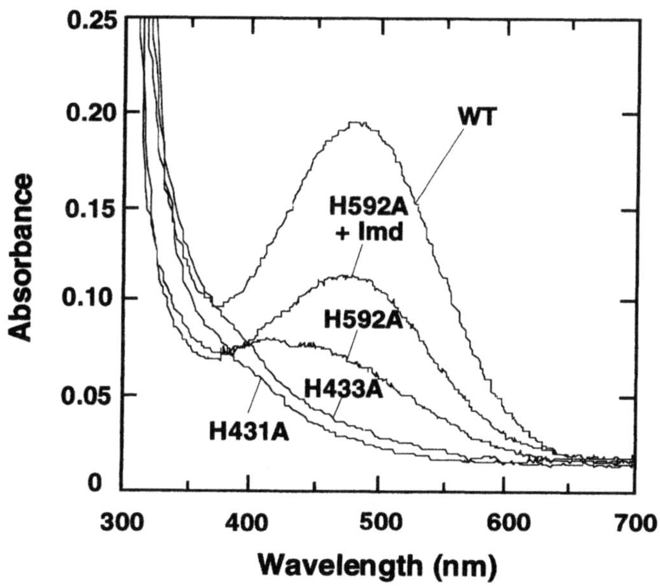

Figure 1: Absorption spectra of WT and mutant enzymes after 24-h incubation with Cu^{2+} ion.

Although the H433A mutant could not form TPQ upon aerobic incubation with Cu^{2+} ion, we have observed transient formation of a ligand-to-metal charge-transfer (LMCT) complex upon addition of Cu^{2+} ions to H433A. Its UV-vis and RR spectra were very similar to those of a tyrosinate-to-Cu^{2+} ion CT complex reported for galactose oxidase (13), suggesting that the phenol ring of Tyr382 is positioned slightly closer to the bound Cu atom in H433A than in the presumed initial state of TPQ biogenesis in the WT enzyme.

Single crystals of the Cu/TPQ-containing H592A and Cu-reconstituted H433A mutants both diffracted to 1.8-2.0 Å resolution in the X-ray crystallographic analysis, and their structures were refined by molecular replacement using the WT enzyme structure (9). The *Fo —Fc* difference Fourier map at a 3.5-σ level of the H592A mutant clearly indicated a positive electron density peak, which was ascribed to an imidazole molecule bound to the enzyme; only the metal-binding side of the imidazole ring appeared to be fixed. Moreover, the TPQ ring in the imidazole-rescued H592A mutant was markedly close to the Cu atom; this could be a reason for its very low catalytic activity, together with the absence of a hydrogen bond with the 4-hydroxyl group of Tyr284, which is probably important as a pivot for the TPQ ring flipping during catalysis (14). The phenol ring of Tyr382 in the H433A mutant was in an orientation very similar to that seen in the precursor WT enzyme (9), and this structure was thought to mimic the Cu coordination geometry (tetrahedral) in the initial stage of TPQ biogenesis.

Altogether, the results reported here demonstrate the importance of the Cu coordination structure constructed by the three Cu-binding His residues for TPQ biogenesis and suggest that precise placement of the Cu atom relative to the precursor Tyr382 is essential for the efficient ring oxidation leading to the TPQ cofactor in copper amine oxidase.

References

1. Janes, S. M., et al. (1990) *Science 248:* 981-987.
2. Matsuzaki, R., et al. (1994) *FEBS Lett. 351:* 360-364.
3. Cai, D., and Klinman, J. P. (1994) *J. Biol. Chem. 269:* 32039-32042.
4. Parsons, M. R., et al. (1995) *Structure 3:* 1171-1184.
5. Wilmot, C. M., et al. (1997) *Biochemistry 36:* 1608-1620.
6. Murray, J. M., et al. (1999) *Biochemistry 38:* 8217-8227.
7. Wilmot, C. M., et al. (1999) *Science 286:* 1724-1728.
8. Kumar, V., et al. (1996) *Structure 4:* 943-955.
9. Wilce, M. C. J., et al. (1997) *Biochemistry 36:* 16116-16133.
10. Li, R., Klinman, J. P., and Mathews, F. S. (1998) *Structure 6:* 293-307.
11. Tanizawa, K., et al. (1994) *Biochem. Biophys. Res. Commun. 199:* 1096-1102.
12. Ruggiero, C. E., et al. (1997) *Biochemistry 36:* 1953-1959.
13. Whittaker, M. M., et al. (1989) *J. Biol. Chem. 264:* 7104-7106.
14. Schwartz, B., et al. (1998) *Biochemistry 37:* 16591-16600.

Biochemistry and Molecular Biology of Vitamin B_6
and PQQ-dependent Proteins
ed. by A. Iriarte, H.M. Kagan and M. Martinez-Carrion

LYSYL OXIDASE ACTIVATES THE TRANSCRIPTION ACTIVITY OF COLLAGEN III PROMOTER: POSSIBLE INVOLVEMENT OF KU ANTIGEN

Giampuzzi M., Gusmano R.[†], Di Donato A.[¶]
Department of Nephrology ; [†] Fondo Malattie Renali del Bambino.
[¶] Corresponding author: e-mail <a-dido@usa.net>
Gaslini Children's Hospital – Largo G. Gaslini, 5 - 16147 Genova, Italy

SUMMARY

Lysyl oxidase is an extracellular enzyme controlling the maturation of the collagen. Because the similar behavior of lysyl oxidase and collagen III expressions in fibrotic tissues, we investigated the influence of lysyl oxidase over-expression on the promoter activity of COL3A1 gene. Our results showed that when COS-7 cells over-expressed the mature form of lysyl oxidase, COL3A1 promoter activity was significantly increased. Electrophoretic Mobility Shift Assay showed a binding activity in the region from –100 to -76, that was significantly increased by lysyl oxidase over-expression. We identified the binding activity as Ku antigen. The study suggests a new co-ordinated mechanisms that might be critical for the development of fibrosis.

INTRODUCTION

Lysyl Oxidase (LOX) (protein-6-oxidase; EC 1.4.3.13) is the key-enzyme that controls collagen and elastin cross-linking and consequent maturation (Pinnel and Martin). Several reports have recently suggested an association between fibrosis and increased LOX activity (Di Donato, et al., Murawaki, et al., Sommer, et al.). The localization of the enzyme is mainly extracellular where its acknowledged substrates are located, although, it has been shown that LOX is localized intracellularly and inside the nucleus as well (Di Donato, et al., Li, et al.). Considering the critical role of collagen III in the organ fibrosis and that its expression level changes are often preceded or paralleled by a similar modifications of LOX activity, we investigated the effects of LOX over-expression on the activity of human collagen III alpha 1 chain (COL3A1) promoter. Our results showed a dramatic increase in COL3A1 promoter activity when the recombinant LOX was over-expressed. Furthermore we defined the region of COL3A1 promoter most likely involved in this LOX dependent activation and some of the possible mechanisms involved.

MATERIALS AND METHODS

Cell culture

Monkey renal fibroblast COS-7 cell line was obtained from American Type Culture Collection (ATCC, Rockville, MD) and grown according to ATCC indication.

Expression vectors

The different fragments of COL3A1 promoter described in the text were derived from the IdF8 clone kindly provided by dr. F. Ramirez (Mt. Sinai School of Medicine, New York, NY) both by endonuclease restriction and PCR amplification. The fragments were then subcloned into pGL2-basic vector. Lysyl oxidase coding sequence for the mature 32 kDa protein, obtained by PCR, was previously cloned into pTrc-His vector (Di Donato, et al.). The fragment NcoI-Bgl II including the poly-His tag together with LOX coding sequence was subcloned into pSG5.

Transient transfections

COS-7 cells were transfected by DEAE dextran method (Vaheri and Pagano). In order to normalize each transfection, 2-4 μg of pCMV-lacZ, expressing the ß-galactosidase enzyme was always co-transfected.

Luciferase Assay Luciferase Assay

Luciferase activity was determined by measuring luminescence in a TD-20/20 luminometer (Turner Designs, USA), according to the Luciferase Assay Kit (Promega Inc., USA) directions.

Gel Retardation Assay

The assay was performed as previously described (Barresi, et al.), using 2-5 μg of nuclear extracts in a total volume of 20 μl..

Protein Analysis

Approximately 30 μg of proteins from the indicated cell fractions were separated by electrophoresis on a 10% polyacrylamide gel and blotted to a Hybond-Super C nitrocellulose membrane (Amersham, UK). The blots were probed with Omni-Probe/M21, an anti-$(His)_6$ tag polyclonal antibody to detect recombinant LOX. To detect Ku80 and Ku70 in the indicated cell extracts, anti-Ku80 polyclonal (M-20) anti-Ku70 polyclonal (C-19) (Santa Cruz Biotechnology Inc., CA, USA) were used. The blots were developed using an alkaline phosphatase conjugated second antibody.

RESULTS AND DISCUSSION

Several fragments of COL3A1 promoter, ranging with their 5' from -1375 to -35 bp from the transcription start site and extending throughout +61 bp, were cloned into pGL2-basic vector. In length order the constructs are: pRUP4, pRUP6, pFV1, pDD1and pDD2, respectively starting at -1375, –257, -117, -55 and -35 (fig. 1). COS-7 cells were transfected with pSG5-LOX, a mammalian expression vector that expresses the mature form of LOX tagged with $poly(His)_6$ at the NH_2 terminus. Interestingly, in our conditions, the recombinant LOX protein was largely expressed also in the nuclear compartment, in agreement with recent findings (Li, et al.) (data not shown). Any attempts to detect the recombinant LOX in the extracellular compartment did not give any positive results (data not shown).

```
        FV1 →
-117                              Consensus 31
        TATTTATCAT TTCTTTTACT GCTGAGGGGAT GGGTGCGGCT CTCATATTTC AGAAAGGGGC

        DD1 →                   DD2 →
- 56        Consensus 11                               Consensus 414   +1
       TGGAAAGTGA GGGAAGCCAA ACTTTTTCCTA TTTAAGGCCA AAGCAAAGGA ATCTCAGTGG

+ 4            Start Codon                                          +60
       CTGAGTTTTA TGACGGGCCC GGTGCCTGAAG GGCAGGGAAC AACTTGATGG TGCTA
```

Fig 1 - The nucleotide sequence of COL3A1 promoter fragments FV1 and its derivatives (DD1 and DD2) used in the study. Homology among oligonucleotide 31, 11 and 414 is underlined (see text).

When COS-7 cells were co-transfected with pSG5-LOX the activity of the promoter was increased up to an average of 12 times, depending on the construct. In fig 2A a summary of the responses of the different promoter fragments to LOX expression is shown.

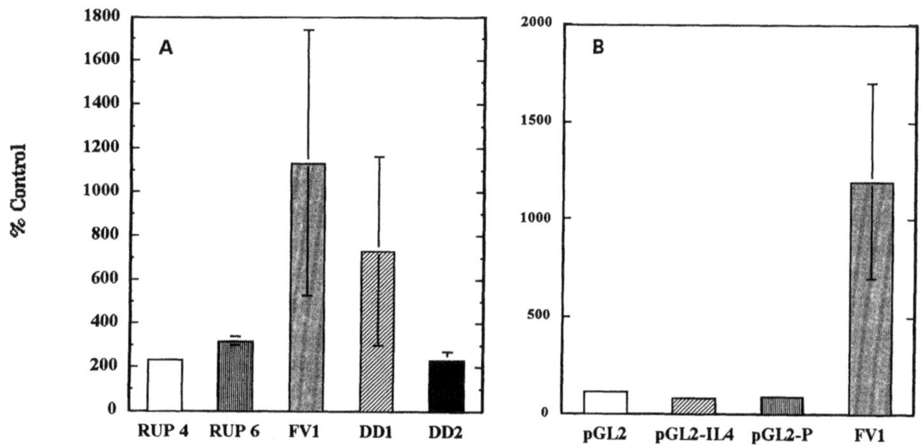

Fig 2 A: Luciferase activity of the indicated pGL2 constructs carrying different fragments of COL3A1 promoter in presence of recombinant LOX expression. B: pGL2: pGL2 basic vector; pGL2-IL4: pGL2 carrying interleukin 4 minimal promoter; pGL2-P: pGL2 carrying SV40 promoter; pFV1: pGL2 carrying pFV1 COL3A1 promoter fragment. The results are the average of at the least three independent experiments performed in triplicate ± s.e.m.

FV1 and DD1 regions are the most responsive, although DD2 retains a significant 2-3 time inducibility upon LOX expression. Then we tested the specificity of LOX effect for COL3A1 promoter by co-transfecting pSG5-LOX together with pGL2-control vector, deprived of any promoter region or on heterologous promoters, like the SV40 contained in pGL2-promoter vector and interleukine 4 minimal promoter, also subcloned into pGL2Fig. 2B shows that in all cases, but with pFV1, LOX expression did not affected the promoter activity. Moreover LOX induction of COL3A1 promoter is dependent on its catalytic activity, being completely suppressed by its inhibitor, the ß-amino-propionitrile (data not shown). Based on these functional findings (fig 2), we guessed that a LOX-responsive sequence must be present at the least in three copies in the FV1 fragment, since we observed a response to LOX expression in all three tested fragments, FV1, DD1 and DD2. Therefore we analyzed the homology among the three fragments, aligning DD2 with the extra nucleotides present in DD1 and in FV1. The result of the alignment showed in fact an overlapping consensus sequence: (A/G)CT(G/A) A(G/A)GG(A/G)A. This sequence doesn't seem to match the consensus for any of the known transcription factors or DNA binding. EMSA experiments, using a probe containing the consensus sequence specific for the fragment FV1 (–100 to –76), indicated as oligonucleotide 31, showed that this sequence binds a proteic complex. Fig. 3 shows that LOX over-expression increased the binding of this complex by 3-4 times.

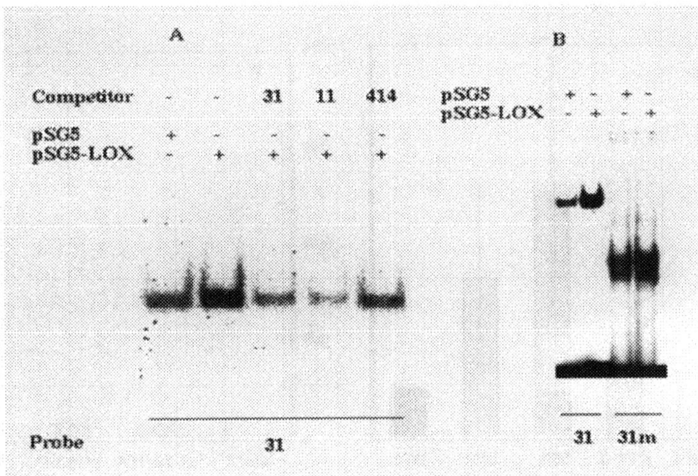

Fig.3 –A: EMSA using oligonucleotide 31 as [^{32}P] labelled probe in presence of: .pSG5: nuclear extracts from COS 7 cells transfected with pSG5 alone or pSG5-LOX: expressing mature Competitor: as above but in presence of a 100 time molar excess of indicated cold oligonucleotide. B: as above but also using a mutated version of oligonucleotide 31 (31m), as radiolabeled probe.

In the same figure the specific competition of the DNA-protein complex by an excess of the cold oligonucleotide 31 or the other two consensus sequences, 11 and 414, is shown. The mutation of oligonucleotide 31 (31m) resulted also in a completely different EMSA pattern (fig 3B) and the pFV1 construct carrying the same mutation showed a dramatic impairment in the response to LOX over-expression (data not shown). An homology search against the EMBL DNA data base returned NRE1 (nuclear regulatory element), among the many possible nucleotide matching sequences. Recent investigations (Torrance, et al.) showed that NRE1 is able to bind the Ku antigen heterodimer complex (Ku80 and Ku70). Ku antigen usually binds to DNA free ends or nicked double strand DNA and mediates the binding of DNA-dependent protein kinase (DNA-PK). The whole complex is involved in some of the main DNA repair and recombination processes (Dynan and Yoo). Therefore, we tested the presence of Ku in the DNA-protein complex shown in our EMSA. Fig 4 shows indeed that both antibodies were able to induce a typical super-shift. In the case of anti-Ku80 the appearance of the super-shift even abolished the retarded band. We obviously tested whether the increased binding of Ku was the result of an higher protein level induced by LOX. That was not the case as tested by Western Blot (data not shown). Experiments are in process in order to assess a possible physiological link between increased binding of Ku by COL3A1 promoter and its enhanced activity in presence of LOX over-expression.

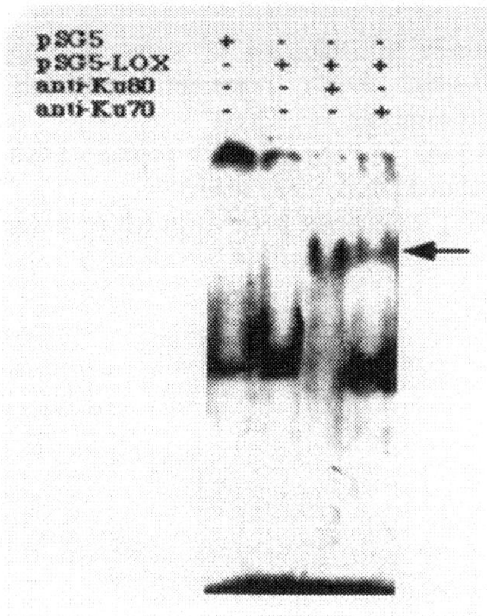

Fig. 4 – EMSA using oligonucleotide 31 as [^{32}P] labelled probe in presence of: .pSG5: vector alone; pSG5-LOX: same vector expressing the coding sequence for mature LOX. Anti-Ku80 and anti-Ku70 indicates the presence in the incubation mixture of the specific antibodies. The arrow points to the super-shift.

Further studies are also needed to completely clarify the regulatory mechanism that might link LOX to Ku. At the moment it is intriguing to think that some of LOX tumor suppressor activities might be achieved through an improved Ku DNA repair activity.

REFERENCES

R. Barresi, M. Sirito, G. Karsenty and R. Ravazzolo, (1994), A negative cis-acting G-fer element partecipates in the regulation of expression of the H-ferritin-encoding gene (FERH), 140, 140-

A. Di Donato, E. Jivotenko, M. Di Duca, G. Ghiggeri, R. Acinni, J. Campolo, F. Ginevri and R. Gusmano, (1997), Lysyl oxidase expression and collagen cross-linking during chronic adriamycin nephropathy., 76, 192-200

A. Di Donato, J. Lacal, M. Di Duca, M. Giampuzzi, G. Ghiggeri and R. Gusmano, (1997), Micro-injection of recombinant lysyl oxidase blocks oncogenic p21-Ha-Ras and progesterone effects on Xenopus laevis oocyte maturation, 419, 63-68

W. S. Dynan and S. Yoo, (1998), Interaction of Ku protein and DNA-dependent protein kinase catalytic subunit with nucleic acids, 26, 1551-9

W. Li, K. Nellaiappan, T. Strassmaier, L. Graham, K. Thomas and H. Kagan, (1997), Localization and activity of lysyl oxidase within nuclei of fibrogenic cells., 94, 12817-12822

Y. Murawaki, Y. Kusakabe and C. Hirayama, (1991), Serum lysyl oxidase activity in chronic liver disease in comparison with serum levels of prolyl hydroxylase and laminin, 14, 1167-1173

S. R. Pinnel and G. R. Martin, (1968), The cross-linking of collagen and elastin: enzymatic conversion of lysine in peptide linkage to α-aminoadipic-δ-semialdheide (allysine) by an extract from bone, 61, 708-716

P. Sommer, C. Gleyzal, M. Raccurt, M. Delbourg, M. Serrar, P. Joazeiro, S. Peyrol, H. Kagan, P. C. Trackman and J. A. Grimaud, (1993), Transient expression of lysyl oxidase by liver myofibroblasts in murine schistosomiasis, 69, 460-470

H. Torrance, W. Giffin, D. J. Rodda, L. Pope and R. J. Hache, (1998), Sequence-specific binding of Ku autoantigen to single-stranded DNA, 273, 20810-9

A. Vaheri and J. Pagano, (1965), Infectious poliovirus RNA: a sensitive method of assay, 77, 4216

Biochemistry and Molecular Biology of Vitamin B_6
and PQQ-dependent Proteins
ed. by A. Iriarte, H.M. Kagan and M. Martinez-Carrion

TGF-β1 Regulation of Gingival Lysyl Oxidase and Connective Tissue Growth Factor

Mehmet Ilhan Uzel, Hsiang Hsi-Hong, Michael C. Sheff, and Philip C. Trackman
Boston University Goldman School of Dental Medicine
Boston, MA, USA 02118

TGF-β is a regulator of lysyl oxidase in several tissues and cell types. The present study investigates the regulation of lysyl oxidase by TGF-β1 in human gingival fibroblasts. Results indicate that lysyl oxidase activity and mRNA levels are slowly increased compared to previous studies performed in osteoblastic cells. A possible role for connective tissue growth factor in mediating TGF-β regulation was assessed. TGF-β stimulates CTGF production, but TGF-β and CTGF appear to promote collagen accumulation by mostly independent mechanisms in human gingival cells. In addition, preliminary studies demonstrating the presence of CTGF in oral tissues are discussed.

Introduction

Gingival overgrowth results from a variety of pathologic conditions including an undesired consequence of therapy with certain antihypertensive calcium channel blockers, the immunosuppressive drug cyclosporin A, and the anti-seizure drug dilantin (Hassell and Hefti, 1991). Recently, increased levels of several cytokines have been found in drug-induced gingival overgrowth tissues (Dill *et al.*, 1993; Iacopino *et al.*, 1997; Plemons *et al.*, 1996; Saito *et al.*, 1996; Williamson *et al.*, 1994). Evidence for a contributing role for TGF-β-mediated stimulation of extracellular matrix production by human gingival fibroblasts in familial gingival fibromatosis has been reported (Tipton and Dabbous, 1998).

Among the genes stimulated by TGF-β1 is connective tissue growth factor (CTGF) (Igarashi *et al.*, 1993). CTGF has been found at elevated levels in several different hyperplastic or fibrotic tissues including skin, and aorta (Frazier *et al.*, 1996; Igarashi *et al.*, 1996; Igarashi *et al.*, 1995;

Oemar *et al.*, 1997). Some of the biological effects of TGF-β, including stimulation of extracellular matrix accumulation, may be mediated by CTGF (Frazier *et al.*, 1996).

The present study determined in human gingival fibroblasts the regulation of lysyl oxidase by TGF-β1. In addition, changes in α-1 type I collagen mRNA levels were measured. As regulation by TGF-β1 was surprisingly slow in these cells, we determined whether CTGF was regulated by TGF-β1, and assessed for a possible role of CTGF in mediating TGF-β1 regulatory effects in human gingival fibroblasts. The results support the concept that TGF-β1 and CTGF stimulate collagenous extracellular matrix production by mostly parallel pathways.

Materials and Methods

Preparation of fibroblastic cultures. Non-hyperplastic gingival tissue from different healthy adult subjects undergoing routine periodontal procedures were obtained from the Boston University School of Dental Medicine Clinic after informed consent according to Boston University Institutional Review Board approved procedures. Fibroblastic cells were obtained from explants (Piche *et al.*, 1989).

Cell culture. Human gingival fibroblasts were cultured at 37 °C in a 5% CO_2 atmosphere in DMEM containing 10% NBS, 0.1 mM non-essential amino acids and antibiotics at passage 3 from frozen stocks in 100 mm tissue culture plates under standard conditions and were passaged at a ratio of 1:7. Cells were grown for three days to about 75 - 85% confluence, and were placed in serum-free medium and cultured for an additional 24 hours. Cells were then re-fed with this medium either containing or lacking the cytokine being tested. Cells were harvested at intervals.

Lysyl oxidase enzyme activity. Lysyl oxidase enzyme activity was determined utilizing a tritium release assay with human tritiated recombinant tropoelastin as substrate (Bedell-Hogan *et al.*, 1993) except that incubations were carried out at 55° C for 60 min. Enzyme activity determinations were normalized to the number of cells counted from the same cultures.

Collagen accumulation. Human gingival fibroblasts were cultured to confluence in 100 mm cell culture dishes. Cells were then cultured in 15 ml of medium containing in addition 50 μg/ml ascorbate either in the presence or absence of 50 ng/ml CTGF. Cells were re-fed fresh medium containing ascorbate +/- 50 ng/ml CTGF once every 48 hours for the duration of the experiment.

Cells were harvested after 0, 4, 1, and 18 days. Aliquots of cell suspensions were analyzed for DNA content (Vytasek, 1982) and additional aliquots were utilized for amino acid analysis. Per cent collagen was calculated based on the hydroxyproline content of 13% by weight.

Results

Regulation by TGF-β1. The regulation of lysyl oxidase enzyme activity, and the regulation of mRNA levels of lysyl oxidase, α-1-type I collagen, and elastin by TGF-β1 was determined in human gingival fibroblasts. TGF-β1 (400 pM) treatment of N6 human gingival fibroblasts resulted in time-dependent stimulation of lysyl oxidase enzyme activity. Fold-increases in lysyl oxidase activity were significantly higher after 48 hours of treatment (2.77 +/ 0.45), but were not consistently higher after 4 and 24 hours. Time-dependent changes in collagen, lysyl oxidase, and elastin steady state mRNA levels caused by 400 pM TGF-β1 treatment of gingival fibroblasts indicate that α-1-type I collagen and lysyl oxidase mRNA levels increase no earlier than after 48 hours of treatment (not shown). Similar results were obtained in experiments performed with gingival fibroblasts from three different donors, indicating that the results are generally true for cells cultured from adult human gingiva (Table 1).

Table 1. Regulation of lysyl oxidase enzyme activity, and steady state mRNA levels for lysyl oxidase, α-1-type I collagen, and elastin in human gingival fibroblast cultures treated with 400 pM TGF-β1 for 48 hours.

	N4 cells	N5 cells	N7 cells	Mean +/- SE
Determination	Fold Change in TGF-β1-treated Cultures Compared to Untreated Controls			
L. O. Activity	2.66	2.17	1.65	2.16 +/- 0.25
L.O. mRNA	0.96	2.09	2.15	1.73 +/- 0.39
COL1A1 mRNA	1.8	2.16	2.80	2.25 +/- 0.29
Elastin mRNA	1.26	1.10	1.38	1.25 +/- 0.08

Fold changes in lysyl oxidase enzyme activity were each statistically significant by unpaired t-test ($p < 0.05$). COL1A1: α-1-type I collagen. Permission granted: Hong *et al.* (1999) *Lab. Invest.* 79, 1655-67.

Regulation of CTGF by TGF-β1. Time-dependent regulation of CTGF mRNA levels by 400 pM TGF-β1 in human gingival fibroblasts was determined. As shown in Figure 1, 2.4 kb CTGF mRNA levels increased dramatically from nearly undetectable levels in untreated controls, to easily detectable levels after four, 24, and 48 hours of treatment. Scanning laser densitometry indicates that the stimulation of CTGF mRNA levels was at least 20-fold, and was maximally

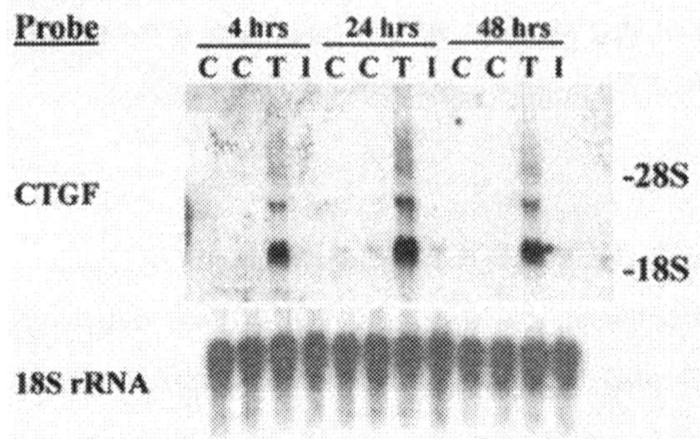

Figure 1. Northern blot analyzing CTGF mRNA levels after treatment of human gingival fibroblasts with TGF-β1 or IL-6. (T), 400 pM TGF-β1, (I), 20 ng/ml IL-6, (C), no cytokine. Permission granted: Hong *et al.* (1999) *Lab. Invest.* 79, 1655-67.

stimulated at 24 hours of treatment. CTGF was not induced by IL-6 (Figure 1). TGF-β1-dose dependent increases in CTGF mRNA levels were found (not shown). CTGF protein levels (38 kDa and 20 kDa molecular species) were also found to be elevated in media of TGF-β1 treated cells after 4, 24, and 48 hours of treatment in Western blotting studies (data not shown).

Regulation by exogenous CTGF. Next, we evaluated whether CTGF exogenously added to human gingival fibroblasts regulates lysyl oxidase enzyme activity, and mRNA levels of lysyl oxidase and α-1 type I collagen. Zero, 1 ng/ml, 10 ng/ml and 50 ng/ml CTGF was added to human gingival fibroblasts in serum-free medium, and lysyl oxidase enzyme activity in media samples was determined after 24 and 48 hours normalized to cell number. Results from several experiments show 1.3-fold stimulation of lysyl oxidase enzyme activity and mRNA levels by 50 ng/ml CTGF requiring 48 hours of treatment. Taken together, results indicate that CTGF is a weak stimulator of lysyl oxidase enzyme activity.

CTGF-induced collagen accumulation. It is possible that small increases in lysyl oxidase enzyme activity over time may be functionally important, and might result in increased insoluble collagen accumulation. Lysyl oxidase catalyzes the final enzymatic step required for cross-linking of collagens and elastin and contributes to the efficiency of accumulation of insoluble collagen. Thus, human gingival fibroblasts were grown in the presence and absence of 50 ng/ml CTGF and cell layers were analyzed for total DNA, and for total protein, and hydroxyproline levels. At day zero there was little collagen accumulation in both the control and CTGF treated cultures. Cell layer collagen/DNA increased after 4, 11, and 18 days in control cells, as expected for cultures accumulating insoluble collagen. CTGF further increased cell layer collagen by 1.5, 2.0, and 1.5-fold on day 4, 11, and 18, respectively.

Discussion

The present study evaluated the hypothesis that CTGF may be induced by TGF-β1 in gingival cells and tissues, and that CTGF itself might stimulate insoluble collagen accumulation. Data show that CTGF mRNA levels are strongly induced in human gingival fibroblasts by TGF-β1. Treatment of human gingival fibroblasts with intact CTGF resulted in modest and slow stimulation of lysyl oxidase activity, and an increase in insoluble collagen accumulation by cultured human gingival fibroblasts. Thus, CTGF-induced increased lysyl oxidase activity may contribute to the observed increased insoluble collagen accumulation.

It is interesting that CTGF stimulates lysyl oxidase enzyme activity without increasing lysyl oxidase mRNA levels. This finding predicts that CTGF increases the amount of active lysyl oxidase by influencing either lysyl oxidase stability, or by stimulating secretion or processing of the inactive lysyl oxidase proenzyme. Pro-lysyl oxidase is processed extracellularly principally by procollagen C-proteinase (Panchenko *et al.*, 1996), and extracellular processing of lysyl oxidase may contribute to the regulation of the level of active enzyme (Feres-Filho *et al.*, 1995).

Preliminary findings suggest that CTGF is highly expressed in human dilantin gingival hyperplasia tissues, and not in non-hyperplastic gingival tissues (data not shown). Studies in progress are evaluating expression of CTGF protein in other forms of gingival hyperplasia. Such detailed clinical characterization coupled to histological and immunohistochemical analyses of human gingival overgrowth tissues may result in new identification and classification of differences among these tissues that may provide insight into the etiology and treatment of different forms of gingival overgrowth.

Acknowledgements

We thank Fibrogen, Inc. South San Francisco, CA for the generous gifts of hrCTGF, anti-CTGF antibodies, and for a cDNA CTGF probe. We thank the Boston University Goldman School of Dental Medicine Clinical Research Center, and the Franciscan Children's Hospital and Rehabilitation Center for supporting these studies by facilitating gingival tissue collection.

References

Bedell-Hogan, D., Trackman, P., Abrams, W., Rosenbloom, J. and Kagan, H. (1993) Oxidation, cross-linking, and insolubilization of recombinant tropoelastin by purified lysyl oxidase. *J. Biol. Chem.*, 268, 10345 - 10350.

Dammeier, J., Brauchle, M., Falk, W., Grotendorst, G.R. and Werner, S. (1998) Connective tissue growth factor: a novel regulator of mucosal repair and fibrosis in inflammatory bowel disease? *Int J Biochem Cell Biol*, 30, 909-22.

Dill, R.E., Miller, E.K., Weil, T., Lesley, S., Farmer, G.R. and Iacopino, A.M. (1993) Phenytoin increases gene expression for platelet-derived growth factor B chain in macrophages and monocytes. *J Periodontol*, 64, 169-73.
Feres-Filho, E.J., Choi, Y.J., Han, X., Takala, T.E.S. and Trackman, P.C. (1995) Pre- and post-translational regulation of lysyl oxidase by transforming growth factor β1 in osteoblastic MC3T3-E1 cells. *J. Biol. Chem.*, 270, 30797 - 30803.
Frazier, K., Williams, S., Kothapalli, D., Klapper, H. and Grotendorst, G.R. (1996) Stimulation of fibroblast cell growth, matrix production, and granulation tissue formation by connective tissue growth factor. *J Invest Dermatol*, 107, 404-11.
Hassell, T. and Hefti, A.F. (1991) Drug-induced gingival overgrowth: Old problem, new problem. *Crit. Rev. Oral Biol. Med.*, 2, 103 - 137.
Iacopino, A.M., Doxey, D., Cutler, C.W., Nares, S., Stoever, K., Fojt, J., Gonzales, A. and Dill, R.E. (1997) Phenytoin and cyclosporine A specifically regulate macrophage phenotype and expression of platelet-derived growth factor and interleukin-1 in vitro and in vivo: possible molecular mechanism of drug-induced gingival hyperplasia. *J Periodontol*, 68, 73-83.
Igarashi, A., Nashiro, K., Kikuchi, K., Sato, S., Ihn, H., Fujimoto, M., Grotendorst, G.R. and Takehara, K. (1996) Connective tissue growth factor gene expression in tissue sections from localized scleroderma, keloid, and other fibrotic skin disorders. *J Invest Dermatol*, 106, 729-33.
Igarashi, A., Nashiro, K., Kikuchi, K., Sato, S., Ihn, H., Grotendorst, G.R. and Takehara, K. (1995) Significant correlation between connective tissue growth factor gene expression and skin sclerosis in tissue sections from patients with systemic sclerosis. *J Invest Dermatol*, 105, 280-4.
Igarashi, A., Okochi, H., Bradham, D.M. and Grotendorst, G.R. (1993) Regulation of connective tissue growth factor gene expression in human skin fibroblasts and during wound repair. *Mol Biol Cell*, 4, 637-45.
Nishida, T., Nakanishi, T., Shimo, T., Asano, M., Hattori, T., Tamatani, T., Tezuka, K. and Takigawa, M. (1998) Demonstration of receptors specific for connective tissue growth factor on a human chondrocytic cell line (HCS-2/8). *Biochem Biophys Res Commun*, 247, 905-9.
Oemar, B.S., Werner, A., Garnier, J.M., Do, D.D., Godoy, N., Nauck, M., Marz, W., Rupp, J., Pech, M. and Luscher, T.F. (1997) Human connective tissue growth factor is expressed in advanced atherosclerotic lesions. *Circulation*, 95, 831-9.
Panchenko, M.V., Stetler-Stevenson, W.G., Trubetskoy, O.V., Gacheru, S.N. and Kagan, H.M. (1996) Metalloproteinase activity secreted by fibrogenic cells in the processing of pro-lysyl oxidase. *J. Biol. Chem.*, 271, 7113 - 7119.
Piche, J.E., Carnes, D.L., Jr. and Graves, D.T. (1989) Initial characterization of cells derived from human periodontia. *J Dent Res*, 68, 761-7.
Plemons, J.M., Dill, R.E., Rees, T.D., Dyer, B.J., Ng, M.C. and Iacopino, A.M. (1996) PDGF-B producing cells and PDGF-B gene expression in normal gingival and cyclosporine A-induced gingival overgrowth. *J Periodontol*, 67, 264-70.
Saito, K., Mori, S., Iwakura, M. and Sakamoto, S. (1996) Immunohistochemical localization of transforming growth factor β, basic fibroblast growth factor and heparan sulfate glycosaminoglycan in givgival hyperplasia induced by nifedipine and phenytoin. *J. Periodont. Res.*, 31, 545 - 555.
Tipton, D.A. and Dabbous, M.K. (1998) Autocrine transforming growth factor β stimulation of extracellular matrix production by fibroblasts from fibrotic human gingiva. *J. Periodontol.*, 69, 609 - 619.
Vytasek, R. (1982) A sensitive fluorometric assay for the determination of DNA. *Anal. Biochem.*, 120, 243 - 248.
Williamson, M.S., Miller, E.K., Plemons, J., Rees, T. and Iacopino, A.M. (1994) Cyclosporine A upregulates interleukin-6 gene expression in human gingiva: possible mechanism for gingival overgrowth. *J Periodontol*, 65, 895-903.

Biochemistry and Molecular Biology of Vitamin B_6
and PQQ-dependent Proteins
ed. by A. Iriarte, H.M. Kagan and M. Martinez-Carrion

Pyridoxine, dopa decarboxylase, and tetrahydroisoquinoline derivatives in Parkinson's disease

M. Ebadi, P. Govitrapong, and J.R. Haselton

Department of Pharmacology, Physiology, and Therapeutics, University of North Dakota School of Medicine and Health Sciences, 501 North Columbia Road, Grand Forks, ND 58203, USA

Summary

The activity of pyridoxal phosphate-dependent dopa decarboxylase in the corpus striatum of patients with Parkinson's disease is low. However, **pyridoxine** which nullifies the beneficial therapeutic effects of **levodopa** is contraindicated in patients with Parkinson's disease. This problem is obviated by **carbidopa**, a peripheral dopa decarboxylase inhibitor which does not penetrate the blood brain barrier and hence does not affect the formation of dopamine in the brain. Administration of pyridoxine enhances the synthesis of cysteine and dopamine in the brain, and oxidation of dopamine in the presence of cysteine generates neurotoxic products such as **dihydrobenzothiazines**. Moreover, the situation is aggravated by the fact that the concentration of **glutathione**, a cysteine containing peptide, is low in the striatum of Parkinsonian patients. In addition, pyridoxal-containing **tetrahydroisoquinoline** derivatives, alter mitochondrial functions by inhibiting the activity of complex I. Therefore, pyridoxine should be used cautiously in patients with Parkinson's disease.

Introduction

Parkinson's disease is treated with dopamine-function enhancing substances. Birkmeyer and Hornykiewicz (1961, 1962) reported that 50-150 mg of L-DOPA, (-)-3(3,4-dihydroxyphenyl)-L-alanine, administered intravenously produced a decrease in akinesia in patients with Parkinson's disease. The effects were maximal at 2-3 hr and lasted up to 24 hr. Administration of monoamine oxidase inhibitors intensified the effects, whereas D-dopa, dopamine, or 5-hydroxytryptophan, the precursor of serotonin did not produce these beneficial effects. Cotzias (1967) reported that 3-16 gm of oral DL-DOPA given over long periods of time could produce a marked and persistent improvement in the symptoms and signs of Parkinson's disease.

L-DOPA synthesized from tyrosine is rapidly metabolized within the cytoplasm to dopamine by the enzyme aromatic amino acid decarboxylase (AADC). In pathological states with reduced TH activity, the rate of synthesis of L-DOPA is low. When exogenous L-DOPA is administered, blood becomes the source of L-DOPA for catecholamine neurons in the brain. L-DOPA, like tyrosine, is transported across the brain capillary endothelial cell and nerve cell membrane by the L-system, a carrier-mediated transport system for neutral amino acids with branched or ringed side chains. L-DOPA entering cells containing AADC is thought to be metabolized nonselectively to dopamine. AADC is found almost exclusively within catecholaminergic and serotonergic neurons in the brain (Lovenberg et al., 1962; Lloyd and Hornykiewicz, 1972), and mostly in the caudate nucleus and putamen (Sourkes, 1966; Lloyd and Hornykiewicz, 1970).

Subtypes of dopaminergic receptors

Two subtypes of dopamine receptors, D1 and D2, were identified on the basis of pharmacological and biochemical criteria (Kebabian and Calne, 1979). The D1 and the D2 dopamine receptors are pharmacologically distinct, displaying differing affinities not only for endogenous ligand dopamine, but also for several other compounds (see Seeman and Grigoriadis, 1987). Benzazepines, including SCH 23390, bind to D1 receptors with high affinity, and benzamides, such as sulpiride or raclopride, and butyrophenones, such as spiroperidol, selectively label D2 receptors. At least five genes encoding subtypes of dopamine receptors have been isolated and categorized as D1-like or D2-like according to their nucleotide sequences and the pharmacological profile of the expressed proteins. The D1-like receptors include the D1 and the D5 receptors, whereas the D2-like receptors include the two isoforms of the D2 receptor, differing in the length of their predicted third cytoplasmic loop, dubbed *D2 short* (D2S) and *D2 long* (D2L), and the D3 and D4 receptors. The D1 and D5 receptors activate adenylyl cyclase. The D2 receptors couple to multiple effector systems, including the inhibition of adenylyl cyclase activity, suppression of Ca^{2+} currents, and activation of K^{+} currents. The effector systems to which the D3 and D4 receptors couple have not been unequivocally defined.

Neuropharmacology of peripheral dopa decarboxylase inhibitors

Carbidopa, (-)-L-*a*-hydrazino-3,4-dihydroxy-*a*-methylbenzene propionic acid, and **benserazide,** (+-)-DL-seryl-2-(2,3,4-trihydroxybenzyl) hydrazine (fig. 1) are reversible inhibitors of dopa decarboxylase (Heubert et al., 1983; Baldessarini and Greiner, 1973). They inhibit dopa decarboxylase within capillary endothelial cells of the cerebral vessels (Burkard et al., 1964) but do not enter the brain, and therefore, are devoid of central activity. When used in combination with L-DOPA, they inhibit the metabolism in peripheral tissues. Carbidopa and benserazide are concentrated in the kidneys, lungs, small intestine, and liver, with lower concentrations found in the heart. These compounds do not enter the brain to any significant extent (Schwartz et al.,1974; Porter, 1973).

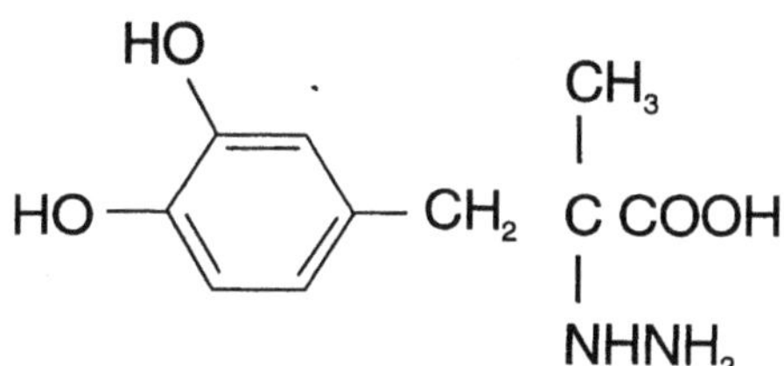

Carbidopa

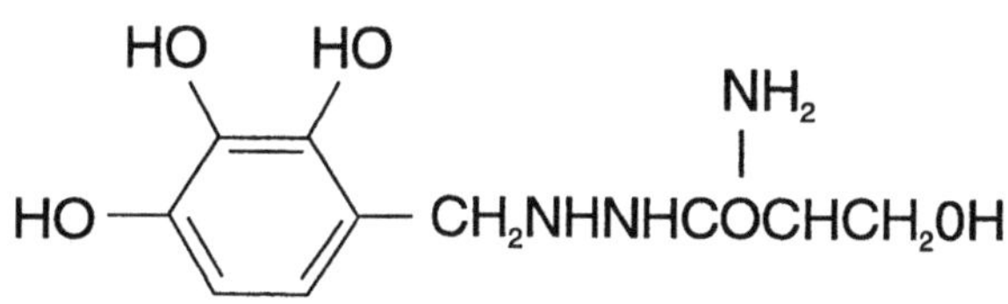

Benserazide

Figure 1: Peripheral dopa decarboxylase inhibitors.

Cysteine neurotoxicity in Parkinson's disease

Six vitamers have vitamin B_6 metabolic activity: pyridoxine, pyridoxal and pyridoxamine, and their 5′-phosphates. The metabolically active coenzyme is pyridoxal 5′-phosphate (PLP). In the liver there is rapid oxidation of the other vitamers to pyridoxal, and rapid phosphorylation to PLP, which is the main circulating vitamer exported from the liver bound to albumin. Uptake into peripheral tissues is by extracellular dephosphorylation, followed by metabolic trapping intracellularly as PLP. PLP that is not bound to enzymes is rapidly dephosphorylated, and surplus pyridoxal in tissues is oxidized to pyridoxic acid, which is the main urinary metabolite of the vitamin.

In amino acid metabolism, PLP reacts with the α-amino group of the substrate; and the reactions of PLP-dependent enzymes include:

(a) decarboxylation of amino acids to yield amines, which are neurotransmitters or hormones, e.g., γ-aminobutyrate, histamine, norepinephrine (and hence epinephrine), serotonin;

(b) transamination of amino acids to yield their keto-acids (oxo-acids), which are then oxidized as metabolic fuels;

(c) a variety of reactions involving the side-chains of amino acids, including kynureninase (*EC* 3.7.1.3), cystathionine synthetase (*EC* 4.2.1.22) and cystathionase (*EC* 4.4.1.1);

(d) decarboxylation of phosphatidylserine to phosphatidylethanolamine in phospholipid synthesis (see Bender, 1999).

Homocysteine is an intermediate in methionine metabolism, and may undergo one of two metabolic fates (as shown in Fig. 2) remethylation to methionine (a reaction that is dependent on vitamin B_{12} and folic acid), or onward metabolism leading to the synthesis of cysteine (trans-sulfuration). The trans-sulfuration pathway has two PLP-dependent enzymes: **cystathionine synthetase** and **cystathionase**. Glutathione (GSH) is synthesized by the activity of two ATP-requiring GSH synthesizing enzymes. Gamma-glutamylcysteine synthetase (γ-GCS) is the rate limiting enzyme for the GSH synthesis. γ-GCS is a heterodimer of heavy, catalytic subunit and light, regulatory subunit and responsive to many stresses, such as heat shock, oxidative stress, or cytokines (Kondo et al., 1999).

Glutathione serves as an essential cofactor for a number of enzymes including formaldehyde dehydrogenase, glyoxylase, maleylacetoacetate isomerase, dehydrochlorinase, and prostaglandin endoperoxidase isomerase. In addition, glutathione, whose level is reduced in the striatum of patients with Parkinson's disease (see Ebadi et al., 1996), possesses a neuroprotective action by transporting cysteine. Elevated cysteine/cystine is excitotoxic by interfering with the N-methyl-D-aspartate (NMDA) receptors. Cysteine forming hemithioketals and hemithioacetals generate substances such as thiazolidinone, which tend to inhibit glutamate decarboxylase. Furthermore,

Puka-Sundvall, et al. (1998) have shown that subtoxic doses of cysteine administered before or after hypoxia-ischemia enhances brain injury. Carbidopa enhances the formation of dopamine in the brain, pyridoxine enhances the formation of cysteine, and oxidation of dopamine in the presence of cysteine generates neurotoxic products such as dihydrobenzothiazines (Shen et al., 1997).

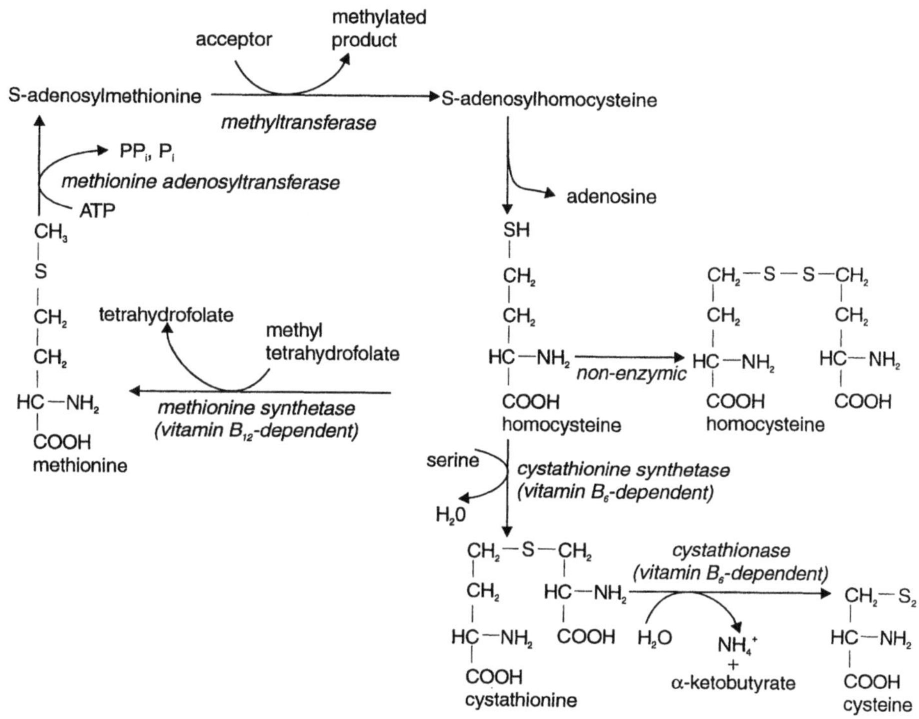

Figure 2: The synthesis of cysteine from cystathionine requires a vitamin-dependent cystathionase.

Pyridoxine, tetrahydroisoquinolines, and Parkinson's disease

By using the Pictet Spengler reaction (Ebadi and Govitrapong, 1981) and cyclizing pyridoxal- or pyridoxal phosphate with biogenic amines, we have synthesized tetrahydroisoquinoline derivatives from pyridoxal and dopamine (1-(2-methyl-3-hydroxy-5-hydroxymethyl-4-pyridyl)-6, 7-dihydroxy-1,2,3,4-tetrahydroiso-quinoline), from pyridoxal phosphate and dopamine (1-(2-methyl-3-hydroxy-5-

phosphinyloxymethyl-4-pyridyl)-6,7-dihydroxy-1,2,3,4-tetrahydroisoquinoline), from pyridoxal and dopa (1-(2-methyl-3-hydroxymethyl-4-pyridyl)-3-carboxyl-6,7-dihydroxy-6,7-dihydroxy-1,2,3,4-tetrahydroisoquinoline), or from pyridoxal phosphate and dopa (1-(2-methyl-3-hydroxyphosphinyloxy-methyl-4-pyridyl)-3-carboxyl 6, 7-dihydorxy-1,2,3,4-tetrahydroisoquinoline). These native compounds did not inhibit dopa decarboxylase. However, when incubated with monoamine oxidase B and, converted to charged molecules, they then became inhibitory not only to dopa decarboxylase, but also to mitochondrial functions by inhibiting the activity of complex I (Ebadi et al., 1998). These studies suggest that endogenously produced tetrahydroisoquinoline derivatives may predispose to Parkinsonism.

Acknowledgments

The authors gratefully acknowledge the excellent secretarial skills of Mrs. KayLynn Bergland, and the editorial assistance of Mrs. Carol Haselton. The studies cited in this paper have been supported from a grant from USPHS No. R02 NS34566-06.

References

[1] Birkmayer W. and Hornykiewicz O. (1961) *Wiener Klinische Wochenschrift* 73, 787-788.
[2] Brikmayer W. and Hornykiewicz O. (1962) *Arch. Psychiat. Nervenkr.* 203, 560-574.
[3] Cotzias G.C., Van Woert M.H., and Schiffer L.M. (1967) *N. Engl. J. Med.* 276, 374-379.
[4] Lovenberg W., Wiessbach H., and Udenfriend S. (1962) *J. Biol. Chem.* 237, 89-93.
[5] Lloyd K.G. and Hornykiewicz O. (1972) *J. Neurochem.* 19, 1549-1559.
[6] Sourkes T.L. (1966) *Pharmacol. Rev.* 18, 53-60.
[7] Lloyd K.G. and Hornykiewicz O. (1970) *Science* 170, 1212-1213.
[8] Kebabian J.W. and Calne D.B. (1979) *Nature* 277, 93-96.
[9] Seeman P. and Grigoriadis D. (1987) *Neurochem. Int.* 10, 1-25.
[10] Heubert N.D., Palfreyman M.G., and Haegele K.D. (1983) *Drug. Metab.* 11, 195-200.
[11] Baldessarini R.J. and Greiner E. (1973) *Biochem. Pharmacol.* 22, 247-256.
[12] Burkard W.P., Gey K.F. and Pletscher A. (1964) *Arch. Biochem. Biophys.* 107, 187-196.
[13] Schwartz E.D., Jordan J.C. and Ziegler, W.H. (1974) *Eur. J. Clin. Pharmacol.* 7, 39-45.
[14] Bender D.A. (1999) *Br. J. Nutr.* 81, 7-20.
[15] Kondo T., Higashiyama Y., Goto S., Iida T., Cho S., Iwanaga M., Mori K., Tani M. and Urata Y. (1999) *Free Rad. Res.* 31, 325-334.
[16] Ebadi M., Srinivasan S.K. and Baxi, M. (1996) *Prog. Neurobiol.* 48, 1-19.
[17] Puka-Sundvall M., Sandberg M. and Hagberg H. (1998) *Brain Res.* 797, 328-332.
[18] Shen X.Y., Zhang F. and Dryhurst G. (1997) *Chem. Res. Toxicol.* 10, 147-155.
[19] Ebadi M. and Govitrapong P. (1981) *Drug. Develop. Res.* 1, 129-136.
[20] Ebadi M., Pfeiffer R.F., Burke W.J., Nelson F.A., Rasmussen N., and Govitrapong, P. (1998) *Soc. Neurosci. Abst.* 24, 1007.

LYSYL OXIDASE

Biochemistry and Molecular Biology of Vitamin B_6
and PQQ-dependent Proteins
ed. by A. Iriarte, H.M. Kagan and M. Martinez-Carrion

Lysyl oxidase: a family of multifunctional proteins

K. Csiszar , C. Jourdan-Le Saux , S. F. T. Fong , K.S.K.Fong, C. D. Boyd

Pacific Biomedical Research Center, University of Hawaii, Honolulu, Hawaii 96822

Summary

Lysyl oxidase (LOX) is a copper containing amine oxidase traditionally described as a secreted enzyme that crosslinks the extracellular matrix proteins elastin and fibrillar collagens. Recently, multiple novel biological functions have been attributed to LOX and experimental evidence indicates that LOX may have other, intra-cellular and intra-nuclear substrates involved in these functions. Our results suggest that several genetically different lysyl oxidases exist that individually function to perform these roles. We report here the gene structure, chromosomal localization, protein domain structure, evolutionary relationship, tissue specific expression and distribution of these novel lysyl oxidases, LOX, LOXL, LOXL2 and LOXL3 that could serve as a family of proteins present in distinct cellular and tissue locations, each with a related but different function.

Introduction

Lysyl oxidase (LOX) belongs to a heterogeneous family of copper dependent amine oxidases that oxidize primary amine substrates to reactive aldehydes (Janes et al.,1992, Lyles, 1996, Dove et al., 1996). From the time of the discovery of lysyl oxidase (Pinell and Martin, 1968) most studies have focused on the specific cross-linking activity and catalytic mechanism of action of this enzyme on collagen and elastin substrates, essential to the biogenesis of connective tissue. LOX oxidizes peptidyl lysine in these proteins to ß-aminoadipic-ß-semialdehyde or allysine. This peptidyl aldehyde can then spontaneously condense with neighboring amino groups or other peptidyl aldehydes to form covalent cross-links in several fibrillar collagen types and desmosines and isodesmosines in elastin (Eyre et al., 1984, Kagan et al., 1994, Smith-Mungo and Kagan, 1998). Recently, multiple novel biological functions have been attributed to lysyl oxidase. Evidence from several laboratories demonstrated that lysyl oxidase may have other intracellular or intranuclear substrates involved in these functions. The range of these novel activities attributed to LOX cover a spectrum of biological functions including developmental regulation, tumor suppression, senescence, cell growth control and chemotaxis (Contente et al., 1990, Lazarus et al., 1994, Mello et al., 1995, Csiszar et al., 1996, Saito et al., 1997, DiDonato et al., 1997). Our hypothesis is that several different lysyl oxidases exist that individually function to perform these roles. In this work we report three novel proteins that fulfill the requirements of being fully functional but genetically distinct lysyl oxidases that could serve as a family of proteins present in distinct cellular and tissue locations, each with a related but different function.

Materials and Methods

A database protein homology search (Human Genome Sciences Inc.) identified several human cDNA clones which had homology to the copper binding and catalytic domains of lysyl oxidase. Using PCR primers derived from these cDNA sequences, 16 additional cDNAs were isolated from a spleen cDNA library and two PAC recombinants, containing overlapping human genomic DNA, were obtained. These clones were used to further characterize new putative lysyl oxidase-like proteins. A full length cDNA sequence was established by DNA sequence analysis of several overlapping cDNA clones. In addition an EST (expressed sequence tag) database (http://www. ncbi. nlm. nih. gov/dbEST/ index. html) was screened for sequence homology to both the highly conserved copper binding site and the catalytic domains observed in LOX, LOXL and LOXL2. More than 1,000 lysyl oxidase-homologous EST entries were identified. A BLAST search of these EST sequences identified 21 unique and novel cDNAs. We have sequenced several of these human cDNA clones. Two of these cDNAs clearly corresponded to LOXL2 cDNAs. In addition, five EST cDNAs contained overlapping sequences and were obtained from a mixed cDNA library of fetal heart, pregnant uterus and melanocytes and cDNA libraries derived from prostate and brain mRNA. Sequence analysis of these cDNAs identified a copper binding site and the characteristic catalytic domain of lysyl oxidase. A full length cDNA sequence was generated from sequenced 5'-end clones that were isolated from brain and placenta cDNA libraries. These cDNAs encode a new lysyl oxidase variant that we have referred to as LOXL3, which represents the fourth member of the lysyl oxidase family. Expression of these novel genes was examined by Northern blot analysis. Chromosomal localization was determined by radiation hybrid mapping. Immunohistochemistry was performed using specific antibodies designed against regions of minimal homology among the LOX proteins.

Results and Discussion

The first protein related to lysyl oxidase (LOX) on chromosome 5q23, that we identified, was the lysyl oxidase-like protein (LOXL) on chromosome 15q23. More recently, a new lysyl oxidase-related protein was reported in senescent fibroblasts that contained the same carboxy-terminal sequence conservation noted in LOX and LOXL (Saito et al., 1997). Our recent results extended the characterization of this new LOXL2 protein, identified the corresponding gene, mapped to chromosome 8p21 and confirmed that LOXL2 is indeed another genetically distinct copper-binding protein closely related to LOX (Jourdan Le-Saux et al., 1998). Structural comparison of the LOX, LOXL and LOXL2 genes provides further convincing evidence for the existence of a gene family. In addition to LOXL and LOXL2, we have recently identified an additional protein, LOXL3 and the corresponding gene on chromosome 2p13, with sequence homology to LOX that permits the classification of this new member as lysyl oxidase-like protein. Five of the exons that encode the conserved copper-binding, carbonyl cofactor and cytokine receptor-like domains are present in the LOX, LOLX and LOXL2 genes. Only three of the thirteen LOXL3 exons encode the conserved domains and there is significant sequence and structure divergence within the eight 5' - end exons of the LOLX2 and LOXL3 genes (Jourdan-Le Saux et al., 1999).

A striking conservation in amino acid sequence within the carboxy-terminal end of all four proteins suggests a functional similarity between LOX, LOXL, LOXL2 and LOXL3. These domains include the copper-binding site containing 4 histidines, two metal-binding domains, a cytokine receptor-like motif, ten cysteines and the catalytic domain. It is very likely therefore, that the novel LOX-like proteins share a functional similarity with the copper-dependent oxidative deamination activity characteristic of lysyl oxidase.

The discovery of the LOX-like proteins and the possibility that additional lysyl oxidase-like proteins may exist, prompted us to re-consider the definition of a lysyl oxidase. A database search for homology to conserved LOX and LOXL copper-binding domains revealed that none of the other copper-binding proteins contain this particular consensus sequence (Gacheru et al., 1988, Krebs and Krawetz, 1993).

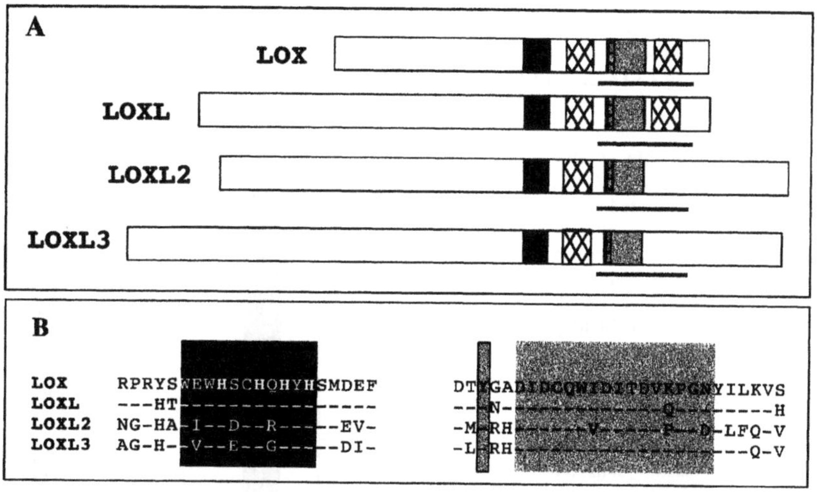

Figure 1. A comparison of conserved domains within lysyl oxidase and lysyl oxidase-related proteins. Panel A illustrates the various domains in lysyl oxidase (LOX) and the presence of these conserved sequences in LOXL, LOXL2 and LOXL3. The lengths of the un-shaded bars represent the relative sizes of the individual proteins. Dark grey boxes: copper binding sites, hatched boxes : metal binding sites, grey boxes: catalytic domains and underlined are the cytokine receptor-like domains. **Panel B** presents the amino acid sequence of LOX, LOXL, LOX2 and LOXL3 proteins within and immediately surrounding the conserved copper-binding and catalytic domains. Amino acid sequences were derived from cDNA sequences.

Similarly, the catalytic domain is also highly conserved within all lysyl oxidases in every species studied so far (human, mouse, rat and chicken; Kim et al., 1995) but was not detected in other known proteins. In contrast, the less defined metal- binding domains and the cytokine receptor-like domain, present in several lysyl oxidases, share homology with proteins unrelated to lysyl oxidases. We, therefore, define lysyl oxidase or lysyl oxidase-related proteins as proteins that contain the copper-binding domain WEW**H**SC**H**Q**H**Y**H** with four histidines and the catalytic domain DIDCQWWIDITDVXPGNY (Janes et al., 1992, Krebs and Krawetz, 1993, Smith-Mungo and Kagan, 1998). A comparison of these domains within different lysyl oxidases is summarized in **Figure 1.**

In addition to the copper binding site, two domains are conserved within all four lysyl oxidases. These are the catalytic domain and the cytokine receptor-like domain. While the catalytic function of LOX is well understood, at least within the extracellular matrix, the role of the cytokine receptor-like domain within the LOX proteins is unknown. In our initial functional analysis of these conserved domains, we have tested whether the catalytic domain present in the

LOXL proteins reflects a function similar to the cross-linking activity of LOX. While our previous results indicated differences in substrate specificity (Kim et al., 1999) it was not clear from these initial data if the catalytic activity of LOXL is different to LOX.

To explore the possibility that LOXL will catalyze the oxidative deamination of lysyl residues, we have tested the cross-linking activity of the LOXL protein on an elastin substrate. We have detected significant cross-linking activity for a bovine aorta-derived LOXL, previously described for LOX only and this activity was inhibited by ßAPN, a specific inhibitor of LOX cross-linking activity (**Figure 2**). These in vitro results indicate that both LOX and LOXL will catalyze the oxidative deamination of lysine residues in an elastin substrate. While this finding does not address whether LOX and LOXL specifically utilize an elastin substrate, it does suggest that all lysyl oxidases, including LOXL2 and LOXL3, will probably be subject to ßAPN inhibition.

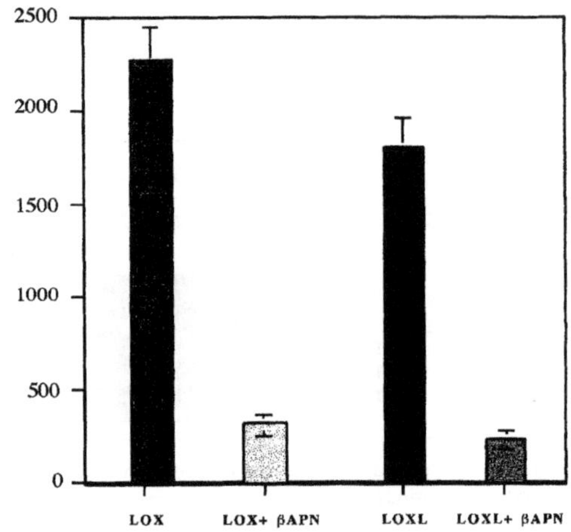

Figure 2. β APN inhibitable amine oxidase activity of LOX and LOXL. LOX and LOXL were isolated from extracts of calf aorta by DEAE-cellulose chromatography as previously described (Decitre et al., 1998). Eluates containing both immunoreactive LOX and LOXL proteins were shown by Western blotting to lack any immuno-crossreactivity with each other. These separated fractions were then assayed for lysyl oxidase activity using a previously published ultra-filtration assay that measures the release of ^{3}H-H_2O from an elastin substrate labeled with L-[4-5-^{3}H] lysine (Shackleton, et al., 1990). ßAPN: ß-amino proprionitrile

The C-terminal regions of human LOX and LOX-like proteins contain sequence homology with the N-terminal extracellular domain of the growth factor and cytokine receptor superfamily that overlaps with the catalytic site (**Figure 3**). We previously reported that the consensus sequence found in the N-terminal modules of Class 1 cytokine receptors, C-x_9-C-x-W-$x_{26\text{-}32}$-C-$x_{10\text{-}13}$-C (where C is cysteine, W is tryptophan and x_n is a defined number of any amino acid) is conserved in human LOX and LOX-like proteins (Kim et al., 1995). Furthermore, the first 13 residues fit the Prosite pattern PS00241, C-[LVDYR]-x(7,8)-[STIVDN]-C-x-W, found in cytokine receptors, where amino acid residues in [ ] indicate the presence of one such residue and numbers in ( ) indicate the number of consecutive residues. Examples of proteins in this class include erythropoietin receptor, granulocyte colony-stimulating factor receptor, granulocyte-macrophage colony-stimulating factor receptor alpha chain, growth hormone receptor, thrombopoeitin receptor, and various interleukin receptor chains (Miyajima et al., 1992). Structural analysis of such receptors indicate that they are related to immunoglobulin constant domains and evolved from primitive fibronectin type III sequence modules that are also common to various adhesion molecules (Bazan, 1990).

Three-dimensional crystallographic analysis of the cytokine receptors suggest that the N-terminal extracellular domain forms two barrel-shaped modules, each consisting of six or seven β-strands

(Bazan, 1990). Cytokines are presumed to bind into the groove created by a kink between these two modules. Based on this model, the C-terminal region of human LOX and LOX-like proteins lacks sequence corresponding to the C-terminal "barrel" module. The LOX and LOX-like proteins contain cysteine and tryptophan residues at conserved positions within the first five ß-strands that form the N-terminal "barrel" module. Furthermore, secondary structure analysis predicts that ß-strands are formed (Figure 3). Therefore, this region of the LOX and LOX-like proteins may fold in a similar manner.

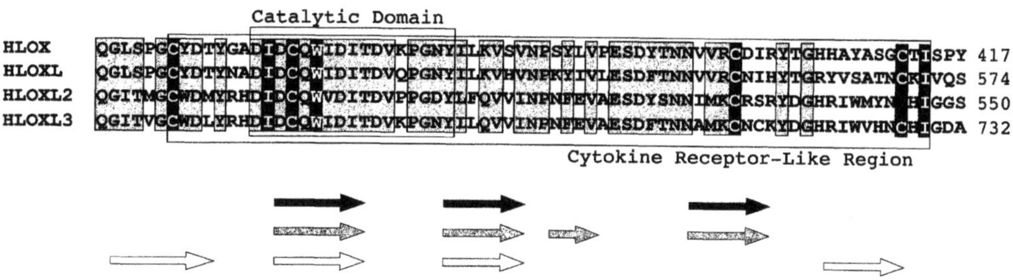

Figure 3 : Amino acid alignment of the C-terminal region of human LOX and LOX-like proteins. Conserved amino acids in each protein are shaded. Residues highlighted in white are conserved in the N-terminal region of Class 1 cytokine receptors, important for secondary and tertiary structural folds. The arrows below the sequence correspond to regions predicted to form ß-strands in LOX (black), LOXL (grey), LOXL2 and LOXL3 (white) according to a consensus of programs from the Network Protein Sequence Analysis web site (http://pbil.ibcp.fr/). The C-terminal amino acid residue position is indicated on the right of each peptide.

In summary, the catalytic site is embedded within a larger partial cytokine receptor-like domain in all LOX and LOX-like proteins. Predicted ß-strand structure within this larger domain indicates similar secondary structure for LOX and LOXL that differs only in one ß-strand. This difference may contribute to defining a different catalytic site within these proteins and may result in different substrate specificity for LOX and LOXL. The secondary structure analysis of LOXL2 and LOXL3 predicts four ß-strands in both proteins and consequently a different structure within the catalytic and cytokine domains. The presence of the partial cytokine binding site overlapping the catalytic site raises the possibility that the LOX proteins may fulfill biological roles different from the catalytic function reported for LOX and shown here for LOXL. Evidence for such roles comes from the works of Lazarus (1994) and Saito (1997) who showed a potential role for LOX and LOLX2 in leukocyte motility and cell adhesion, respectively.

Acknowledgment

This work was supported by NIH grants CA76580 and RR03061.

References

Bazan, J.F. (1990). Structural design and molecular evolution of a cytokine receptor superfamily. Proc. Natl. Acad. Sci., 87:6934-6938.

Contente, S., Kenyon, K., Rimoldi, D., Friedman, R.M. (1990) Expression of gene rrg is associated with reversion of NIH3T3 transformed by LTR-c-H-ras. Science, 249: 796-798.

Csiszar, K., Entersz, I., Trackman, P.C., Samid, D., Boyd, C.D. (1996) Functional analysis of the promoter and first intron of the human lysyl oxidase gene. Mol. Biol. Reports, 23: 97-108.

Decitre M., Gleyzal C., Raccurt M., Peyrol S., Aubert-Foucher E., Csiszar K., Sommer P. (1998) Lysyl oxidase-like protein localizes to sites of de novo fibrinogenesis in fibrosis and in the early stromal reaction of ductla breast carcinomas. Lab. Invest. 78:143-151.

DiDonato, A., Lacal, J.C., DiDuca, M., Giampuzzi, M., Ghigheri, G., Gusmano, R. (1997) Micro-injection of recombinant lysyl oxidase blocks oncogenic p21-Ha-Ras and progesterone effects on Xenopus laevis oocyte maturation. FEBS Lett., 419:63-68.

Dove, J.E., Smith, A.J., Kuchar, J., Brown, D.E., Dooley, D.M., Klinman, J.P. (1996) Identification of the quinone cofactor in a lysyl oxidase from Pichia pastoris. FEBS Lett., 398:231-234.

Eyre, D.R., Paz, M.A., Gallop, P.M. (1984) Cross-linking in collagen and elastin. Ann. Rev. Biochem., 53: 717-748.

Gacheru, S.N., Trackman, P.C., Kagan, H.M. (1990) Evidence for a functional role for histidine in lysyl oxidase catalysis. J. Biol. Chem., 263: 16704-16708.

Janes, S.M., Palcic, M.M., Scaman, C.H., Smith, A.J., Brown, D.E., Dooley, D.M., Mure, M., Klinman, J.P. (1992) Identification of topaquinone and its consensus sequence in copper amine oxidases. Biochemistry, 31:12147-12154.

Jourdan-Le Saux, C., Le Saux, O., Donlon, T., Boyd, C.D., Csiszar, K. (1998) The human lysyl oxidase-related gene (LOXL2) maps between markers D8S280 and D8S278 on chromosome 8p21.1-p21.3. Genomics, 2:305-307.

Jourdan-Le Saux, C,Tronecker h., Bogic L., Bryant-Greenwood GD., Boyd CD., Csiszar K (1999) The LOXL2 gene encodes a new lysyl oxidase-like protein and is expressed at high levels in reproductive tissue. J. Biol.Chem. 18:12939-12944.

Kagan, H.M., Reddy, V.B., Narashimhan, N., Csiszar, K. (1994) Catalytic properties and structural components of lysyl oxidase. In: Molecular Biology and Pathology of Elastic Tissue, (Editors: Mecham and Roberts), Ciba Foundation Symposium Series.

Kim,Y., Boyd, C.D., Csiszar, K. (1995) A new gene with sequence and structure similarity to the gene encoding human lysyl oxidase. J. Biol.Chem., 270:7176-718.

Kim, Y., Peyrol, S., So, C.K., Boyd, C.D., Csiszar, K. (1999) Co-expression of the lysyl oxidase-like gene (LOXL) and the gene encoding type III procollagen in induced liver fibrosis. J. Cell. Biochem., 72:181-188

Krebs, C.J., Krawetz, S.A.(1993) Lysyl oxidase copper talon complex. Biochim. Biophys. Acta, 1202:7-12.

Lazarus, H.M., Cruikshank, W.W., Narasimhan, N., Kagan, H.M., Center, D.M. (1994) Induction of human monocyte motility by lysyl oxidase. Matrix Biol., 14:727-731.

Lyles, G.A. (1996) Mammalian plasma and tissue-bound semicarbazide-sensitive amine oxidases: biochemical, pharmacological and toxicological aspects. Int. J. Biochem. Cell Biol., 28:259-274.

Mello, M.L., Contente, S., Vidal, B.C., Planding, W., Schenck, U. (1995) Modulation of ras transformation affecting chromatin supraorganization as assessed by image analysis. Exp. Cell Res., 220: 374-382.

Miyajima, A., Kitamura, T., Harada, N., Yokota, T., Arai, K. (1992). Cytokine receptors and signal transduction. Ann. Rev. Immunol., 10:295-331.

Pinnell, S.R., Martin, G.R. (1968) The cross-linking of collagen and elastin: enzymatic conversion of lysine in peptide linkage to α-aminoadipic-δ-semialdehyde (allysine) by an extract from bone. Proc. Natl. Acad. Sci. USA, 61: 708-714.

Saito, H., Papaconstantinou, J., Sato, H., Goldstein, S. (1997) Regulation of a novel gene encoding a lysyl oxidase-related protein in cellular adhesion and senescence. J. Biol. Chem., 272: 8157- 8160.

Shackleton, DR., Hulmes, DJ (1990) An ultrafiltration assay for lysyl oxidase. Anal. Biochem., 185:359-62.

Smith-Mungo, L.I., Kagan, H.M. (1998) Lysyl oxidase: properties, regulation and multiple functions in biology. Matrix Biol.,16:387-398.

Biochemistry and Molecular Biology of Vitamin B_6
and PQQ-dependent Proteins
ed. by A. Iriarte, H.M. Kagan and M. Martinez-Carrion

Structural Aspects of Lysyl Oxidase

Frederick T. Greenaway, Chunyan Qiu, Faina Ryvkin

Carlson School of Chemistry and Biochemistry, Clark University, Worcester, MA 01610, USA

Summary

Lysyl oxidase is unusual amongst the copper-containing amine oxidases. It is much smaller (29kDa) and has a high cysteine content, with ten conserved cysteines. Quantitative NTSB assays show that lysyl oxidase contains five disulfide links and no free cysteines. Thus lysyl oxidase has six covalent cross-links (including one on the cofactor) that provide constraints for molecular modeling studies directed at elucidating the secondary structure of lysyl oxidase and clarifying the geometry of the active site. Computer modeling of a 34-residue, histidine-rich section of lysyl oxidase implicated as a copper-binding region indicates that this region alone does not account for the known properties of the copper-binding site, so we are now extending modeling to the 140-amino acid C-terminal region.

Introduction

Lysyl oxidase, LO, (EC 1.4.3.13) is a highly-conserved 29kDa copper-containing amine oxidase that catalyzes the oxidation of ε-amino groups of lysine side chains in collagen and elastin (1), ultimately resulting in non-enzymatic formation of the cross-links responsible for the mechanical strength and flexibility of these important structural proteins. LO is one of the most interesting amine oxidases but is also the least characterized and most difficult to treat experimentally because of the difficulty of obtaining large quantities and its limited solubility in the absence of urea. The enzyme contains an interesting new organic cofactor, lysyl tyrosylquinone (LTQ) (2), a close relative of the topaquinone (TPQ) cofactor found in all other copper-containing amine oxidases. LO also contains one Cu(II). There are few data available regarding this copper, but one EPR study has indicated that it exists as Cu(II) in a distorted octahedral environment with a primarily nitrogen ligand environment (3). All known primary structures of LO show very high homology. The metal binding ligands are presumed to be conserved and it is usually supposed, largely by analogy with the larger copper-containing amine oxidases, that three of the eight fully-conserved histidines are bonded

```
Bovine:                                     FPQR  NQGTSDFLPSRPR      HS
Chick:   MSMYNLRCAAEENCLASSAYRADVRDYDNRVLLRFPQRVKNQGTSDFLPSRPRYSWEWHS
Human:   MSMYNLRCAAEENCLASTAYRADVRDYDHRVLLRFPQRVKNQGTSDFLPSRPRYSWEWHS-290
Rat:     MSMYNLRCAAEENCLASSAYRADVRDYDHRVLLRFPQRVKNQGTSDFLPSRPRYSWEWHS
Mouse:   MSMYNLRCAAEENCLASSAYRADVRDYDHRVLLRFPQRVKNQGTSDFLPSRPRYSWEWHS
                                                           #####*****

              MDEFSHY          VAEGHK
         CHQHYHSMDEFSHYDLLDASSHRKVAEGHKASFCLEDTSCDYGYYRRYACTAHTQGLSPG
         CHQHYHSMDEFSHYDLLDANTQRRVAEGHKASFCLEDTSCDYGYHRRFACTAHTQGLSPG-350
         CHQHYHSMDEFSHYDLLDASTQRRVAEGHKASFCLEDTSCDYGYHRRFACTAHTQGLSPG
         CHQHYHSMDEFSHYDLLDANTQRRVAEGHKASFCLEDTSCDYGYHRRFACTAHTQGLSPG
         *********************#####    ↑x-link

   DTYNA                    VSVNPSYLVPESDYSNNIVR     YTGHHAYASGQSISPY
  CYDTYNADIDCQWIDITDVKPGNYILKVSVNPSYLVPESDYSNNIVRCDIRYTGHHAYASGCTISPY
  CYDTYGADIDCQWIDITDVKPGNYILKVSVNPSYLVPESDYTNNVVRCDIRYTGHHAYASGCTISPY-417
  CYDTYAADIDCQWIDITDVQPGNYILKVSVNPSYLVPESDYSNNVVRCEIRYTGHHAYASGCTISPY
  CYDTYAADIDCQWIDITDVQPGNYILKVSVNPSYLVPESDYTNNVVRCDIRYTGHHAYASGCTISPY
      ↑x-link
```

Figure 1: Primary sequence of the C-terminus of LO from various sources. The numbering sequence is that of human LO. **** represents peptide P24; #### represents the added residues in P34. Conserved cysteines and histidines are underlined, and LTQ cross-linked residues are indicated by ↑.

---

to the copper. On the basis of structure-prediction calculations Krebs and Krawetz (4) have suggested His-292, His-294, and His-296 as likely copper ligands.

So far, nothing is known about the 3-D structure of LO, which shows unusual stability in 6 M urea and at high temperatures (*e.g.* it is fully active at 55°C). All other copper-containing amine oxidases, which are much larger with two identical subunits of about 75-90 kDa, have few or no cysteines. However, LO has ten that are fully conserved. We are investigating the hypothesis that the disulfide cross-links play a crucial role in stabilizing the structure of LO.

We set out to test the proposed model for the copper-binding site (4), and to develop a computer model for the 3-D structure of LO utilizing energy minimization methods. An adequate computer model requires that the disulfide links be identified, so we also set out to determine these. When incorporated in our model, these will provide important constraints for the calculation of low energy structures. As a preliminary step, and to test the computer methods, we chose to first work with a 34-residue histidine-rich section of the LO sequence that has been implicated in copper binding (4).

Materials and Methods

LO was isolated from bovine aorta using slight modifications of previously published procedures (1). Free cysteines were determined by DTNB (Ellman's reagent, 5,5'-dithiobis(2-nitrobenzoic acid) (5). Disulfides were determined using NTSB (disodium 2-nitro-5-thiosulfobenzoate) (6). The net absorbance at 416 nm (ε_M = 13600 $M^{-1}cm^{-1}$) was used to calculate the number of reactive thiols. Since both free cysteines and also half pairs of the cleaved disulfide bonds react with NTSB, the number of disulfide bonds can be calculated by subtracting the number of free cysteines from the total number of reactive thiols.

Molecular dynamics and energy minimization studies were carried out in two ways.

(a) Three specific histidines were assigned as copper ligands and Cu-N interatomic vectors were constrained to lengths and angles consistent with crystallographically-determined values (7) for such sites. All possible choices of histidines in P34 were chosen, in threes, and distance geometry calculations were carried out with an in-house program to identify favored structures as inputs for energy minimization and interactive modeling using the Consistent Valence Force Field, CVFF91, calculation program in Discover, with additional copper parameters from Wiesemann *et al.* (7).

(b) We then changed to calculations not based on prior assumptions about the copper ligands using Insight II (Molecular Simulations Inc., San Diego), with Discover-3 for energy minimization calculations. Calculations were first done for the apopeptide, then with Cu(I) or Cu(II) utilizing the ESFF force field and incorporating a 10Å pre-optimized layer of water at pH 7.0.

Results and Discussion

We found no free cysteines using Ellman's reagent, although we are as yet reluctant to definitively rule out the possibility that one free cysteine is present. The average number of NTBs released per mole of enzyme was 3.74 for lysozyme (which has 4 disulfides) and 4.91 for LO, with standard deviations of 0.36 and 0.23. There is some uncertainty in the interpretation of these results as we used bovine aortic LO, for which the primary sequence is not yet completely determined. The reported partial sequence (8) shows very high homology with LO from other sources (Figure 1) except for a C412Q substitution near the C-terminal end, although it now seems likely that this assignment is in error (private communication, G. Offner and P. Trackman). If there are ten cysteines, as seems probable, they are all involved in disulfide linkages.

Using procedure (a) we modeled the specific binding site suggested by Krebs and Krawetz (4), with histidines 292, 294, and 296 (see Figure 1) as the Cu(II) ligands. We conclude that not all of these histidines can bind at the same time, as the resulting structure is of very high energy. Using procedure (b) we investigated the likely Cu(II)-binding site in P24 and P34. Results without copper show that the peptides have little structure, in confirmation of our CD and NMR results (not discussed here). Addition of Cu(II) results in a well defined structure, and the lowest energy structures for Cu(II)P24 and Cu(II)P34 are similar. Both peptides bond to Cu(II) through two histidine nitrogens (His-289δ and His-296δ), one water, and two carbonyl oxygens (from Ser-302 and Gln-293). His-303 is close to the copper but not directly bonded, with the copper ion lying directly below the plane of the ring. A comparison with the metal binding site for Cu(I)P34 indicates that His-303 moves over 3Å upon metal reduction.

As the 34-residue region poorly represents the copper-binding site of LO, we are now modeling a 140-amino acid C-terminal region that contains the organic cofactor and all likely copper ligands.

Acknowledgements

This work was partially supported by grant R15 GM49392-01A1 from the National Institutes of Health. We thank Dale F. Mierke and Donald J. Nelson for their contributions.

References

1. Kagan, H. M., Trackman, P. C. (1991) Properties and function of lysyl oxidase, *Am. J. Respir. Cell Mol. Biol. 5:* 122-132.
2. Wang, S. X., Mure, M., Medzihradszky, K. F., Burlingame, A. L., Brown, D. E., Dooley, D. M., Smith, A. J., Kagan, H. M., Klinman, J. P. (1996) A cross-linked cofactor in lysyl oxidase: redox function for amino acid side chains. *Science 273:* 1078-1084.
3. Gacheru, S. N., Trackman, P. C., Shah, M. A., O'Gara, C. Y., Spacciapoli, P., Greenaway, F. T., Kagan, H. M. (1990) Structural and Catalytic Properties of Copper in Lysyl Oxidase. *J. Biol. Chem. 265:* 19022-19027.
4. Krebs, C. J., Krawetz, S. A. (1993) Lysyl oxidase copper-talon complex: a model. *Biochim. Biophys. Acta 1202:* 7-12.
5. Riddles, P. W., Blakeley, R. L., Zerner, B. (1979) Ellman's reagent: 5,5'-dithiobis(2-nitrobenzoic acid) – a reexamination. *Analytical Biochemistry 94:* 75-81.
6. Thannhauser, T. W., Konishi, Y., Scheraga, H. A. (1984) Sensitive quantitative analysis of disulfide bonds in polypeptides and proteins. *Analytical Biochemistry 138:* 181-188.
7. Wiesemann, F., Tiepel, S., Krebs, B., Howeler, U. (1994) Force field calculations on the structures of transition metal complexes. 1. Application to copper(II) complexes in square planar coordination. *Inorg. Chem. 33:* 1891-1898.
8. Trackman, P. C., Pratt, A. M., Wolanski, A., Tang, S.-S., Offner, G. D., Troxler, R. F., Kagan, H. M. (1990) Cloning of Rat aorta lysyl oxidase cDNA: complete codons and predicted amino acid sequence. *Biochemistry 29:* 4863-4870.

Biochemistry and Molecular Biology of Vitamin B_6
and PQQ-dependent Proteins
ed. by A. Iriarte, H.M. Kagan and M. Martinez-Carrion

Chemotaxis of Vascular Smooth Muscle Cells by Lysyl Oxidase

Wande Li and Herbert M. Kagan, Department of Biochemistry, Boston University School of Medicine, Boston, MA USA 02118

Summary

Purified bovine aorta lysyl oxidase conventionally oxidizes peptidyl lysine in elastin and collagen thus initiating the spontaneous formation of covalent crosslinkages between individual molecules of these fibrous proteins. The growing evidence that this catalyst may have additional functions in biology is enhanced by the present finding that it is a potent chemokine for vascular smooth muscle cells. This chemotactic effect requires the lysyl oxidase-catalyzed formation of hydrogen peroxide accompanying the oxidation of peptidyl lysine in a cell-associated substrate.

Introduction

Lysyl oxidase (LO) is a copper-dependent amine oxidase of connective tissue that plays a critical role in the biogenesis of structural matrices by catalyzing the oxidation of peptidyl lysine to peptidyl α-aminoadipic-δ-semialdehyde. This aldehyde serves as the immediate precursor of the lysine-derived crosslinkages which insolubilize and stabilize the extracellular matrix proteins, collagen and elastin. Levels of LO increase in fibrotic diseases, exemplified by atherosclerosis, scleroderma, and liver and lung fibrosis, while expression of the enzyme is decreased in diseases involving impaired copper metabolism, exemplified by Menke's disease. In addition to copper, lysyl oxidase contains a carbonyl cofactor, lysinetyrosyl quinone which is post-translationally derived from tyrosine 349 and lysine 314 of the enzyme protein, using amino acid sequence positions predicted from the sequence of rat aorta lysyl oxidase cDNA. Evidence indicates that this carbonyl cofactor condenses with the substrate amino group and serves as a transient electron sink during substrate oxidation (1). Lysyl oxidase is synthesized as a preproprotein. Following signal peptide removal, the resulting proenzyme is N-glycosylated, the copper and carbonyl cofactors are inserted or generated, respectively, and the protein is secreted as a 50 kDa proenzyme from which the 32 kDa functional enzyme is released by specific proteolysis in the extracellular space (2).

Recent evidence suggests that LO may have roles in biology in addition to that involving the crosslinking of elastin and collagen. Thus, LO has been found to suppress the oncogenic potential of ras oncogene and recent evidence suggests that this may involve the catalytic expression of the enzyme on a protein signal transduction component acting downstream of ras (3). Of further interest, LO has been shown to be chemotactic for human blood monocytes (4). This property of the enzyme is prevented by the inhibition of LO activity by β-aminopropionitrile (BAPN). While elastin and collagen proteins are conventionally accepted as substrates for this enzyme, these results indicate that other proteins may also be oxidized by this enzyme. We have previously noted that the substrate specificity of LO can be flexible. In fact, the purified bovine enzyme oxidized peptidyl lysine in a variety of basic, globular proteins, including histone H1, although not in proteins with isoelectric points below pH 8 (5), indicating that LO prefers protein substrates which are predominantly cationic.

Materials and Methods

Enzyme Purification and Assay. Calf aorta LO (32 kDa) was isolated as described (6). The specific activity of the enzyme preparation used in this study was 7.5 x 10^5 cpm of [^{3}H]H_2O release/mg protein in 2 h at 37°C, using 1.25 x 10^5 cpm of [^{3}H]tropoelastin substrate per assay (6).

Cell Culture and Preparation of Conditioned Media. VSMC were isolated from the aortas of 2-3 day old rat pups as described (7). Primary cultures were maintained in 10% fetal bovine serum (FBS) in Dulbecco's Modified Eagle's Medium (DMEM) with antibiotics at 37°C in 7.5% CO_2 atmosphere. Cells were used at passages 3-6 for this study. To obtain growth arrested cultures, cells were incubated at 3-4 x 10^3 cells/cm^2 in 10% FBS/DMEM for 48 h and then in 0.3% FBS/DMEM for 72 h. To prepare conditioned media, growth-arrested cells were cultured in serum-free DMEM in the absence or presence of 10 nM LO or 10 nM LO plus 300 units/ml catalase. After 6 h of incubation, media were collected and assayed for chemotactic activity.

Chemotaxis Assay. Chemotaxis assays were performed using multiwell cluster plates (Corning Costar Co., Cambridge. MA) containing collagen or fibronectin-coated polycarbonate filters (8 μm micropores) as described (8). Primary cultures were refed with 10% FBS/DMEM 24 h before experiments. Following trypsinization, cells were resuspended in 0.1% bovine serum albumin (BSA)/DMEM. Native BSA is not a substrate for LO (5) and was selected as the protein component of this medium on that basis. Agents tested as chemoattractants were added to the lower compartments of the wells. Cells (1 x 10^5) in 1 ml of 0.1% BSA/DMEM were placed in the upper compartments of wells. When included in the experiments, BAPN, an irreversible inhibitor of LO activity, or catalase were added to both the upper and the lower compartments at the indicated concentrations. The assembled chambers were incubated for 6 h, the microporous filters were then removed for Giemsa staining. The numbers of cells migrating from the upper to the lower surfaces of the membrane filters were counted in a total of 10 fields using a Diaphot Nikon microscope at high magnification.

Results and Discussion

As noted, LO chemotactically attracts human blood monocytes, raising the possibility that it may exert a pro-inflammatory effect in response to injury or infection. It has long been known that normally quiescent, contractile smooth muscle cells of the arterial media are activated in early stages of atherosclerosis and then migrate to the subendothelial region of the artery where they set up residence, undergo rounds of proliferation, and then secrete and, by virtue of LO activity also produced by these cells, insolubilize quantities of elastin and collagen. The deposition of these proteins in the intima contributes to the growing mass of the arterial lesion which can progress to the point that occlusion of arterial blood flow occurs with resultant ischemia of downstream tissue. We have previously shown that lysyl oxidase activity is significantly elevated in the early stages of arterial lesion development (9). In view of these considerations, we explored the possibility that vascular smooth muscle cells (VSMC) might be chemotactically activated by lysyl oxidase in an *in vitro* assay system.

As shown in Table 1, neonatal rat aorta vascular smooth muscle cells (VSMC), added as a suspension above the microporous filter of the modified Boyden chamber, are chemotactically attracted to lysyl oxidase present in solution at different concentrations below the microporous filter. The maximum effect (3.5-fold over the control) is seen at 10 nM LO (0.32 $\mu g\ ml^{-1}$).

The possibility that the catalytic activity of LO is required for this chemotactic effect was assessed by inhibiting the enzyme with BAPN and/or by assessing for the effect of the H_2O_2 product of LO action on peptidyl lysine by including catalase under selected incubation conditions. A series of experiments were performed probing for the role of LO catalysis and, if catalysis proved to be essential, for the cellular and/or media source of potential substrates of LO in the chemotactic

Table 1

Chemotaxis of VSMC by LO

LO, nM	Cells per Field	% Control	*p* value
0	4 ± 2.2[a]	(100)	
0.3	5.9 ± 2.1	148 ± 36	
1	10.4 ± 2.5	250 ± 25	<0.001[b]
3	11.4 ± 2.9	285 ± 25	<0.001
10	14 ± 3.4	350 ± 24	<0.001
30	9.3 ± 3	233 ± 32	<0.01
100	7.5 ± 2.5	188 ± 33	<0.05

[a]Data are presented as means ± standard deviations and are representative of three experiments in each of which each control or variable was assessed in triplicate incubations. [b]Values significantly different from the control were assessed from *p* values determined by ANOVA analysis.

effect. The possibility was initially assessed that an LO substrate may pre-exist within the non-conditioned DMEM growth medium by preincubating the freshly prepared, DMEM medium for 6 hours at 37 °C in the presence and absence of 10 nM LO, with each of these incubation conditions including the presence or absence of 300 units of catalase. The incubated samples were added to the bottom wells of the chemotactic chambers to assay for their chemotactic activities against the VSMC present above the filters, as described. Chemotactic assays were performed in the presence and absence of 100 μM BAPN added to suspended cells and to the test solutions at the initiation of the chemotaxis assay incubations. As shown (Figure 2, Panel A), non-conditioned medium incubated in the absence of LO did not stimulate VSMC chemotaxis (bars 1 and 2). The media preincubated with LO was strongly chemotactic (bar 3), but this response was eliminated either by inclusion of BAPN in the chemotaxis assay (bar 4) or of catalase in the preincubation (bar 5). These results indicate that the activity of LO is essential and that it is the H_2O_2 product of enzyme action that is the chemotactic mediator. If the H_2O_2 had been generated during the preincubation of LO with the non-conditioned medium, the subsequent addition of BAPN would not have inhibited the chemotactic response. Since the addition of BAPN only during the chemotaxis assay did inhibit this effect, this argues that the chemotactic response required the expression of catalytic function (and, thus, the formation of H_2O_2) during the chemotactic assay period when the carry-over, functional LO had potential access to the VSMC.

The possibility was then tested that substrates synthesized and secreted by the VSMC during culture might be oxidized in the freely soluble condition by LO added to the growth medium after the conditioning period and incubated for 6 hours with the conditioned medium only after its separation from the VSMC. As shown (Figure 2, Panel B), the conditioned medium not subsequently incubated with LO did not exhibit chemotactic activity above background (bars 6 and 7). Incubation of the conditioned medium with added LO resulted in a strong chemotactic response (bar 8) which was prevented from developing by the simultaneous presence of added catalase during the preincubation with LO (bar 9) or by the subsequent inclusion of BAPN in the chemotaxis chamber assay (bar 10). The conclusions applicable to the results of Panel A also apply here. Thus, the chemotaxis reflects the catalytic function of carry-over enzyme during the chemotaxis assay period, only. There does not appear to be evidence that soluble substrates secreted by the VSMC during the culture period are the sources of enzyme-dependent generation of H_2O_2.

The possibility was then considered that the presence of added LO during the period of cell culture, as the cells were conditioning the medium, would influnce the chemotactic response. As shown (Figure 2, Panel C) once again, conditioned medium, alone, not incubated with added LO, was not chemotactic (bars 11 and 12). The medium which had been cultured in the presence of 10 nM LO was strongly chemotactic (bar 13). This LO-supplemented, conditioned medium retained approximately 50% of its chemotactic potential when BAPN was subsequently included in the chemotaxis assay period (bar 14). Inclusion of catalase during the conditioning period together with LO completely prevented the chemotactic response (bar 15). These data indicate that the presence of LO during the conditioning period results in the LO-dependent formation of H_2O_2 to elicit a chemotactic response which, as would be expected, is not prevented by the addition of BAPN only during the chemotactic assay period. Since that addition of BAPN reduced the total chemotactic response by about 50%, this indicates that 50% of the response stems from the expression of LO function during the chemotaxis assay period and 50% from the carry-over of the H_2O_2 previously generated.

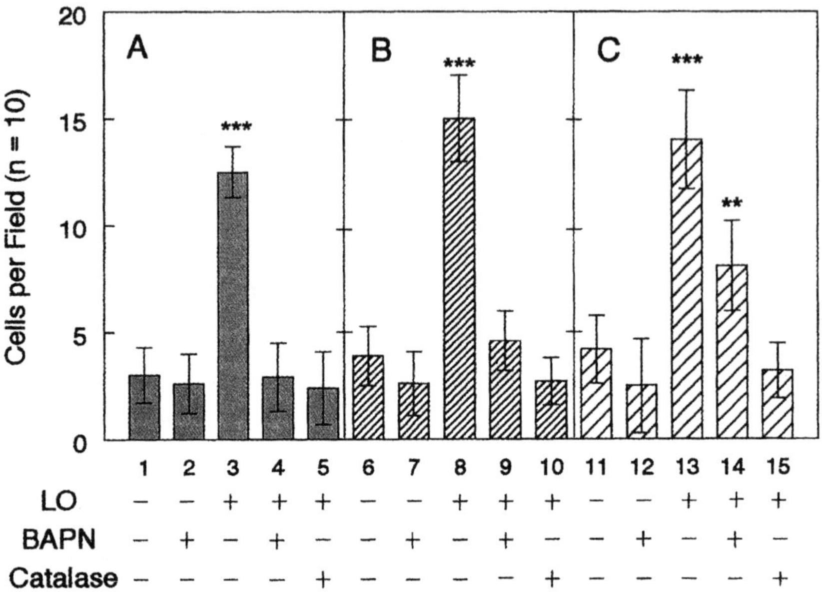

Figure 1. Conditioned and non-conditioned cell culture media as substrate sources for the production of mediators of LO-dependent chemotaxis. Panel A, non-conditioned DMEM preincubated with LO. Panel B, conditioned medium separated from cells and preincubated with LO. Panel C, DMEM incubated with LO during conditioning by VSMC. LO, BAPN and/or catalase were added as described in the text.

In toto, these data suggest that the added LO must be in contact with the cells or with proteins tightly associated with the cells for LO to generate the chemotactic H_2O_2 product. Such protein substrates might include conventional, cell-associated LO substrates such as fibrillar forms of collagen or elastin, or, in view of the potentially flexible substrate specificity of LO, other cell- or cell-membrane associated proteins. Future work will assess these possibilities.

The present data reveal that LO, which is a secreted product of vascular smooth muscle cells, can stimulate the migration of these cells to extracellular sites of this enzyme. The migration of VSMC from the medial region to the sub-endothelial, intimal region of arterial walls occurs as an early event in atherosclerosis (10). These cells contribute to the development of potentially occluding vascular lesions in this disease by proliferating, secreting and crosslinking abundant quantities of the elastin and collagen substrates of LO which they also secrete at these sites. The resulting release of H_2O_2 within the developing intimal lesion may stimulate the subsequent, continuing migration of additional medial cells into the lesion, thus contributing to the chronic nature of atherosclerotic disease. Indeed, the oxidation of lysine in collagen and elastin by LO in wound repair may be an important source of chemotactic H_2O_2 in many instances of inflammation and fibrogenesis.

Acknowledgements
The authors gratefully acknowledge the support of this research by NIH grant HL 13262.

References

1. Smith-Mungo, L. I., and Kagan, H. M. (1998) *Matrix Biol.* **16**, 387-398.
2. Panchenko M.. V., Stetler-Stevenson, W. G., Trubetskoy, O. V., Gacheru, S. N. and Kagan, H. M. (1996) *J. Biol. Chem.* **271**, 7113-7119.
3. Di Donato, A., Lacal, J. C., Di Duca, M., Giampuzzi, M., Ghiggeri, G., and Gusmano, R. (1997) *FEBS Lett.* **419**, 63-68.
4. Lazarus, H. M., Cruikshank, W. W., Narasimhan, N., Kagan, H. M., and Center, D. M..(1995) *Matrix Biol.* **14**, 727-731.
5. Kagan H. M., Williams M. A., Williamson, P. R., and Anderson, J. M. (1984) *J. Biol. Chem.* **259**, 11203-11207.
6. Bedell-Hogan, D., Trackman, P. C., Abrams, W., Rosenbloom, J., and Kagan, H. M.. (1993) *J. Biol Chem.* **268**,10345-10350.
7. Oakes, B. W., Batty, A. C., Handley, C. J., and Sandberg, L. B. (1982) *Eur. J. Cell Biol.* **27**, 34-46.
8. Kundra, V., Anand-Apt,e B., Feig, L. A., and Zetter, B. R. (1995) *J. Cell Biol.* **130**, 725-731.
9. Kagan, H. M., Raghavan, J., and Hollander, W. (1981) *Arteriosclerosis* **1**, 287-291.
10. Ross, R. (1993) *Nature* **362**, 801-809

EVOLUTION AND BIOLOGICAL IMPLICATIONS

Biochemistry and Molecular Biology of Vitamin B_6
and PQQ-dependent Proteins
ed. by A. Iriarte, H.M. Kagan and M. Martinez-Carrion

Biological Implications of the Different Hsp70 Binding Properties of Mitochondrial and Cytosolic Aspartate Aminotransferase.

A. Artigues, M.T. Bengoechea-Alonso, D.L. Crawford, A. Iriarte, and M. Martinez-Carrion.
Division of Molecular Biology and Biochemistry, School of Biological Sciences, University of Missouri- Kansas City, Kansas City, MO 64110.

Summary

The cytosolic molecular chaperone Hsp70 discriminates between the cytosolic (cAAT) and mitochondrial (mAAT) aspartate aminotransferase isozymes, recognizing and binding exclusively mAAT. By screening a library of synthetic peptides, we have identified six putative Hsp70-binding sites in mAAT. Phylogenetic analyses indicate that these Hsp70-binding sequences show less variability in mAAT than cAAT and contain more fixed differences between the two isozymes than the rest of the sequence. Thus, sequence variation between cAAT and mAAT might direct their selective interaction with molecular chaperones and thereby contribute to their correct localization in the cell.

Introduction

Members of the Hsp70 family of molecular chaperones are involved in a broad spectrum of cellular processes including protein folding and translocation across membranes, both in prokaryotes and in the different compartments of eukaryotic cells (1, 2). These functions involve the ATP-dependent interaction of the chaperone with aggregation-prone hydrophobic regions exposed in the surface of unfolded polypeptides either under conditions of stress or during synthesis of new chains on the ribosome (2). Earlier studies suggested that Hsp70 interacts with most nascent chains (3). Yet, this association appeared to be transient for some proteins while others remained bound to Hsp70 for longer periods of time. In addition, we have shown that Hsp70 binds rat liver mAAT newly synthesized in cell-free extracts, but not its cytosolic counterpart, cAAT (4). Since mAAT resides in the mitochondrial matrix but it is synthesized in the cytosol, its association with Hsp70 may contribute to the efficient targeting and translocation of this polypeptide into mitochondria, perhaps by preventing premature folding of the passenger protein in the cytosol. However, the structural features of the non-native polypeptides, which allow Hsp70 to discriminate between different potential unfolded substrates, are largely unknown. We have explored this aspect of the Hsp70 function using the mAAT/cAAT isozyme system. Our results suggest that the selective recognition of mAAT by Hsp70 might be explained by the presence of specific recognition sequences in this polypeptide, although the slower folding kinetics of mAAT may also contribute to this selectivity.

Materials and Methods

Published procedures were used in the purification of recombinant rat liver mAAT and cAAT (5) and Hsp70 from bovine brain (6). The AAT proteins were unfolded by incubation in 2 mM TrisHCl, pH 2.0 for 30 minutes at room temperature. In manual mixing experiments, refolding was initiated by the dilution of aliquots of unfolded proteins in 10 volumes of refolding buffer (100 mM Hepes, pH 7.5, 0.1 mM EDTA, 1 mM dithiothreitol, 10 μM pyridoxal 5'-phosphate) at 10°C (7). The yield of refolding was estimated by monitoring the recovery of enzyme activity as described (8). For stopped-flow fluorescence studies, samples of the enzymes at pH 2.0 were rapidly diluted (Bio-Logic SM-F4 stopped-flow spectrometer, Molecular Kinetics) 2-fold in refolding buffer at 10°C. Excitation was at 280 nm and emission was monitored using a cutoff filter of 309 nm. The dead time of the instrument was determined to be 1-2 ms. The final concentration of AAT monomer was 0.6 μM. The kinetic refolding traces were analyzed using the Bio-Kine (v. 2.1) software supplied with the stopped-flow instrument or Sigma Plot (v. 5.0; Jandel Corp.). Sedimentation velocity coefficients were measured in a Beckmann XL-A analytical ultracentrifuge using an An-60 Ti rotor at 20°C and 30,000 rpm or 15,000 rpm for the native mAAT or mAAT;Hsp70 complexes, respectively. Sedimentation coefficients were calculated using the Optmima XL-A Data Analysis software.

Results and Discussion

Both mAAT and cAAT can be refolded after denaturation at low pH (7). However, the two isozymes display very different behavior when refolding is performed in the presence of the molecular chaperone Hsp70. Addition of Hsp70 to the refolding reaction has no effect on the

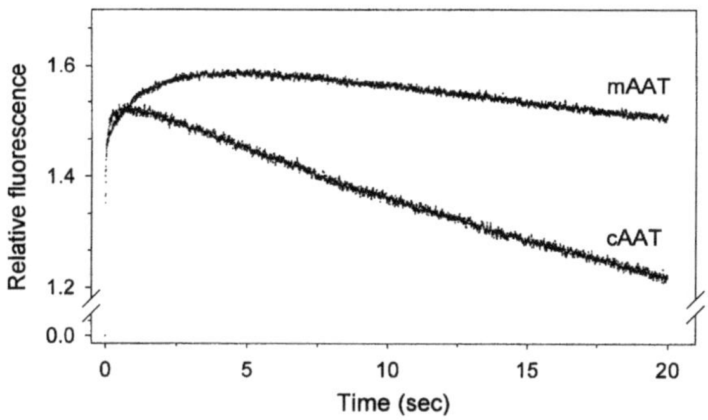

Figure 1. Kinetics of refolding of acid-unfolded AATs monitored by following intrinsic fluorescence (λ_{ex}=280 nm). The final protein concentration was 0.6 μM monomer for both proteins and the temperature was maintained at 10 °C. The refolding reaction was performed in the presence of an excess (10μM) of the coenzyme pyridoxal 5'-phosphate. The curves represent the average of 6 kinetic traces.

reactivation of cAAT but decreases dramatically (from 80 to 20%) the yield of refolding of mAAT. At low concentrations of the chaperone both mAAT and Hsp70 were recovered in insoluble aggregates (7), but they remained in solution when a large excess of Hsp70 over mAAT (100-fold molar excess or higher) was added. Co-inmunoprecipitation experiments using mAAT or Hsp70 antibodies confirmed the presence of soluble complexes between Hsp70 and mAAT. The lack of enzymatic activity and its susceptibility to trypsin hydrolysis indicate that mAAT bound to Hsp70 remains largely in an unfolded conformation. Previous studies also showed that the affinity of mAAT for binding to Hsp70 is progressively lost as the protein folds to its native state (7). Thus, Hsp70 could discriminate between substrates such as mAAT and cAAT according to their rate of folding to an intermediate it can no longer recognize. In fact, preliminary stopped-flow fluorescence analysis of the refolding reaction of the two acid-unfolded isozymes shows that cAAT fold significantly faster than mAAT (Figure 1). For both proteins, the kinetic traces were multiphasic and included an initial raise in fluorescence followed by a subsequent decrease to the native protein value. The initial rapid burst phase was completed over the first 10 ms. For mAAT, this was followed by an intermediate phase of further increase in fluorescence (rate constant, 0.8 s^{-1}, about 10% of the total increase) which was not detected in the cAAT reaction (Figure 1). The rate constant for the slow final phase of fluorescence decay is approximately 10-fold faster for cAAT (rate constant, 2.5 min^{-1}) than for mAAT (0.3 min^{-1}). The faster refolding kinetics of cAAT may explain, at least in part, the inability of Hsp70 to bind and arrest refolding of this isozyme.

The binding of Hsp70 to a particular protein might be also determined by the presence of specific recognition sequences in the polypeptide substrate. Indeed, these two factors (folding kinetics and presence of binding sequences) may in unison contribute to the ability of chaperones to discriminate between closely related substrates. The large sedimentation coefficient of the Hsp70:mAAT complex (11.3 S compared to 5.3 S for the native mAAT dimer) is compatible with the presence of various Hsp70 molecules associated with a single mAAT polypeptide chain and therefore suggest that the mAAT sequence may contain multiple binding sites for Hsp70.

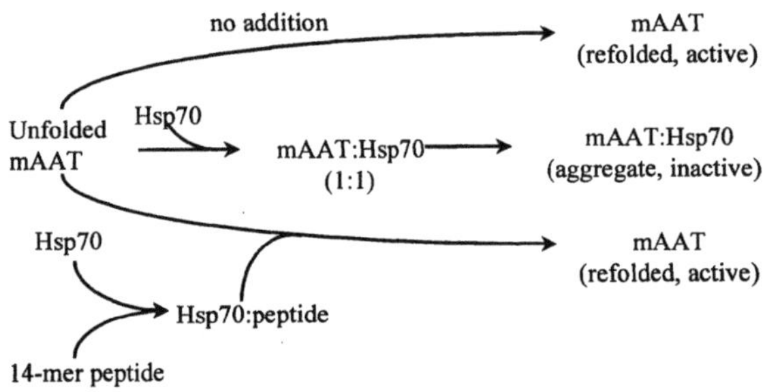

Scheme 1

Using a classical competition assay (Scheme 1) and a collection of synthetic peptides corresponding to the entire sequence of mAAT, we identified six putative Hsp70-binding sequences in the primary structure of mAAT (9). Binding to Hsp70 was confirmed by the ability of these peptides to stimulate the weak ATPase activity of Hsp70 (2). These Hsp70-recognition sites are not clustered in any particular section of the polypeptide chain and localize to regions of the chain which in the native state adopt regular secondary structure conformations, mainly α-helices, but are still mostly exposed to the solvent (Figure 2). Due to the characteristics of the assay used in these studies, additional binding sites may exist in the mAAT polypeptide that we were unable to detect because they were spliced between two consecutive 14-mer peptides. By contrast, several cAAT synthetic peptides corresponding to the Hsp70-binding regions found in mAAT were also analyzed, but none of them competed with unfolded mAAT for binding to Hsp70 or stimulated the chaperone's ATPase activity. Based on this limited screening, it appears that cAAT actually lacks recognition sites for Hsp70, at least in regions analogous to those identified for mAAT.

A structural comparison between the cytosolic and mitochondrial rat liver isozymes reveals that

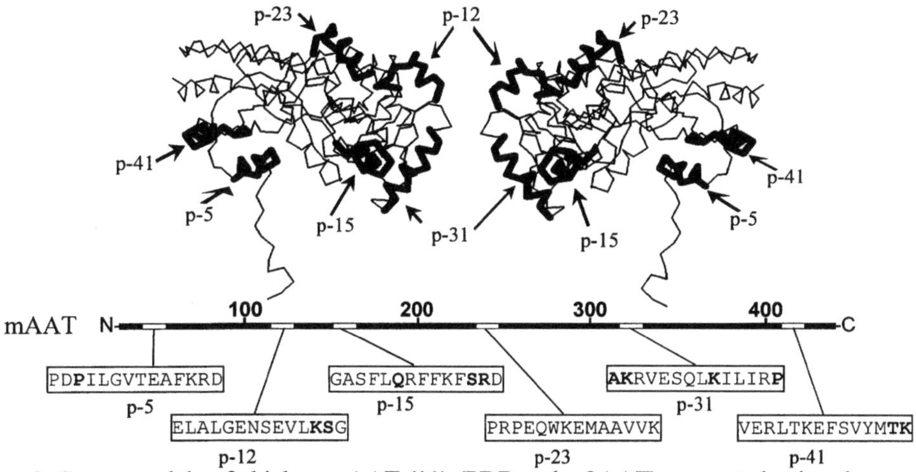

Figure 2: Space models of chicken mAAT (11) (PDB code, 8AAT) generated using the program Rasmol (RasWin Molecular Graphics). The two images represent 180°C rotated views of the same subunit. The six Hsp70-binding regions are shown in a thicker black trace. Their amino acid sequence and position within the primary structure of mAAT (*bar*) are indicated below. The m-residues (10) have been highlighted in *bold*.

these potential Hsp70-binding sites are on regions with the lowest similarity score between the two isozymes (9). Furthermore, phylogenetic analysis of different AAT isozymes using either the complete amino acid sequence or only the 84 amino acid residues from the six Hsp70-binding sequences led us to conclude that the chaperone binding sequences are evolving differently than the rest of the protein sequence. This is further supported by the fact that the amino acid variation for the six Hsp70-binding peptides among vertebrate AATs is significantly lower in mAAT (12%) than the same residues in cAAT (34%). In addition, there are more fixed differences between the two isozymes in the portions of the sequence comprising the Hsp70-binding peptides (51%) than in the remaining of the protein (39%). In a previous report of a comparative analysis of the mAAT sequences from chicken, turkey and pig with cAAT from chicken and pig, 36 residues (m-residues) were identified as unique to the mitochondrial isozymes (10). Of these, 12 (33%) are located in the six Hsp70-binding regions identified in this study (Fig 2). In fact, with the exception of p-23, all of these peptides contain one or more m-residue, which is not surprising since our more extensive phylogenetic analysis clearly shows that these Hsp70-binding sequences contain more fixed differences between the two isozymes than the rest of the polypetide. Clearly, these two isozymes are more likely to have different amino acid residues in the Hsp70-

binding regions than in any other section of the protein. In summary, based on these results, we can hypothesize that the sequence divergence observed between cytosolic and mitochondrial isozymes might has evolved to promote efficient translocation of the mitochondrial form into mitochondria and rapid folding of the cytosolic protein in the cytosol.

Acknowledgments

We thank B. Conti-Fine for her generous gift of the peptide library and J. Bollin for the purification of m- and cAAT.

References

1. Bukau, B., and Horwich, A. L. (1998) The Hsp70 and Hsp60 chaperone machines. *Cell* **92,** 351–366
2. McKay, D. B. (1993) Structure and mechanism of 70-kDa heat-shock-related proteins. *Adv. Protein Chem.* **44,** 67–98
3. Beckman R. P., Mizzen, L.A., and Welch, W.J. (1990) Interaction oh Hsp70 with newly synthesized proteins: implications for protein folding and assembly. *Science* **248,** 850-854
4. Lain, B., Iriarte, A., Mattingly, J. R., Jr., Moreno, J. I., and Martinez-Carrion, M. (1995) Structural features of the precursor to mitochondrial aspartate aminotransferase responsible for binding to Hsp70. *J. Biol. Chem.* **270,** 24732–24739
5. Mattingly, J. R., Jr., Iriarte, A., and Martinez-Carrion, M. (1995) Homologous proteins with different affinities for groEL. The refolding of the aspartate aminotransferase isozymes at varying temperatures. *J. Biol. Chem.* **270,** 1138–1148
6. Welch, W. J., and Feramisco, J. R. (1985) Rapid purification of mammalian 70,000-dalton stress proteins: affinity oft he proteins for nucleotides. *Mol. Cell. Biol.* **5,** 1229–1237
7. Artigues, A., Iriarte, A., and Martinez-Carrion, M. (1997) Refolding intermediates of acid-unfolded mitochondrial aspartate aminotransferase bind to Hsp70. *J. Biol. Chem.* **272,** 16852–16861
8. Martinez-Carrion, M., Turano, C., Chiancone, E., Bossa, F., Giartossio, A., Riva, F., and Fasella, P. (1967) Isolation and characterization of multiple forms of glutamate-aspartate aminotransferase from pig heart. *J. Biol. Chem.* **242,** 2397–2409
9. Artigues, A. Crawford, D.L. Iriarte, A., and Martinez-Carrion, M. (1998) Divergent Hsc70 binding properties of mitochondrial and cytosolic aspartate aminotransferase. *J. Biol. Chem.* **273,** 33130-33134
10. 10 Graf-Hauser, U., Wilo, K.J., and Christen, P. (1983) The covalent structure of mitochondrial aspartate aminotransferase from chicken. Identification of segments of the polypeptide chain invariant specifically in the mitochondrial isoenzyme. *J. Biol. Chem.* **258,** 813-8826
11. McPhalen C.A., Vincent M.G., and Jansonius J.N. (1992). X-ray structure refinement and comparison of three forms of mitochondrial aspartate aminotransferase. *J. Mol. Biol.* **225,** 495-517

Biochemistry and Molecular Biology of Vitamin B_6
and PQQ-dependent Proteins
ed. by A. Iriarte, H.M. Kagan and M. Martinez-Carrion

Molecular Evolution of Alanine:Glyoxylate Aminotransferase Intracellular Targeting

Joanna D. Holbrook & Christopher J. Danpure,
MRC Laboratory for Molecular Cell Biology and Department of Biology,
University College London, Gower Street, London WC1E 6BT, United Kingdom

Under the influence of dietary selection pressure, the subcellular distribution of the pyridoxal-phosphate-dependent enzyme alanine:glyoxylate aminotransferase (AGT) has changed on numerous occasions throughout the evolution of mammals. Thus, AGT is mitochondrial in carnivores, peroxisomal in herbivores, and both mitochondrial and peroxisomal in omnivores. The variable distribution of AGT results from the variable use of two alternative transcription and translation start sites, such that an N-terminal mitochondrial targeting sequence is either included or excluded from the open reading frame.

AGT is targeted to different organelles in different species.

As a result of natural selection for optimised function, the vast majority of enzymes in the eukaryotic cell have a single and unvarying location irrespective of cell type or species. However, a remarkable exception to this general rule is alanine:glyoxylate aminotransferase (AGT), an intermediary-metabolic pyridoxal-phosphate-dependent enzyme that has different subcellular distributions in different species. For example, AGT is peroxisomal in the livers of some animals, such as humans, great apes, Old World monkeys, rabbits, guinea pigs, fruit bats, koalas and wallabies, but it is mainly mitochondrial in others, such as cats, dogs, ferrets, shrews, moles, hedgehogs, frogs, newts and terrapins. In yet other animals, such as most New World monkeys, rats, mice, squirrels, and opossums, AGT is found in both peroxisomes and mitochondria. In addition to this variable organellar distribution, significant amounts of AGT can also be found in the cytosol in some animals, such as guinea pigs and frogs. The intracellular location of AGT appears to have changed on numerous occasions during the evolution of mammals and, in most cases, it appears that the changes have resulted from the loss of mitochondrial, rather than peroxisomal, AGT.

AGT distribution is related to diet and metabolic function.

The subcellular distribution of AGT in mammals appears to be related to natural diet. Thus, herbivores tend to have peroxisomal AGT, carnivores tend to have mitochondrial AGT, and omnivores tend to have AGT in both peroxisomes and mitochondria. The dual organellar distribution of AGT is thought to reflect its dual metabolic role of gluconeogenesis and glyoxylate detoxification. AGT catalyses at least two reactions of metabolic significance:-

$$\text{alanine} + \text{glyoxylate} \rightarrow \text{pyruvate} + \text{glycine} \qquad 1.$$

$$\text{serine} + \text{pyruvate} \rightarrow \text{hydroxypyruvate} + \text{alanine} \qquad 2.$$

Although both reactions 1 and 2 are predicted to be important in gluconeogenesis, reaction 1 can also be considered as glyoxylate detoxification, because a failure to transaminate glyoxylate to glycine allows it to be oxidised to oxalate instead, catalysed either by the peroxisomal enzyme glycolate oxidase (reaction 3) or the cytosolic enzyme lactate dehydrogenase (reaction 4):-

$$\text{glyoxylate} + H_2O + O_2 \rightarrow \text{oxalate} + H_2O_2 \qquad 3.$$

$$\text{glyoxylate} + H_2O + NAD^+ \rightarrow \text{oxalate} + NADH + H^+ \qquad 4.$$

Reactions 3 and 4 are potentially life-threatening because the calcium salt of oxalate is highly-insoluble and can crystallise out in regions of high concentration, such as the kidney and urinary tract, to form stones. This happens in the hereditary disorder primary hyperoxaluria type 1 (PH1), a progressive condition, caused by a functional deficiency of AGT, that can eventually lead to chronic kidney failure and death.

Although for its gluconeogenic function, it is predicted that reaction 1 would be most efficiently located in the mitochondria or cytosol, for its glyoxylate detoxification function it would be better placed in the peroxisomes, which is the major site of glyoxylate synthesis. Proof of the latter is amply shown by the observation that one third of PH1 patients have disease, not because they have no AGT, but because it is in the wrong place (i.e. it is mistargeted from the peroxisomes to the mitochondria).

The relative importance of these two putative roles for AGT depends on diet. Thus, gluconeogenesis is the more important function of AGT in carnivores, while glyoxylate detoxification is more important in herbivores. Both functions are presumed to be equally important in omnivores. The observation that normal human AGT has a 'herbivorous' (i.e. peroxisomal) distribution, despite the fact that we are usually considered to be omnivores, is likely to be a reflection our herbivorous ancestry.

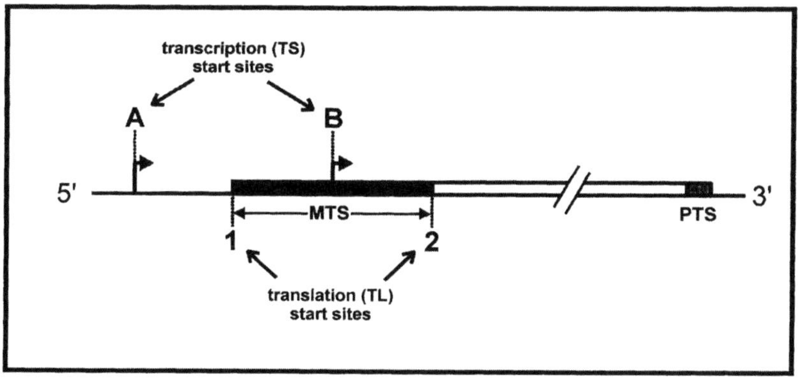

Figure 1. Simplified structure of the mammalian AGT gene

The ancestral transcription (TS) start sites (A & B) and translation start sites (1 & 2) are described in the text. The N-terminal mitochondrial targeting sequence (MTS) is encoded by the region between 1 and 2. The peroxisomal targeting sequence (PTS) is a tripeptide located at the C-terminus.

The AGT gene can encode two targeting sequences.

The archetypal mammalian AGT gene (Figure 1) has the potential to encode an N-terminal mitochondrial targeting sequence (MTS) and a C-terminal type 1 peroxisomal targeting sequence (PTS1). The final intracellular distribution of AGT is dependent upon the expression of the MTS rather than the PTS1, as the former is functionally dominant over the latter. Thus, a decrease in mitochondrial targeting can automatically lead to an increase in peroxisomal targeting. The twenty-two amino acid MTS can be included or excluded from the open reading frame as a result of the variable use of two transcription start sites (A and B), and two in-frame translation start sites (1 and 2) (Figures 1 & 2).

Species	TS start	TL start	N-terminal MTS (residues -22 to -1)	C-terminal PTS (residues 390 to 392)	Subcell. distrib.
Human	A	2		KKL	P
Marmoset	A+B	1+2	MFQALAKASAALGPRAAGWVRT	KKL	M+P
Rabbit	A	2		SQL	P
Cat	A	1+(2)	MFRALARASATLGPQVAGWART	NKL	M+(P)
Rat	A+B	1+2	MFRMLAKASVTLGSRAASWVRN	NKL	M+P
Guinea Pig	B	2		HRL	P

Figure 2. Effect of transcription (TS) and translation (TL) start site usage on the targeting sequences included in the open reading frame and the subcellular distribution of AGT.

Abbreviations are the same as those used in Figure 1. Subcellular distribution = mitochondrial (M), peroxisomal (P), or both mitochondrial and peroxisomal (M+P). In the cat, although most translation starts from site 1, there is some readthrough to site 2 due to leaky ribosome scanning. Thus, although most AGT is targeted to mitochondria a small amount is also targeted to peroxisomes.

In what we presume to be the more primitive condition, all sites (i.e. A and B, 1 and 2) are present. In this situation, currently found in animals such as the rat and marmoset, two proteins of different sizes are synthesised. The longer form which contains the MTS is targeted to the mitochondria, whereas the shorter form is targeted to the peroxisomes. However if one or more of these sites are absent then the subcellular distribution of AGT is different. For example, translation start site 1 has been lost by point mutation independently in the human, rabbit and guinea pig ancestral lines (Figure 2). The resulting exclusion of the MTS from the open reading frame allows all the AGT to be targeted to the peroxisomes. In the cat, both translation start sites are still present but transcription start site B has been lost, consequently most translation begins at site 1 and most AGT is targeted to the mitochondria. A small proportion of the AGT is nevertheless peroxisomal in the cat due to leaky ribosome scanning and some translation from site 2.

The amino acid composition of the MTS and PTS1 of AGT have also changed during the evolution of mammals (Figure 2). Four different C-termini have so far been identified, none of which match completely the consensus PTS1 sequence S/A/C-K/R/H-L/M. The PTS1 of AGT is highly context specific insofar as it is able to target only AGT to peroxisomes, not reporter proteins. Thus, as yet unidentified ancillary peroxisomal topogenic sequences are likely to be present in other parts of the AGT molecule. There is some evidence that AGT PTS1s are less efficient than the consensus PTS1 SKL. The evolutionary significance of this is unclear, but it is possible that in the case of some species, at least, a more efficient PTS1 would compete too strongly with the MTS.

The evolutionary variability of the amino acid composition of the AGT MTS could be due to the well-recognised degeneracy of MTSs or might be a reflection of previous exclusion from the open reading frame (with no pressure to conserve) followed by reinstitution.

Variable AGT targeting in primates is a molecular adaptation to positive selection pressure.

Recent work on the molecular evolution of AGT targeting in primates has contributed more evidence that AGT distribution has varied under the positive selection pressure of dietary changes. Although the distribution of AGT in catarrhines (humans, great apes and Old World monkeys) looks uniform (i.e. it is peroxisomal in all species), there is evidence that the MTS (and hence mitochondrial targeting) has been lost independently on at least four occasions. In addition, the MTS has been lost at least once in the New World monkeys even though in most species it has been retained. Analysis of the relative numbers of synonymous and non-synonymous mutations accumulated in the region of the AGT gene encoding the MTS suggests that there has been recent strong, yet episodic, positive selection pressure to lose, or decrease the efficiency of, mitochondrial AGT targeting in anthropoid primates. This selection pressure could be associated with a recent shift in diet, from one that is insectivorous/frugivorous to one that is frugivorous/folivorous. Such a dietary shift would be predicted to change the optimal AGT distribution from one that is both mitochondrial and peroxisomal to one that is just peroxisomal.

Unanswered questions.

The variability of AGT targeting during the evolution of mammals is unparalleled. Although much has been learnt about its molecular bases, there are still a number of enigmas and many questions still remain. For example:-

- Is mitochondrial AGT a selective disadvantage to herbivores, as well as being an advantage for carnivores?
- Would up-regulation of AGT expression in both organelles be as adaptive as a change in distribution?
- Why do ruminants, the classic herbivores, have low to non-existent levels of AGT?
- What was the distribution of AGT in the ancestral mammal, ancestral vertebrate, ancestral metazoan and ancestral eukaryote?
- If humans have a non-ideal distribution of AGT, is there any selection pressure to reacquire some mitochondrial enzyme?

Further reading.

Danpure, C.J. Peroxisomal and mitochondrial targeting of alanine:glyoxylate aminotransferase in mammalian evolution and disease. *Bioessays,* 19, 317-326 (1997).

Danpure, C.J., Fryer, P., Jennings, P.R., Allsop, J., Griffiths, S. & Cunningham, A. Evolution of alanine:glyoxylate aminotransferase 1 peroxisomal and mitochondrial targeting. A survey of its subcellular distribution in the livers of various representatives of the classes Mammalia, Aves and Amphibia. *Eur. J. Cell Biol.* 64, 295-313 (1994).

Danpure, C.J. & Purdue, P.E. Primary hyperoxaluria. In "*The Metabolic and Molecular Bases of Inherited Disease*", eds. Scriver, C.R., Beaudet, A.L., Sly, W.S. & Valle, D., 7th edition, chapter 75, pp 2385-2424, McGraw-Hill, New York (1995).

Holbrook, J.D., Birdsey, G.M., Yang, Z., Bruford, M.F. & Danpure, C.J. Molecular adaptation of alanine:glyoxylate aminotransferase targeting in primates. *Mol. Biol. Evol.* 17, 387-400 (2000).

Oatey, P.B., Lumb, M.J. & Danpure, C.J. Molecular basis of the variable mitochondrial and peroxisomal localisation of alanine:glyoxylate aminotransferase. *Eur. J. Biochem.* 241, 374-385 (1996).

Biochemistry and Molecular Biology of Vitamin B_6
and PQQ-dependent Proteins
ed. by A. Iriarte, H.M. Kagan and M. Martinez-Carrion

Common Structural Elements in the Architecture of the Cofactor-Binding Domains in Unrelated Families of Pyridoxal Phosphate-Dependent Enzymes

A.I. Denesyuk, K.A. Denessiouk[*], J.V. Lehtonen[*], T. Korpela and M.S. Johnson[*]

Finnish-Russian Joint Biotechnology Laboratory, University of Turku, BioCity 6A, FIN-20520 Turku, Finland
[*]Department of Biochemistry and Pharmacy, Åbo Akademi University, P.O. Box 66, FIN-20521 Turku and Turku Centre for Biotechnology, Åbo Akademi University and University of Turku, P.O. Box 123, FIN-20521 Turku, Finland

Summary

A detailed comparison of the structures of AspAT, TrpSβ, AlaR, D-AAT and GP has revealed seven common structural segments of the polypeptide chain, which form an extensive common structural organization of the backbone chain responsible for the appropriate disposition of "key" residues interacting with PLPs in these enzymes. Four of the seven fragments in TrpSβ, AlaR, and GP share a similar α-β-α-β motif. In addition, this common α-β-α-β supersecondary structure contains an analogous hydrophobic "minicore" formed from three residues. Despite these similarities, the five enzyme families represented by spatial structures and compared within this study are not considered to be related through a common ancestral protein.

Introduction

An earlier study of three-dimensional structures of pyridoxal-5'-phosphate (PLP)-dependent enzymes has shown that there are at least five distinct folds of the PLP-binding domain (Alexander et al., 1994; Grishin et al., 1995; John, 1995; Jansonius, 1998). These five dissimilar folds are formed: (a) in the large domain of aspartate aminotransferase (AspAT) and in the other members of the α/γ-family of PLP-dependent enzymes; (b) in the carboxy-terminal domain of the β-subunit of tryptophan synthase (TrpSβ) and in the amino-terminal domain of threonine deaminase (TD); (c) in the carboxy-terminal domains of D-amino acid aminotransferase (D-AAT) and branched-chain amino acid aminotransferase (BCAT); d) in the barrel domains of alanine racemase (AlaR) and the mouse ornithine decarboxylase (mODC), and in the yeast hypothetical protein; and e) in the carboxy-terminal domain of glycogen phosphorylase (GP).

Although the topologies of these five type folds do not resemble each other, there are definite local structural similarities in the construction of their PLP-binding sites. The phosphate group of PLP is bound near the amino-terminus of an "anchoring" α-helix and the adjacent edge of the corresponding β-sheet in each of the folds. The present work demonstrates more extensive structural similarities among PLP-binding domains in these folds than was observed previously. The similarities include both the "anchoring" α-helix and an adjacent β-sheet. These structural units play an important role in the recognition of PLP in the five different families of PLP-dependent enzymes represented by solved structures.

Materials and Methods

The coordinates of AspAT from *Escherichia coli* (access code 1ASN), TrpSβ from *Salmonella typhimurium* (1A5A), TD from *Escherichia coli* (1TDJ), D-AAT from *Bacillus sphaericus* (1DAA), AlaR from *Bacillus stearothermophilus* (1BD0), mODC (7ODC), the yeast hypothetical protein (1B54) and GP from rabbit skeletal muscle (1A8I) were obtained from the Brookhaven Protein Data Bank (PDB; Bernstein et al., 1977). The molecular modeling software package SYBYL (Tripos Associates, Inc., St. Louis, MO) and Silicon Graphics ONYX and O2 workstations were used to superimpose the three-dimensional structures. Subsequent comparisons of the three-dimensional structures were made with computer programs described in Lehtonen et al. (1999) using two dual-processor personal computers under the Linux operating system.

Results and Discussion

Overall Multiple Structural Alignment: First of all, a detailed comparison of the structures of AspAT, TrpSβ, D-AAT, AlaR and GP, which are representative proteins of five dissimilar folds, was carried out (Denessiouk et al., 1999). The phosphate group of PLP is bound near the amino-terminus of an "anchoring" α-helix ("H5" in AspAT, "H25" in TrpSβ, "H16" in D-AAT, "H9" in AlaR, and "H31" in GP) and the adjacent edge of the corresponding β-sheet in each of these enzymes (secondary structure assignments are taken according to the X-ray coordinate files from the PDB). Thus, preliminary "hand-made" pairwise superpositions of the PLP-binding domains of the five protein structures were made on the basis of the C_α-atoms from the amino-termini of these α-helices using the "Fit Atoms" command of the computer program SYBYL. As a result, β-strands "S3s1" in AspAT, "S4C" in TrpSβ, "S2D" in D-AAT, "S8B" in AlaR, and "S1G" in GP were found superimposed as well. The protein structures were superimposed again over the C_α-atoms from the "anchoring" α-helix and the β-strand. Equivalent segments of three or more consecutive residues were then identified and extended until the main chains diverged completely (the direction of the polypeptide chain was not taken into consideration in these comparisons). The procedure was repeated several times and, finally, a number of strands, helices and loops totalling from 54 (D-AAT/AlaR pair) to 78 (TrpSβ/GP pair) structurally equivalent C_α-atom pairs (equivalencies) were defined and superimposed.

Independently, the five proteins were compared with our newly developed computer program GENFIT. The GENFIT algorithm is described in Lehtonen et al. (1999), and can make structural comparisons automatically and without any input from the user except for the structural coordinate files. After superposition, equivalencies were found to reside within 7 to 9 segments depending on the pair of proteins compared. Seven segments are common to all five enzymes and contain 43 C_α-atoms.

Equivalent Segments Form Analogous Supersecondary Structures: One of the most noticeable features among all ten possible pairwise comparisons of PLP-binding domains is that AlaR, TrpSβ and GP have analogous super-secondary structures that surround their PLP-binding sites (Fig. 1). Four of the seven common fragments (28 residues) in these enzymes share a similar α-β-α-β motif, although the enzymes have completely different amino acid sequences

	I	II
	H24(207-219)	S4C(226-230)
TrpSβ	209 GEETKAQIL 217	225 DA**V**IA 229
	H9(160-176)	S4A(182-186)
TD	170 ALELLQQDA 178	181 DR**V**FV 185
	H8(176-191)	S8B(198-200)
AlaR	178 FSYQYTRFL 186	199 VH**C**AN 203
	H7(197-213)	S1C(218-220)
H.protein	195 ENRDFATLV 203	219 LS**M**GM 223
	H8(206-225)	S3B(231-233)
mODC	208 FVQAVSDAR 216	232 LD**I**GG 236
	H30(650-659)	S1G(662-665)
GP	650 VSLAEKVIP 658	661 DLSEQ 665

	III	IV
	H25(235-244)	S5C(251-259)
TrpSβ	234 GSNAIG**M**FAD 243	250 VG**L**I 253
	H10(191-203)	S5A(208-214)
TD	190 GGLAAG**V**AVL 199	207 IK**V**I 210
	H9(204-209)	S1B(217-220)
AlaR	204 SAASLR**F**PDR 213	216 NM**V**R 219
	H8(227-232)	S2A(237-240)
H.protein	224 SADFRE**A**IRQ 233	236 AE**V**R 239
	H9(248-262)	S2B(270-273)
mODC	249 EEITSV**I**NPA 258	271 II**A**E 274
	H31(677-683)	S2G(687-690)
GP	675 GTGNMK**F**MLN 684	686 AL**T**I 689

Figure 1: Four of the seven equivalent segments in TrpSβ, TD, AlaR, the yeast hypothetical protein, mODC, and GP form equivalent α-β-α-β supersecondary structures about the cofactor binding sites, which incorporate "anchoring" α-helices (segment III). Secondary structure assignments are shown above the sequences (H = helix; S = strand) and according to the X-ray coordinate files from the PDB (Bernstein et al., 1977). The structurally-equivalent positions containing common hydrophobic amino acids are marked in bold. The single-letter amino acid code is used.

and different functions, and the similarity in their folds does not extend further. The equivalent elements of secondary structure run in the same direction within this α-β-α-β motif among the three proteins.

It is possible to speculate that this α-β-α-β motif was initially exploited to form "PLP-anchoring units" in unrelated families of PLP-dependent enzymes. It is known, for example, that many enzymes with a β/α-fold have a carboxy-terminal phosphate-binding site in common, which involves the glycine-rich region between a β-strand and an α-helix (Farber & Petsko, 1990; Brändén, 1991; Wilmanns et al., 1991; Bork et al., 1995). In order to investigate the relationship between the α-β-α-β motifs of three different folds, we also included comparisons of the structures of TD, the yeast hypothetical protein and mODC. The common α-β-α-β motif and the root-mean-square deviation (RMSD) for the equivalent 28 C_α-atom pairs in all six enzymes are given in Fig. 1 and Table 1 respectively. The overall result for the spatial comparison of α-β-α-β supersecondary structures is that the differences between motifs of TrpSβ, AlaR and GP with the different folds are practically the same as between the motifs of AlaR, the yeast hypothetical protein and mODC with the similar folds.

Table 1. The resulting RMSD (Å) between the common α-β-α-β supersecondary structures (28 equivalences) in six enzymes.

Enzyme	TD	AlaR	H. protein	mODC	GP
TrpSβ	1.14	3.16	2.92	2.90	3.90
TD		2.88	2.95	2.63	3.80
AlaR			2.44	3.35	3.40
H. protein				3.37	3.19
mODC					4.07

Equivalent Hydrophobic Positions: There are no identical amino acids common to all six α-β-α-β motifs after the structure-based alignment (Fig. 1). Only two equivalent positions (marked in bold in Fig. 1) from segments III and IV contain non-polar residues in all six enzymes. One additional position in segment II (also marked in bold) contains a hydrophobic residue in each of these enzymes but not in GP. All of these hydrophobic and non-polar residues are clustered together in space forming a common hydrophobic minicore in the cofactor binding site.

Final Remarks: Despite these similarities, it is generally accepted that these proteins do not share a common heritage, but have arisen on five separate occasions. The common and contiguous α-β-α-β structure accounts for only 28 residues and all five enzymes differ greatly in both the orientation of the PLP pyridoxal rings and their contacts with residues close to the common structural elements.

Acknowledgements

This research was supported from grants from the Academy of Finland, the Technology Development Centre Finland ʹTEKESʹ, the Hasselblad Foundation, and the Graduate School in Informational and Structural Biology.

References

Alexander, F.W., Sandmeier, E., Mehta, P.K. and Christen, P. (1994) Evolutionary relationships among pyridoxal-5'-phosphate-dependent enzymes. *Eur. J. Biochem. 219*: 953-960.

Bernstein, F.C., Koetzle, T.F., Williams, G.J.B., Meyer, E.J., Jr., Brice, M.D., Rodgers, J.K., Kennard, O., Shimanouchi, I. and Tasumi, M. (1977) The Protein Data Bank: a computer-based archival file for macromolecular structures. *J. Mol. Biol. 12*: 535-542.

Bork, P., Gellerich, J., Groth, H., Hooft, R. and Martin, F. (1995) Divergent evolution of a β/α-barrel subclass: detection of numerous phosphate-binding sites by motif search. *Protein Sci. 4*: 268-274.

Brändén, C.I. (1991) The TIM barrel - The most frequently occurring folding motif in proteins. *Curr. Opin. Struct. Biol. 1*: 978-983.

Denessiouk, K.A., Denesyuk, A.I., Lehtonen, J.V., Korpela, T. and Johnson, M.S. (1999) Common structural elements in the architecture of the cofactor-binding domains in unrelated families of pyridoxal phosphate-dependent enzymes. *Proteins 35*: 250-261.

Farber, G.K. and Petsko, G.A. (1990) The evolution of α/β barrel enzymes. *Trends Biochem. Sci. 15*: 228-234.

Grishin, N.V., Phillips, M.A. and Goldsmith, E.J. (1995) Modeling of the spatial structure of eukaryotic ornithine decarboxylases. *Protein Sci. 4*: 1291-1304.

Jansonius, J.N. (1998) Structure, evolution and action of vitamin B_6-dependent enzymes. *Curr. Opin. Struct. Biol. 8*: 759-769.

John, R.A. (1995) Pyridoxal phosphate-dependent enzymes. *Biochim. Biophys. Acta 1248*: 81-96.

Lehtonen, J.V., Denessiouk, K.A., May, A.C.W. and Johnson M.S. (1999) A generic non-linear alignment algorithm for homologous and non-homologous protein structure comparisons. *Proteins 34*: 341-355.

Wilmanns, M., Hyde, C.C., Davies, D.R., Kirschner, K. and Jansonius, J.N. (1991) Structural conservation in parallel β/α-barrel enzymes that catalyze three sequential reactions in the pathway of tryptophan biosynthesis. *Biochemistry 30*: 9161-9169.

Biochemistry and Molecular Biology of Vitamin B_6
and PQQ-dependent Proteins
ed. by A. Iriarte, H.M. Kagan and M. Martinez-Carrion

Mitochondrial Localization of Eukaryotic NifS-Like Proteins

Y. Nakai, Y. Yoshihara,[§] M. Nakai,[¶] H. Hayashi, and H. Kagamiyama

Dept. Biochemistry, Osaka Medical College, Takatsuki, Osaka 569-8686, [§]Laboratoy for Neurobiology of Synapse, RIKEN Brain Science Institute, Saitama 351-0198, and [¶]Division of Enzymology, Institute for Protein Research, Osaka Univ., Suita, Osaka 565-0871, Japan

Summary

We have isolated mouse and yeast nifS-like genes and characterized their products (mNfs1 and yNFS1, respectively). We found that both proteins have an amino-terminal extension, which contains a typical mitochondrial targeting presequence, followed by the several characteristic features common to bacterial NifS proteins. By Western blot analysis, both the mNfs1 and the yNFS1 were found in the mitochondrial fraction. The mNfs1 was shown to distribute in various mouse tissues. We also found that the y*NFS1* gene-disrupted mutant is lethal. These results suggest that both eukaryotic NifS-like proteins may play some significant roles in mitochondrial functions.

Introduction

To date, various iron-sulfur cluster-containing proteins (iron-sulfur proteins) have been known to be involved in various redox reactions of metabolic pathways and in electron transfer of the respiratory chain (Cammack, 1992, Johnson,.1998). In diazotrophic bacterium, the nitrogenase, a complex of the several iron-sulfur proteins of different cluster-types, is synthesized from the products of the nitrogen fixation genes (*nif* genes). In *Azotobacter vinelandii* NifS was shown to possess a PLP-mediated cysteine desulfurase activity that produces L-alanine and elemental sulfur from L-cysteine (Zheng et al., 1993), and its physiological function has been considered to supply an elemental sulfur to the iron-sulfur ([Fe-S]) clusters of the active site of nitrogenase (Zheng et al., 1994, Zheng and Dean, 1994). Recently, another NifS-like protein named IscS has been found in *A. vinelandii* (Zheng et al., 1998), which is very similar to the *Eschelichia coli* NifS-like protein IscS (Flint, 1996), and both IscS proteins are thought to be participate in the cluster assembly of constitutively expressed iron-sulfur proteins (Zheng et al., 1998).

In contrast to various descriptions of prokaryotic *nifS* (and *nifS*-like) genes, few studies have been devoted to the eukaryotic NifS-like proteins. The best-characterized *nifS*-like gene in eukaryote is *Saccharomyces cerevisiae NFS1* (y*NFS1* in the present study, alternatively named as *spl1*) which may be involved in tRNA splicing (Kolman and Söll, 1993). In the present study, we have cloned the mouse *nifS*-like cDNA (named m*Nfs1*) and identified its product, the mNfs1 protein, in various tissues. Mitochondrial localization of both the mNfs1 protein and its yeast counterpart has also been demonstrated.

Materials and Methods

The full-length m*Nfs1* gene was obtained from an unamplified mouse olfactory epithelium λgt10 cDNA library (5×10^5 recombinant phages) using a probe of a partial cDNA fragment of m-*Nfs1* which was amplified with degenerated primers as the template. On the other hands, entire y*NFS1* gene was amplified by PCR from the *Saccharomyces cerevisiae* (strain D273-10B (*MAT* α)) genomic DNA.

Deduced sequences of mNfs1 and yNFS1 proteins were aligned with those of other bacterial NifS/IscS proteins and the human counterpart. Mitochondrial targeting sequences were searched by a protein sorting signal searching program (PSORT). Possible PLP-binding lysine and its neighboring residues are detected by the program PROSITE.

Eleven tissues (brain, heart, lung, thymus, liver, pancreas, stomach, spleen, kidney, skeletal muscle, and testis) from the mouse (ddY, 5 weeks, males) were extracted, homogenized, and suspended in the phosphate-buffered saline. Mouse liver cells were further separated into subcellular (nuclear and cell-debris, mitochondria, microsome, and cytosol) fractions by sucrose density gradient centrifugation.

The yNFS1 proteins are constitutively expressed with a promoter of glycelaldehyde-3-phosphate dehydrogenase in the yeast strain W303-1B (*MAT α ade2-2 his3-11,15 ura3-1 leu2-3,112 trp1-1 can1-100*), or YN101, a W303-1B derivative strain in which chromosomal y*NFS1* was expressed under the galactose-inducible promoter. Subcellular and submitochondrial fractionations of yeast cells were performed essentially according to the published procedure (Nakai et al., 1993).

Western analyses of mNfs1 and yNFS1 proteins were performed with purified anti-mNfs1 polyclonal antibody. The immunoreactive proteins were detected with ECL Western Detection Kit and Hyper ECL film (Amersham, UK).

Results and Discussion

The complete cDNA of 2,051 nucleotides which contained the entire m*Nfs1* cDNA was cloned and sequenced (the EMBL accession number AJ222660). The m*Nfs1* potentially encodes a protein of 459 amino acid residues with a calculated molecular mass of 50,716 daltons (Da). We also obtained the y*NFS1* gene, which encoded a protein of 497 amino acid residues, from the genomic DNA by PCR amplification. The amino acid sequence of mNfs1 shows the highest degree of homology to that of the yNFS1 (68.5% identity) and also exhibits significant similarity to those of diazotrophic bacterial NifS proteins (39.7 – 43.9% identity).

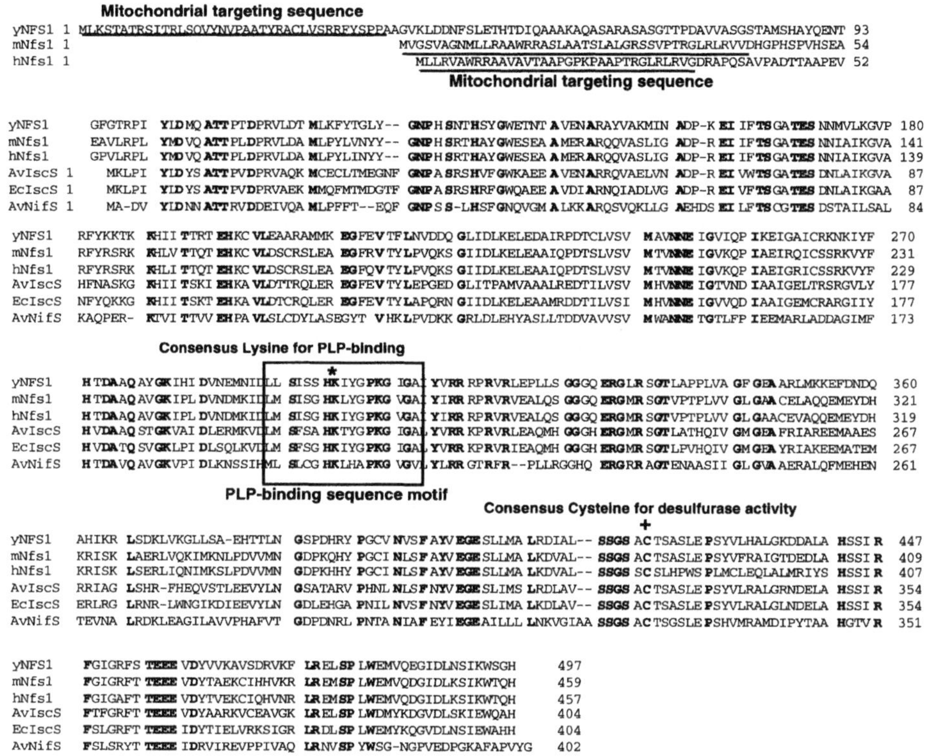

Figure 1. Amino acid sequence alignment of eukaryotic NifS-like proteins and bacterial NifS/IscS proteins. yNFS1; yeast NifS-like protein, mNfs1; mouse NifS-like protein, hNfs1; human NifS-like protein (Land and Rouault,1998), AvIscS; *Azotobacter vinelandii* IscS (Zheng et al., 1998), AvNifS; *Azotobacter vinelandii* NifS, EcIscS; *Escherichia coli* IscS (Flint, 1996). Amino acid residues conserved among them all were described in bold. Mitochondrial targeting presequences were underlined. A putative PLP-binding lysine (*) and its neighborhoods were in a box. A cysteine for the desulfurase activity was indicated with plus (+).

Protein sequences of mNfs1 and yNFS1 were aligned with that of the human NifS-like protein hNfs1 (Land and Rouault,1998) and those of the bacterial (*Escherichia coli* and *Azotobacter vinelandii*) NifS/IscS proteins (Fig. 1). MOTIF–Search analysis identified a putative pyridoxal phosphate (PLP)-binding motif in both the mNfs1 and the yNFS1 sequences; the sequence of the motif is conserved among bacterial NifS/IscS proteins (Fig. 1). It contains a lysine (marked with an star in Fig. 1) which is believed to form a Schiff base with PLP. The cysteine residue essential for the cysteine desulfurase activity of *A. vinelandii* NifS, conserved among the bacterial NifS/IscS proteins, is also found in the mNfs1 protein (Cys383; marked with a plus in Fig. 1).

Both eukaryotic NifS-like proteins have an amino-terminal extension more than fifty residues longer than those of bacterial NifS proteins. Although the amino-terminal 58 residues of mNfs1 have no significant sequence similarities to the corresponding region of yNFS1, PSORT program detected mitochondrial sorting signals in the amino-terminal extensions of both (underlined in Fig. 1).

We found that the m*Nfs1* transcript of the predicted size (2.1-kb) was detected in various mouse tissues, suggesting that the m*Nfs1* is ubiquitously expressed. With anti-recombinant mNfs1-antibody, we also detected the mature-size mNfs1 protein of 47-kDa in various mouse tissues. The 47-kDa immunoreactive protein was detected mainly in the mitochondrial fraction of mouse liver as was the mitochondrial marker protein F1-β (the β-subunit of mitochondrial F1-ATPase). We also showed that the mNfs1 exist in the matrix fraction of mitochondria.

The intracellular localization of yNFS1 was also analyzed by Western blotting using anti-mNfs1 antibody. The immunoreactive 50-kDa protein, which corresponded to the mature-size protein, was associated mainly with the mitochondrial fraction. A small fraction of the 50-kDa protein was also detected in the cell debris fraction which contains unbroken cells and a nucleus. Upon submitochondrial fractionation, the 50-kDa protein was predominantly recovered in the matrix fraction, thus suggesting that the yNFS1 protein also exists in the mitochondrial matrix.

We constructed a yeast strain YN101 in which the chromosomal y*Nfs1* gene was expressed under the inducible promoter. When the chromosomal y*Nfs1* in YN101 was depressed, cells became lethal. However, when a plasmid-borne yNFS1 was expressed under a constitutive promoter, the cell viability was restored under the chromosomal y*NFS1* depression. It indicates that the yNfs1 plays an essential physiological function(s) in yeast cells. Kolman and Söll have also shown that the *S. cerevisiae NFS1* gene to be essential for cell viability because the *NFS1* disruptant exhibits haploid lethality (Kolman and Söll, 1993). they have suggested that the *S. cerevisiae NFS1* gene is involved in tRNA splicing based on the fact that its mutated dominant allele (*NFS1-1*) could complement the splicing-deficiency of an inactivated supressor tRNA gene (Kolman and Söll, 1993). But the functional relationship between these enzymes and the yNFS1 has not yet been clarified.

A. vinelandii NifS possesses a cysteine desulfuration activity (Zheng et al., 1993) and is believed to support the [Fe-S] cluster formation in nitrogenase maturation *in vivo*. However, it can also participate in the construction of [Fe-S] clusters of other iron-sulfur proteins, SoxR (Hidalgo and Demple, 1996) and FNR (Green et al., 1996) *in vitro*. *E. coli* IscS also have desulfuration activity similar to *A. vinelandii* NifS, and it has been shown to contribute to the formation of the [Fe-S] cluster in dihydroxy-acid dehydratase (Flint, 1996). These findings indicate that the prokaryotic NifS proteins can supply an elemental sulfur for the [Fe-S] cluster of iron-sulfur protein(s). Significant sequence homology found between the prokaryotic NifS/IscS proteins and eukaryotic NifS-like proteins, including the PLP-binding site and the cysteine residue essential for the cysteine desulfuration activity of the *A. vinelandii* NifS protein(Zheng et al., 1993, Zheng et al., 1994), suggests that the eukaryotic counterparts also participate in the formation of the [Fe-S] cluster of some iron-sulfur proteins in mitochondria *in vivo*. Mitochondria contain a variety of iron-sulfur proteins such as adrenodoxin, aconitase and Rieske protein of the respiratory complex III. We are now investigating to ascertain whether the eukaryotic NifS-like protein functions in the [Fe-S] cluster formation *in vivo*.

References

Cammack, R. (1992) Iron-sulfur clusters in enzymes: themes and variations. In Cammack, R. (ed.), *Iron-Sulfur Proteins. 38*. Academic Press, San Diego, CA: 281-322.

Flint, D. H. (1996) *Escherichia coli* contains a protein that is homologous in function and N-terminal sequence to the protein encoded by the *nifS* gene of *Azotobacter vinelandii* and that can participate in the synthesis of the Fe-S cluster of dihydroxy-acid dehydratase. *J. Biol. Chem. 271*: 16068-16074.

Green, J., Bennett, B., Jordan, P., Ralph, E. T., Thomson, A. J., and Guest, J. R. (1996) Reconstitution of the [4Fe-4S] cluster in FNR and demonstration of the aerobic-anaerobic transcription switch *in vitro*. *Biochem. J. 316*: 887-892.

Hidalgo, E. and Demple, B. (1996) Activation of SoxR-dependent transcription in vitro by noncatalytic or NifS-mediated assembly of [2Fe-2S] cluster into apo-SoxR. *J. Biol. Chem. 271*: 7269-7272.

Johnson, M. K. (1998) Iron-sulfur proteins: new roles for old clusters. *Curr. Opin. Chem. Biol., 2*: 173-181.

Kolman, C., and Söll, D. (1993) SPL1-1, a *Saccharomyces cerevisiae* Mutation Affecting tRNA Splicing. *J. Bacteriol. 175*: 1433-1442.

Land, T. and Rouault A. (1998) Targeting of a Human Iron-Sulfur Cluster Assembly Enzyme, *nifs*, to Different Subcellular Compartments Is Regulated through Alternative AUG Utilization. *Molecular Cell 2*: 807-815.

Nakai, M., Endo, T., Hase, T., and Matsubara, H. (1993) Intermitochondrial Protein Sorting. *J. Biol. Chem. 268*: 24262-24269.

Zheng., L., Cash, V. L., Flint, D. H., and Dean, D. R. (1998) Assembly of iron-sulfur clusters. Identification of an *iscSUA-hscBA-fdx* gene cluster from *Azotobacter vinelandii*. *J. Biol. Chem. ,273*: 13264-13272.

Zheng., L. and Dean, D. R. (1994) Catalytic formation of a nitrogenase iron-sulfur cluster. *J. Biol. Chem. 269*: 18723-18726.

Zheng, L., White, R. H., Cash, V. L., and Dean, D. R. (1994) *Mechanism for the desulfurization of L-Cysteine Catalyzed by the nifS Gene Product. Biochem. 33*: 4714-4720.

Zheng, L., White, R. H., Cash, V. L., Jack, R. F., and Dean, D. R. (1993) Cysteine desulfurase activity indicates a role for NIFS in metallocluster biosynthesis. *Proc. Natl. Acad. Sci. USA* ***90***: 2754-2758.

Biochemistry and Molecular Biology of Vitamin B_6
and PQQ-dependent Proteins
ed. by A. Iriarte, H.M. Kagan and M. Martinez-Carrion

Mechanistic Studies Of 8-Amino-7-Oxononanoate Synthase.

Scott P. Webster[1]*, Dmitriy Alexeev[2], Dominic J. Campopiano[1], Lisa J. Mullan[1], Lindsay Sawyer[2], Robert. L. Baxter[1].
[1]Edinburgh Centre for Protein Technology, Department of Chemistry, University of Edinburgh, West Mains Road, Edinburgh EH9 3JJ, UK and [2]Structural Biochemistry Group, University of Edinburgh, Swann Building, Mayfield Road, Edinburgh, EH9 3JR, UK.

Summary
8-amino-7-oxononanoate synthase is a pyridoxal 5'-phosphate-dependent enzyme which catalyses the decarboxylative condensation of L-alanine with pimeloyl-CoA to form 8-amino-7-oxononanoate. Individual steps in the reaction mechanism of *Escherichia coli* AONS have been probed by spectroscopy, pre-steady state kinetics and crystallography. Our results suggest that conformational transitions, induced by substrate and product binding, play an important role in catalysis.

Introduction.

The biotin biosynthetic pathway is an essential metabolic pathway found only in plants and microorganisms, making it an attractive target for herbicide and antibiotic development. In *Escherichia coli,* the first committed step in the biosynthesis of biotin is catalysed by the pyridoxal 5'-phosphate-dependent enzyme 8-amino-7-oxononanoate synthase (AONS) (1). This involves the decarboxylative condensation of pimeloyl-CoA and L-alanine to produce 8-amino-7-oxononanoate (AON), coenzyme A and carbon dioxide (Figure 1).

FIGURE 1. Reaction catalysed by AONS.

The structure and overall reaction mechanism of AONS places it in the α-family of PLP-dependent enzymes (2). Alignment of the amino acid sequence of AONS with the enzymes of the α-oxoamine synthase family - 5-aminolevulinate synthase, serine palmitoyltransferase and 2-amino-3-oxobutyrate CoA ligase - shows that almost all of the active site residues are conserved within this family of enzymes (3). The overall tertiary fold of AONS is closest to that of dialkylglycine decarboxylase although they have negligible amino acid sequence identity (4). The PLP cofactor is covalently bound to Lys236 through an imine linkage and interacts with two other basic residues, His133 and His207, which may play an important role in substrate binding and catalysis (3).

TABLE 1: Kinetic parameters for AONS.

pimeloyl-CoA			L-alanine		
k_{cat} (s^{-1})	K_m (μM)	k_{cat}/K_m ($M^{-1}s^{-1}$)	K_m (mM)	k_{cat}/K_m ($M^{-1}s^{-1}$)	k_1 ($M^{-1}s^{-1}$)
0.06(0.01)	25(2)	2400(430)	0.50(0.04)	120(21)	20000(6000)

The steady state kinetic parameters for AONS exhibit no metal ion dependence, and are comparable to those of other α-oxoamine synthases (Table 1) (5). Although the kinetic mechanism of AONS has not been investigated in detail, it is thought to adhere to an ordered bi-bi model in which L-alanine binds to the enzyme active site before pimeloyl-CoA (6). In common with the mechanisms of other PLP-dependent enzymes, the initial step of the AONS catalysed reaction involves displacement of the internal aldimine complex by the incoming amino acid substrate to form an external aldimine. This is followed by heterolytic cleavage of the Cα–H bond of the amino acid leading to a resonance stabilised quinonoid intermediate (7). This quinonoid is thought to react subsequently with the other substrate, pimeloyl-CoA, in a Claisen-type condensation to form a putative β-ketoacid aldimine intermediate. Decarboxylation of the resultant β-ketoacid presumably occurs to generate a second quinonoid intermediate, which is then protonated at C8 to form a product external aldimine. Finally, product release leads to regeneration of the internal aldimine with Lys236.

Here we report a study of the interaction of AONS with its substrates and product, and suggest a mechanistic pathway in which conformational transitions of the enzyme play a critical catalytic role. This mechanism may prove to be common to all four enzymes of the α-oxoamine synthase family.

Materials and Methods

Protein characterisation. Purification of AONS and determination of steady state kinetic parameters were carried out as described previously (3).

Spectroscopic methods. All UV-visible spectra were recorded on a Hewlett-Packard 8452A diode array spectrophotometer. The AONS concentration in all analyses was 10μM in 20mM potassium phosphate (pH 7.5). Reference cuvettes contained all other components except AONS.

Pre-steady state kinetics. Transient absorption kinetics were determined under pseudo-first order conditions using an Applied Photophysics SX17 stopped-flow spectrophotometer, thermostatically controlled at 30°C. For the determination of rates of formation of external aldimine, syringes contained 20μM AONS and 0-100mM L- or D-alanine. Spectra were recorded at 425nm and fitted to single exponential saturation curves. Rate constants were determined from the slope of a best-fit line produced from a plot of observed rate constant versus alanine concentration. For analysis of quinonoid formation, one syringe contained 25μM AONS pre-equilibrated with 10mM L-alanine and the other 0-300μM pimeloyl-CoA. Spectra were recorded at 486nm and the separate phases fitted to single exponential curves.

Determination of dissociation constants. Enzyme samples were dialysed overnight in 20mM potassium phosphate (pH 7.5) containing 100μM PLP. Non-enzyme bound PLP was removed by multiple washes with 20mM potassium phosphate through a Centrex UF2 concentrator (Schleicher & Schuell). Assays typically contained 10-50mM AONS in 20mM potassium phosphate (pH 7.5) with varying amounts of L- or D-alanine (0-80mM). Pimeloyl-CoA (143mM) was included where required. After addition of substrate, the reactants were mixed and allowed to equilibrate for 20mins at 30°C. Spectra were recorded and changes in absorbance at 425nm plotted against L- or D-alanine concentration. Data were fitted to a hyberbolic saturation curve using the equation, $\Delta A_{obs}=\Delta A_{max}.S/K_d+S$ where ΔA_{max} is the maximal absorbance change, S is the L- or D-alanine concentration, and K_d is the dissociation constant.

Crystallography. Crystals of AONS were grown as previously described (3). X-ray diffraction data of crystals containing AON was collected to a resolution of 2.0Å (D. Alexeev, personal communication).

Results and Discussion.

Substrate external aldimine formation. The absorbance spectrum of AONS holo-enzyme shows peaks at 334nm and 425nm which we assign to the non-coplanar and coplanar forms of the AONS-PLP internal aldimine (Figure 2). Addition of L- or D-alanine to AONS leads to an increase in the absorbance of the 425nm band indicating formation of an external aldimine. The pre-steady state rate constant, k_1, for L-alanine external aldimine formation is $2x10^4M^{-1}s^{-1}$ and $125M^{-1}s^{-1}$ for D-alanine, which is not a substrate. If an ordered bi-bi mechanism is assumed we would expect k_1 to approximate k_{cat}/K_m for L-alanine (6). The fact that k_{cat}/K_m is less than k_1 for L-alanine suggests that there may be a slow step following external aldimine formation that is necessary for productive binding of pimeloyl-CoA.

FIGURE 2: Changes in cofactor absorbance spectra. (A) AONS (solid line), AONS + L-alanine (dashed line) and (B) AONS + pimeloyl-CoA (solid line), AONS +AON (dashed line).

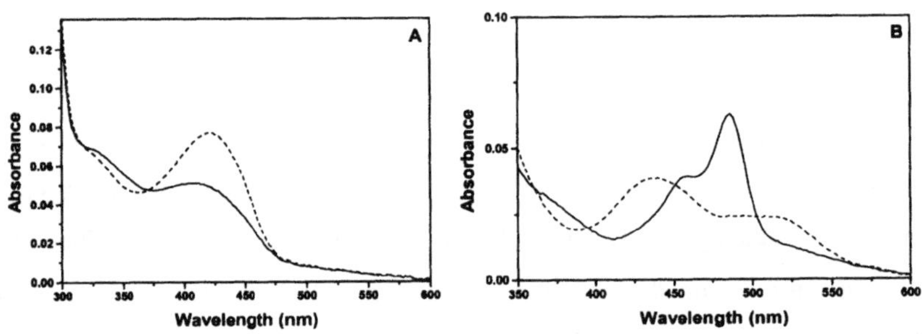

The equilibrium between internal and external aldimine forms of AONS is affected by the presence of pimeloyl-CoA. A plot of absorption increase at 425nm versus L-alanine concentration gives a curve with dissociation constant, K_d=0.9mM, while titration of AONS with D-alanine gives a K_d=9.8mM (Figure 3). In the presence of saturating amounts of pimeloyl-CoA, titration of AONS with D-alanine leads to a more than two-fold decrease in the dissociation constant (K_d=4.5mM). A decrease in the dissociation constant (K_d=0.3mM) of L-alanine in the presence of pimeloyl-CoA is also observed (data obtained from reciprocal plots of steady state kinetic data). Although we have no direct structural evidence, we speculate that the dissociation constant is reduced by binding of pimeloyl-CoA, which blocks the exit path of the amino acid substrate from the bottom of the binding site. Additionally, it is also conceivable that pimeloyl-CoA induces a conformational change which leads to a more tightly bound external aldimine species.

FIGURE 3: Binding of L-alanine to AONS. (A) Absorbance spectra in presence of L-alanine (from bottom to top 0, 0.1, 0.5, 1, 2, 5, 10, 15 and 20mM), (B) plot of absorbance versus titrant concentration.

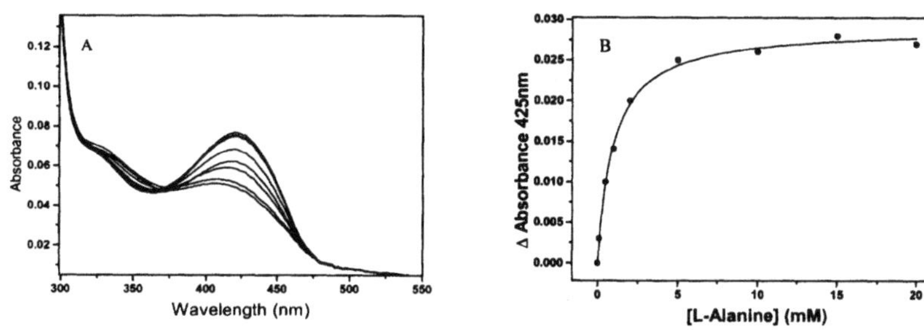

Addition of L- or D-alanine to AONS does not lead to quinonoid formation as indicated by the absence of a band at longer wavelength in the absorbance spectrum (Figure 2). It therefore seems likely that binding of L-alanine alone is insufficient to promote significant proton abstraction at Cα and leads to a benign form of the external aldimine-enzyme complex. The L-alanine external aldimine complex has not yet been crystallised, but has been modelled by analogy with the product aldimine structure (Figure 4). A probable interaction (analogous to the C7 carbonyl interaction in the product aldimine structure) between a carboxylate oxygen of the L-alanine-PLP complex and His133 places the Cα hydrogen away from the catalytic Lys236, making reaction with Lys236 unlikely. Proton abstraction would thus require a rotation around the N-Cα bond in the pyridine plane, which would direct the Cα hyrogen towards Lys236. Such a rotation would place the Cα carboxylate, previously interacting with His133, within hydrogen bonding distance of the side chain of Asn47, and thus generating a stable external aldimine complex.

Reaction with pimeloyl-CoA. In addition to enhancing the binding of L-alanine to AONS, addition of pimeloyl-CoA to the pre-formed external aldimine complex leads to formation of a quinonoid intermediate (Figure 2). Pre-steady state kinetic experiments revealed that the formation of this intermediate (k_{obs}=45s^{-1}) is preceded by a distinct lag phase (~30ms). It is likely that this lag represents conformational transitions of the enzyme which place the Cα hydrogen of the L-alanine-PLP complex and the catalytic lysine in a favourable orientation for catalysis. Modelling suggests that docking of pimeloyl-CoA in the active site may lead to displacement of the carboxylate oxygen of L-alanine from His133 which then forms a hydrogen bond with Asn47. This new orientation of the external aldimine places the Cα proton close to Lys236 in a position perpendicular to the pyridine ring leading to cleavage of the Cα-H bond in accordance with Dunathan's hypothesis (8). Following this pimeloyl-CoA-induced Cα deprotonation, the quinonoid is free to attack the thioester carbonyl of pimeloyl-CoA liberating coenzymeA and producing a β-ketoacid-aldimine complex. At present, little is known about the mechanism of decarboxylation of the β-ketoacid, but our results suggest it is likely to lead to formation of a second quinonoid intermediate (Figure 2).

Interaction with AON. In solution, incubation of AONS holo-enzyme with the product of the reaction, AON, gives rise to a spectrum with maxima at 436nm and 520nm (Figure 2). It is likely that these bands reflect the equilibrium between the external aldimine and quinonoid forms of the product complex respectively. The quinonoid form has an absorption maximum at a substantially longer wavelength than the substrate-PLP quinonoid formed earlier in the catalytic pathway and probably arises from the decarboxylation of the putative β-ketoacid described above.

The structure of AONS incubated with its product, AON, has been determined to 2.0Å resolution. In contrast to the equilibrium between quinonoid and aldimine forms existing in solution, only the aldimine form is observed in the crystal. The N8=C4A double bond of the AON-PLP complex is coplanar with the pyridine ring while the C7 carbonyl oxygen of the complex is hydrogen bonded to His133. The carboxylate of AON is bound to Arg21 from the N-terminal domain and Arg349 from the C-terminal domain of the enzyme. A magnesium ion is coordinated to O3, N8, the hydroxyl of Ser179 and two water molecules. Since the enzymatic activity of AONS exhibits no magnesium dependence it is unlikely that the ion plays a role in catalysis and may have arisen from trace ions present in the crystallisation buffer.

FIGURE 4: Cofactor binding site. PLP internal aldimine (left) and AON-PLP external aldimine (right).

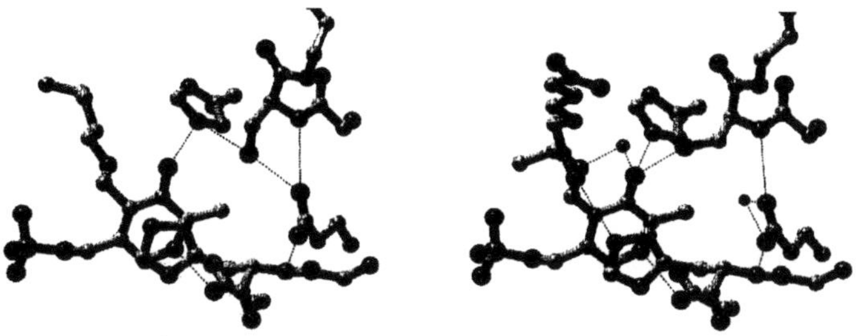

C-terminal domain movement. The overall refined protein structure of the AON-enzyme complex is close to that of the holo-enzyme. However, the β-sheet of the C-terminal domain undergoes a conformational transition whereby three β-strands of this flexible domain fold towards the active site of the AONS catalytic dimer (Figure 5). This movement narrows the access to the binding site and may be analogous to the apparent transition which occurs upon binding of pimeloyl-CoA to the pre-formed substrate external aldimine complex. The release of this structural cap over the active site must be reversed upon product release and concomitant reformation of the internal aldimine. The energy barrier for this process is likely to be large and it may represent the rate determining step in the enzymatic reaction, since neither substrate external aldimine formation or quinonoid formation approximate k_{cat}.

FIGURE 5: Conformational changes in AONS structure.

Conclusion. These results suggest a mechanistically reasonable pathway for the catalytic action of AONS which may prove common for other members of the α-oxoamine family of enzymes. Further studies will shed light on the role of active site residues in the catalytic reaction and may explain the involvement of conformational changes in the protein during catalysis.

References

1. Eisenberg, M. (1987) *Biosynthesis of biotin and lipoic acid*, Vol. 1, American Society of Microbiology, Washington DC.
2. Alexander, F. W., Sandmeier, E., Mehta, P. K., and Christen, P. (1994) *Eur. J. Biochem.* **219**, 953-960.
3. Alexeev, D., Alexeeva, M., Baxter, R. L., Campopiano, D. J., Webster, S. P., and Sawyer, L. (1998) *J. Mol. Biol.* **284**, 401-419.
4. Toney, M. D., Hohenester, E., Cowan, S. W., and Jansonius, J. N. (1993) *Science* **261**, 756-759.
5. Ferreira, G. C., and Dailey, H. A. (1993) *J. Biol. Chem.* **268**, 584-590.
6. Segel, I. H. (1975) *Enzyme Kinetics*, John Wiley & Sons, New York.
7. Ploux, O., and Marquet, A. (1996) *Eur. J. Biochem.* **236**, 301-308.
8. Dunathan, H. C. (1966) *Proc. Natl. Acad. Sci. U.S.A.* **55**, 712-716.

Acknowledgements.

We are grateful to the BBSRC for funding. Thanks also to Marina Alexeeva, Mhairi Brunton and Dr Rory M. Watt for technical assistance.

*Current address: PanTherix Ltd, Unit 6.03 Kelvin Campus, West of Scotland Science Park, Maryhill Road, Glasgow G20 0SP, United Kingdom. E-mail: scott.webster@pantherix.co.uk

TRYPTOPHAN SYNTHASE, TRYPTOPHANASE AND SERINE TRANSHYDROXYMETHYLASE

Biochemistry and Molecular Biology of Vitamin B_6
and PQQ-dependent Proteins
ed. by A. Iriarte, H.M. Kagan and M. Martinez-Carrion

Structure and Function of Tryptophan Synthase

Edith Wilson Miles, Ying-Xin Fan, Kwang-Hwan Jhee, Hyeon-Su Ro, Peter McPhie, Sangkee Rhee, and David R. Davies

National Institutes of Health, Bethesda, MD 20892, USA

Summary

Crystal structures of the tryptophan synthase $\alpha_2\beta_2$ complex show binding sites in the β subunit for pyridoxal phosphate, for substrates, and for a monovalent cation. Cations, including guanidine at low concentration, activate tryptophan synthase. Molecular modeling indicates that guanidine could bind to the cation site in the β subunit. Our results demonstrate that the conformational states of the tryptophan synthase $\alpha_2\beta_2$ complex are regulated by monovalent cations, chaotropic agents, pH, α subunit ligands, and temperature.

Introduction

Bacterial tryptophan synthase is composed of α and β subunits that combine reversibly to form an $\alpha_2\beta_2$ complex [1]. The β subunit was the first pyridoxal phosphate (PLP) enzyme in Fold II [2] to be analyzed by crystallography in the structure of the tryptophan synthase $\alpha_2\beta_2$ complex from *Salmonella typhimurium* [3]. Crystal structures of wild-type and mutant forms of the tryptophan synthase $\alpha_2\beta_2$ complex reveal a unique indole tunnel, binding sites for a monovalent cation and other ligands, and ligand-induced conformational changes [3-5].

Results and Discussion

Activation by Monovalent Cations Including Guanidinium—A low concentration (0.02 M) of guanidine hydrochloride (GuHCl) activates the tryptophan synthase $\alpha_2\beta_2$ complex 1.4-fold, whereas 0.1 M NaCl activates 3.2-fold [6] (Figure 1). Higher concentrations of GuHCl inhibit. The results suggest that GuH^+ activates tryptophan synthase by binding to the monovalent cation site in the β subunit [4].

Cation	Relative Activity
None	1.0
NaCl (0.1 M)	3.2
GuHCl (0.02 M)	1.4
GuHCl (0.1 M)	0.4

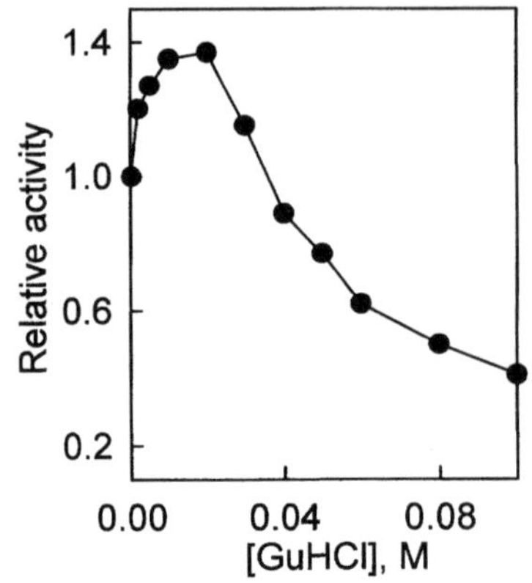

Figure 1. Effects of GuHCl and of NaCl on the relative activity of the $\alpha_2\beta_2$ complex in the reaction of L-serine and indole in the absence of other cations (modified from [6]).

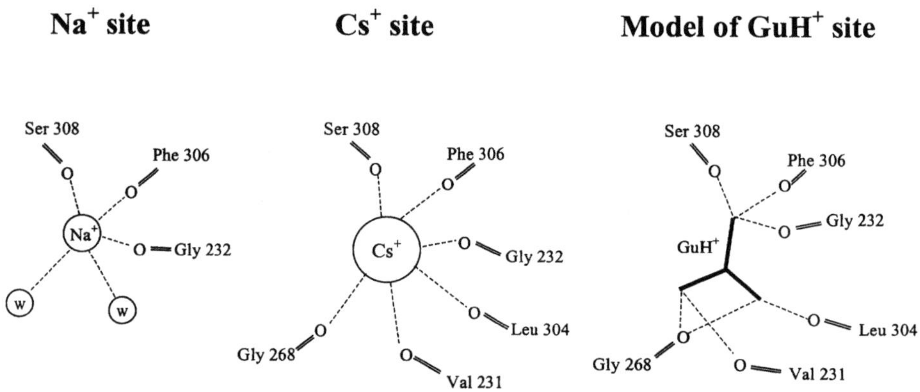

Figure 2. Cartoons of the Na^+ and Cs^+ binding sites in the β subunit based on the X-ray coordinates [4] and of the GuH^+ binding site based on molecular modeling (modified from [6]).

Cartoons of the binding sites for Na^+ and Cs^+ in the β subunit [4] show that Na^+ interacts with the peptide carbonyl oxygens of 3 residues and with 2 waters (W), whereas the larger Cs^+ interacts with the peptide carbonyl oxygens of 6 residues (Figure 2). Molecular modeling of GuH^+ into the cation site shows favorable interactions between GuH^+ (dark lines) and the peptide carbonyl oxygens of the 6 residues that interact with Cs^+. Two of the nitrogens of GuH^+

are located in the model close to the positions of the two waters in the Na^+ site. Thus, binding of GuH^+ to the cation site appears feasible.

Modulation of the Active Site Conformation by GuHCl—Addition of increasing GuHCl concentrations has strikingly different effects on the rates of the two different β-replacement reactions of indole with L-serine or β-chloro-L-alanine catalyzed by the tryptophan synthase $\alpha_2\beta_2$ complex [6] (Figure 3A). Low concentrations of GuHCl stimulate the rate of the reaction with β-chloro-L-alanine more than that with L-serine. Furthermore, the decrease in activity at higher GuHCl concentrations is much less sharp with β-chloro-L-alanine than with L-serine. The results suggest that addition of GuHCl results in an altered conformation of the $\alpha_2\beta_2$ complex having higher activity with β-chloro-L-alanine than with L-serine.

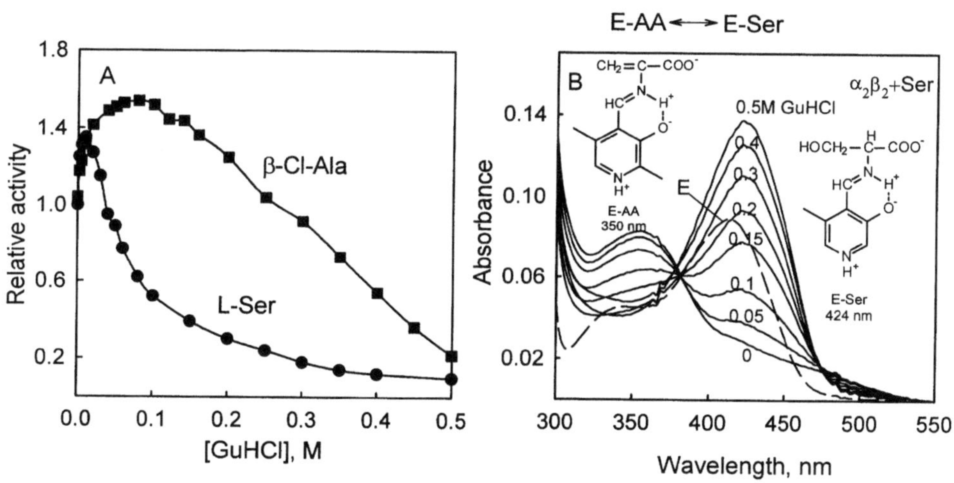

Figure 3. Effects of GuHCl concentration on the activities of the $\alpha_2\beta_2$ complex in the reactions with indole and either L-serine or β-chloro-L-alanine (A) and on the absorption spectra of the $\alpha_2\beta_2$ complex in the presence of 50 mM L-serine (B) (modified from [6]).

Table 1. Conformational States of the Tryptophan Synthase $\alpha_2\beta_2$ Complex

	Open	Closed
Activity (Ser + Ind -> Trp)	Low	High
Preferred Substrate	β-Chloro-L-alanine	L-Serine
Major Serine Intermediate	E-Ser	E-AA
Stabilizers	Lower °C	Higher °C
	Higher pH	Lower pH
	Na^+	Cs^+
	GuHCl or Urea	

To explain our finding that some mutant $\alpha_2\beta_2$ complexes exhibit high activity with β-chloro-L-alanine but low activity with L-serine, we have proposed that the wild-type, but not the mutant, enzymes are converted from a low activity open conformation to a high activity, closed conformation [7]. The closed conformation facilitates protonation of the weak OH^- leaving group of L-serine but is not needed for protonation of the strong Cl^- leaving group of β-chloro-L-alanine [7] (Table 1).

The proposal that GuHCl (>0.1 M) stabilizes the open form of the tryptophan synthase $\alpha_2\beta_2$ complex is supported by the observation that addition of increasing GuHCl concentrations in the presence of L-serine results in increased absorbance at 424 nm and decreased absorbance at 350 nm [6] (Figure 3B). The 424 nm band is attributed to the E-Ser intermediate that accumulates with the low activity, open form of the enzyme, whereas the 350 nm band is attributed the E-AA intermediate that accumulates with the high activity, closed form [7-9] (Table 1).

Modulation of the Active Site Conformation by pH and by Temperature—The pH profile of the activity of the tryptophan synthase $\alpha_2\beta_2$ complex in the reaction with indole and L-serine exhibits a bell shaped curve with maximum activity at pH 7.5 (Figure 4A). In contrast, activity in the reaction with β-chloro-L-alanine and indole increases sharply with increasing pH, suggesting that increasing pH converts the $\alpha_2\beta_2$ complex to the open conformer that has higher activity with β-chloro-L-alanine than with L-serine (Table 1). These results are consistent with the earlier finding that increasing pH stabilizes the open conformer that accumulates E-Ser [9].

We have recently examined the effects of temperature and of cations on the activities of the $\alpha_2\beta_2$ complex in the reaction of L-serine with indole to form tryptophan [10]. Arrhenius plots of the activity data are nonlinear in presence of NaCl or GuHCl but are linear in the presence of CsCl or NaCl + GP, an α subunit ligand. At low temperature, the activity in NaCl is much lower than in CsCl, whereas at high temperature the activities are the same. The conditions that yield nonlinear Arrhenius plots also yield temperature-dependent changes in the equilibrium distribution of the E-Ser and E-AA intermediates in the reaction with L-serine (Figure 4B). E-Ser is the predominant intermediate at low temperature in the presence of NaCl, whereas E-AA is the predominant intermediate at high temperature. E-AA is the predominant intermediate at all temperatures in the presence of CsCl or NaCl + GP. The results support the proposal that increasing temperature stabilizes the active closed conformation of the $\alpha_2\beta_2$ complex. Data analysis yields values of the changes in enthalpy (ΔH_{eq}) and entropy (ΔS_{eq}) for the temperature-dependent conversion. Importantly, the ΔH_{eq} is equal to the difference between the activation energies at low and high temperatures. The large positive entropy change in the open to closed conformational conversion suggests that the closed conformation results from tightened hydrophobic interactions that exclude water from the active site.

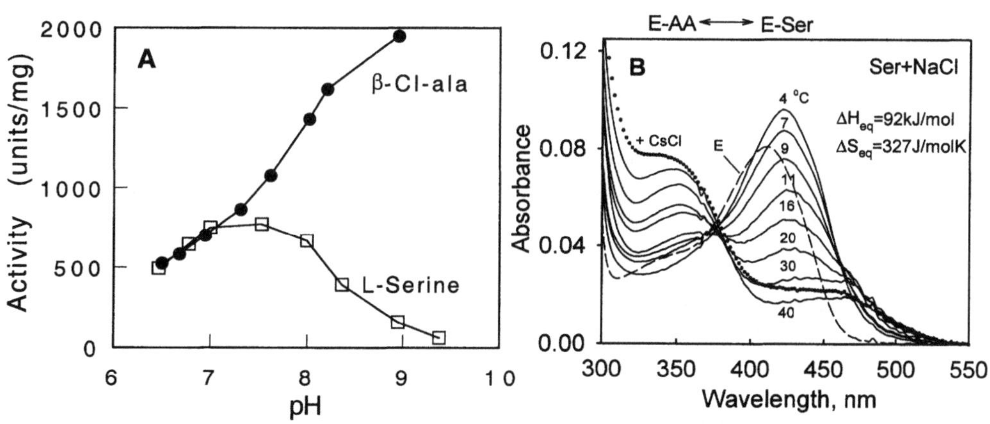

Figure 4. Modulation of the active site conformation by pH (A) and temperature (B). A. pH dependence of the specific activity of the $\alpha_2\beta_2$ complex in the reaction of indole with either L-serine (data from [11]) or β-chloro-L-alanine. B. Effects of temperature on the absorption spectra of the $\alpha_2\beta_2$ complex in the presence of 50 mM L-serine and 0.1 M NaCl (—) or 0.1 M CsCl (••). The spectra in the presence of CsCl were not affected by temperature (modified from [10]).

Conclusions—Our results and the results of others [8,9] provide evidence that higher temperatures, lower pH, α subunit ligands, and Cs^+ stabilize the active, closed conformation, whereas high pH and GuHCl stabilize the open form (Table 1).

References

1. Miles, E. W. (1991) Structural basis for catalysis by tryptophan synthase. *Adv. Enzymol. Relat. Areas Mol. Biol.* **64,** 93-172.
2. Grishin, N. V., Phillips, M. A., and Goldsmith, E. J. (1995) Modeling of the spatial structure of eukaryotic ornithine decarboxylase. *Protein Science* **4,** 1291-1304.
3. Hyde, C. C., Ahmed, S. A., Padlan, E. A., Miles, E. W., and Davies, D. R. (1988) Three-dimensional structure of the tryptophan synthase $\alpha_2\beta_2$ multienzyme complex from *Salmonella typhimurium. J. Biol. Chem.* **263,** 17857-17871.
4. Rhee, S., Parris, K. D., Ahmed, S. A., Miles, E. W., and Davies, D. R. (1996) Exchange of K^+ or Cs^+ for Na^+ induces local and long-range changes in the three-dimensional structure of the tryptophan synthase $\alpha_2\beta_2$ complex. *Biochemistry* **35,** 4211-4221.
5. Rhee, S., Parris, K. D., Hyde, C. C., Ahmed, S. A., Miles, E. W., and Davies, D. R. (1997) Crystal structures of a mutant (βK87T) tryptophan synthase $\alpha_2\beta_2$ complex with ligands bound to the active sites of the α and β subunits reveal ligand-induced conformational changes. *Biochemistry* **36,** 7664-7680.
6. Fan, Y. X., McPhie, P., and Miles, E. W. (1999) Guanidine hydrochloride exerts dual effects on the tryptophan synthase $\alpha_2\beta_2$ complex as a cation activator and as a modulator of the active site conformation. *Biochemistry* **38,** 7881-7890.
7. Ahmed, S. A., Ruvinov, S. B., Kayastha, A. M., and Miles, E. W. (1991) Mechanism of mutual activation of the tryptophan synthase α and β subunits. Analysis of the reaction specificity and substrate-induced inactivation of active site and tunnel mutants of the β subunit. *J. Biol. Chem.* **266,** 21540-21557.
8. Brzovic', P. S., Ngo, K., and Dunn, M. F. (1992) Allosteric interactions coordinate catalytic activity between successive metabolic enzymes in the tryptophan synthase bienzyme complex. *Biochemistry* **31,** 3831-3839.
9. Peracchi, A., Bettati, S., Mozzarelli, A., Rossi, G. L., Miles, E. W., and Dunn, M. F. (1996) Allosteric regulation of tryptophan synthase: effects of pH, temperature, and α subunit ligands on the equilibrium distribution of pyridoxal 5'-phosphate-L-serine intermediates. *Biochemistry* **35,** 1872-1880.
10. Fan, Y. X., McPhie, P., and Miles, E. W. (in press) Regulation of tryptophan synthase by temperature, monovalent cations, and an allosteric ligand. Evidence from Arrhenius plots, absorption spectra, and primary kinetic isotope effects. *Biochemistry.*
11. Ro, H.-S., and Miles, E. W. (1999) Catalytic mechanism of the tryptophan synthase $\alpha_2\beta_2$ complex: Effects of pH, isotopic substitution, and allosteric ligands. *J. Biol. Chem.* **274,** 31189-31194.

Biochemistry and Molecular Biology of Vitamin B_6
and PQQ-dependent Proteins
ed. by A. Iriarte, H.M. Kagan and M. Martinez-Carrion

Salt Bridging and Movalent Cation Binding Regulate Catalysis and Channeling in Tryptophan Synthase

M. F. Dunn*, E. U. Woehl*, D. Ferrari*, O. Hur*, U. Banik§, L.-H. Yang§, and E. W. Miles§.

*Department of Biochemistry, University of California, Riverside, CA 92521, and §National Insititutes of Health, Bethesda, MD 20892, USA

Summary

Control of indole channeling between the α- and β-sites of tryptophan synthase involves allosteric signals that switch αβ-dimeric units between open and closed conformations of low and high activity. Signaling is triggered by covalent bonding at the β-site and ligand binding to the α-site, and is postulated to be transmitted between sites via a scaffolding that includes a monovalent cation (MVC) site and the side chain of βLys 167 in alternating salt bridges with βAsp 305 or αAsp 56. Via site-directed mutation and MVC-substitutions, we have explored the roles of the MVC site and salt bridging in the regulation of channeling. From these experiments, we conclude the βK167-αD56 salt bridge plays an important role in substrate channeling.

Introduction

Channeling of indole between the α- and β-sites of tryptophan synthase occurs via a 25-Å long tunnel (1, 2). Preliminary studies (3-5) indicate that the salt-bridging interactions which link the α- and β-sites mediate the transmission of allosteric signals between the two sites, and modulate catalytic and regulatory properties of the enzyme. Binding of MVCs to an allosteric site located 8 Å from the β-active site also alters catalysis and the regulation of channeling (6-10). Solution studies and x-ray structures show that the α-aminoacrylate and quinonoidal intermediates stabilize closed structures, while the gem diamine and and aldimine species stabilize open or partially open conformations (Scheme 1) (1, 2). The switch between open and closed states is critically important for the regulation of channeling (1, 2, 11). To achieve a more detailed understanding of the relationship of the salt bridge interactions to the MVC site, herein we investigate the catalytic and regulatory properties of the αD56A, βK167T, βD305N and βD305A mutants, both in the presence and in the absence of MVCs.

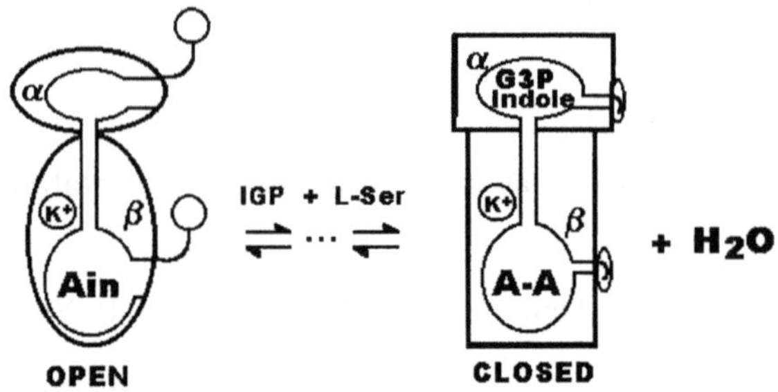

Scheme 1. Cartoons of an αβ-dimeric unit of tetrameric tryptophan synthase depicting the α- and β-sites, the lids that close over the active sites, the interconnecting tunnel, and the MVC site (occupied by K^+). The interconversion of the open and closed conformations is shown occuring as the internal aldimine, Ain, reacts with 3-indolyl-D-glycerol 3'-phosphate (IGP) and L-Ser to form D-glyceraldehyde 3-phosphate (G3P) and indole at the α-site and the α-aminoacrylate (A-A) at the β-site. Redrawn from (8) with permission.

Materials and Methods

The materials and methods used have been previously described (3-9).

Results and Discussion

Influence of Allosteric Effectors on Intermediate Distribution. Figure 1A, B shows the effects of Na^+ and NH_4^+ on the reaction of wild-type enzyme with L-Ser, and with L-Ser and indoline. In the L-Ser reaction (Figure 1A), the α-aminoacrylate Schiff base, E(A-A), with λ_{max} = 350 nm and a shoulder extending beyond 500 nm, is the predominant species formed both in the absence of MVCs and in the presence of NH_4^+ (6-9, 12, 13). Comparable amounts of the L-Ser external aldimine, $E(Aex_1)$, with λ_{max} = 425 nm, and E(A-A) are present at equilibrium when the effector is Na^+. Clearly, NH_4^+ binding stabilizes the closed E(A-A) form, while Na^+ tends to stabilize the partially closed $E(Aex_1)$ form (14). In the reaction of wild-type enzyme with L-Ser and indoline (Figure 1B), the E(A-A) intermediate reacts very rapidly with indoline to form a quinonoidal species, $E(Q)_{Indoline}$ (λ_{max} 464 nm), which very slowly converts to product, dihydroiso-L-tryptophan (15). The yield of $E(Q)_{Indoline}$ is remarkably dependent upon the allosteric effector present. Almost no $E(Q)_{Indoline}$ forms in the absence of MVCs. In the presence of NH_4^+, the yield of $E(Q)_{Indoline}$ is small. The distribution of E(A-A) and $E(Q)_{Indoline}$ is shifted strongly toward $E(Q)_{Indoline}$ by the binding of Na^+. Na^+ is more effective than NH_4^+ in stabilizing the closed form of $E(Q)_{Indoline}$.

The effects of Na^+ and NH_4^+ upon reaction of βK167T with L-Ser, and upon reaction with L-Ser and indoline are shown in Figure 1C, D. Na^+ binding stabilizes $E(Aex_1)$, while NH_4^+ binding gives similar amounts of the E(A-A) and $E(Aex_1)$ species (Figure 1C), as seen with the wild

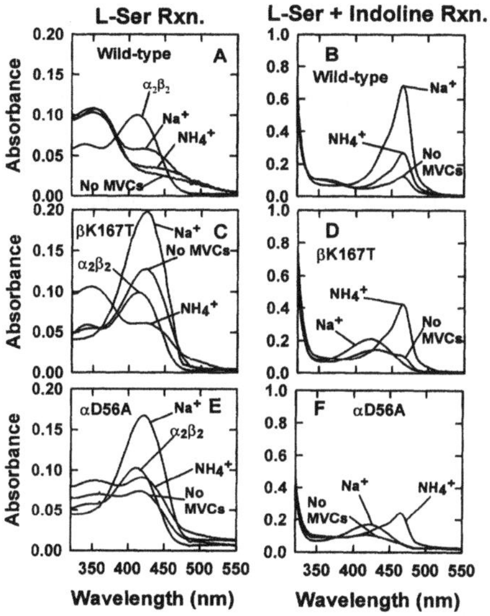

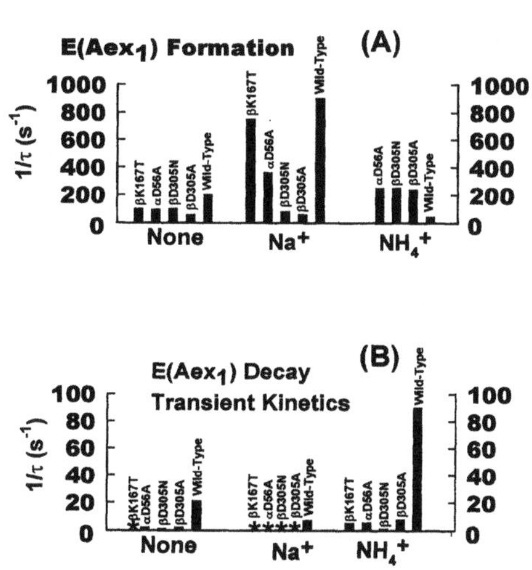

Figure 1. Static spectra for the reaction of wild-type enzyme (A, B) and the βK167T (C, D) and αD56A (E, F) mutants with L-Ser, and with L-Ser and indoline in the absence of effectors, and in presence of Na^+ or NH_4^+. Spectra were measured as follows: in (A), (C), and (E), $\alpha_2\beta_2$, reaction with L-Ser in the absence of effectors (No MVCs), and in the presence of 100 mM NaCl, or 50 mM NH_4Cl. Spectra in (B), (D), and (E) were measured under the same effector conditions as in (A), (C), and (E). Concentrations: [enzyme] =8.3 μM, [L-Ser] = 40 mM and [Indoline] = 4.5 mM when present, all in 50 mM triethanolamine buffer at pH 7.80 and 25°C.

Figure 2. Relaxation rates for the formation (A) and decay (B) of $E(Aex_1)$ measured in the absence of effectors (None), and in the presence of 100 mM Na^+ or 50 mM NH_4^+ for the wild-type, αD56A, βK176T, βD305N, and βD305A enzymes. Conditions: 50 mM triethanolamine buffer at pH 7.80 and 25°C. The * designates mutants that gave no detectable decay.

type enzyme. Reaction of E(A-A) with indoline gives a significant amount of $E(Q)_{Indoline}$ only in the presence of NH_4^+ (Figure 1D). Na^+ binding stabilizes $E(Aex_1)$, and no detectable amount of $E(Q)_{Indoline}$ is formed, indicating Na^+ strongly stabilizes the partially open conformation.

Reaction of L-Ser with MVC-free αD56A gives $E(Aex_1)$. The Na^+ form gives predominantly $E(Aex_1)$, whereas, the NH_4^+ form slightly alters the distribution in favor of E(A-A) (Figure 1E). In the indoline reaction with αD56A (Figure 1F), the MVC-free species gives almost no $E(Q)_{Indoline}$, Na^+ stabilizes $E(Aex_1)$ with no detectable $E(Q)_{Indoline}$, and the NH_4^+ form gives a relatively modest yield of $E(Q)_{Indoline}$. The rates of formation of $E(Q)_{Indoline}$ are noticeably slower than for the wild-type enzyme. Again, Na^+ strongly stabilizes the partially open conformation.

The βD305N mutant (data not shown) shows behavior that is very similar to the wild-type enzyme (see Figure 2A, B). The βD305A mutant was found to show remarkable differences from the wild-type enzyme and the αD56A and βK167T mutants; these differences appear to reflect both an altered role in β-site catalysis and an altered allosteric role involving salt bridging to βR141. Therefore, the behavior of this mutant will be presented elsewhere.

Influence of Allosteric Effectors on Activity. The β-reaction activity of the effector-free wild-type enzyme is strongly activated by the binding of MVCs, whereas the α-reaction activity is somewhat suppressed (6-9). NH_4^+ is a better activator of the β-reaction than is Na^+ (8-fold vs. 5-fold). The α-reaction is inhibited to 33% of the MVC-free value by Na^+ and to 67% by NH_4^+. The extent to which the L-Ser reaction activates the α-site is given by the αβ/α reaction activity ratio (16). In the absence of MVCs, there is almost no α-site activation, but strong activation occurs when Na^+or NH_4^+ bind (30- and 40-fold, respectively). Activation is due to an allosteric interaction between the β- and α-sites that switches the α-site from an inactive (open) state to an active (closed) state triggered by the formation of the E(A-A) intermediate (14, 16, 17).

The β-reaction catalyzed by βK167T is not significantly stimulated by Na^+, while NH_4^+ stimulates by 107-fold (compared to the MVC-free species). The α-activity is suppressed to ~ 10% of the wild-type enzyme activity and is unaffected by MVCs. The β-reaction activates the α-site 12-fold when Na^+ is bound, and 36-fold when NH_4^+ is bound.

In comparison to the MVC-free species, Na^+ binding slightly inhibits the β-reaction activity of αD56A, NH_4^+ binding activates 5-fold, giving activities that are 25-30 % that of the corresponding wild-type species. The α-site is only slightly activated by the L-Ser reaction in the presence of Na^+ (2- fold), whereas the NH_4^+ form actives 61-fold. As measured by the extent of α-site activation due to the β-reaction, NH_4^+ is more effective than Na^+ in restoring this site-site interaction to each enzyme species.

Influence of Effectors on E(Aex$_1$) Formation and Decay. E(Aex_1) formation and decay time courses were determined by fluorescence stopped-flow methods (Figure 2A,B) (7-9, 12). The relaxation rate constants for formation of the E(Aex_1) species of βK167T, αD56A and wild-type enzyme are significantly increased by the binding of Na^+, while the βD305N and βD305A mutants are unaffected (Figure 2A). NH_4^+ also increases the E(Aex_1) formation rates for αD56A, βD305N and βD305A, but inhibits the wild-type enzyme. These data indicate neither the βK167-αD56 salt bridge, nor a closed conformation are essential for E(Aex_1) formation.

Figure 2B shows that, except for MVC-free βK167T, both in the absence of MVCs and in the presence of NH_4^+, all of the E(Aex_1) species decay to an equilibrating mixture of E(Aex_1) and E(A-A). Both in the absence of MVCs and in the presence of NH_4^+, the wild-type enzyme shows the fastest decay. In the presence of Na^+, only the wild-type enzyme shows a detectable decay phase. Figure 2B indicates the βK167-αD56 salt bridge is important for formation of the closed conformation of E(A-A), and Na^+ binding destabilizes the closed conformation.

Implications for Channeling and Catalysis. These studies establish that the NH_4^+ species all exhibit properties that are significantly altered in comparison to the MVC-free and Na^+ species. The distribution of E(Aex_1) and E(A-A) formed in the NH_4^+ species appears similar for wild-type enzyme and all the mutants (viz., Figure 1). Figure 2 shows that Na^+ binding impairs conversion of the mutant E(Aex_1) species to E(A-A), while NH_4^+ does not. Na^+ binding stabilizes wild-type $E(Q)_{Indoline}$, but prevents its accumulation in the αD56A and βK167T mutant enzymes. These data show that NH_4^+ binding gives closely similar wild-type and mutant enzymes where the interconversion of the open and closed states is facile, while Na^+ does not.

The conversion of αβ-dimeric units to the completely closed, activated conformation is postulated to depend upon the covalent transformation of E(Aex_1) to E(A-A) at the β-site and substrate binding to the α-site (1, 16-18). This conformational transition is accompanied by the closing of loops over the α-site and domain movements in the β-subunit that block the entrance from solution to the β-site. Interactions at the MVC site and the switching of βLys 167 between the alternative salt bridges linking the α- and β-sites are essential for triggering the switch between open and closed states (3-10). The extent of activation parallels the extent to which the distribution of species in the L-Ser reaction favors the formation of E(A-A).

Mutation of βK167 to Thr, or αD56 to Ala strongly impairs the regulatory and catalytic activities of tryptophan synthase, while mutation of βD305 to Asn does not. The impaired behavior indicates the βK167-αD56 salt bridge plays a critically important role in the switch between open and closed states, whereas the βK167-αD305 salt bridge is only of minor importance. Unlike Na^+ (10), NH_4^+ complexes are restricted to a tetrahedral geometry dictated by the H-bonding of the four N-H protons to acceptor ligands. The repair of catalytic and allosteric function by NH_4^+ likely is due to the re-establishment of wild-type-like conformations stemming from the tetrahedral geometry of the NH_4^+ complex at the MVC site.

References

1. Pan, P., Woehl, E., & Dunn, M. F. (1997) *Trends in Biochemical Sciences 22*, 22-27.
2. Miles, E. W., Rhee, S., & Davies, D. R. (1999) *J. Biol. Chem. 274*, 12193-12196.
3. Rowlett, R., Yang, L. H., Ahmed, S. A., McPhie, P., Jhee, K. H., & Miles, E. W. (1998) *Biochemistry* 37, 2961-2968).
4. Yang, X-J. & Miles, E. W. (1993) *J. Biol. Chem.* 268, 22269-22272.
5. Ahmed, S. A., Ruvinov, S. B., Kayastha, A. M., & Miles, E. W. (1991) *J. Biol. Chem. 266* 21548-21557.
6. Peracchi, A., Mozzarelli, A., & Rossi, G. L. (1995) *Biochemistry 34*, 945-9465.
7. Woehl, E. U., & Dunn, M. F. (1995) *Biochemistry, 34*, 9466-9476.
8. Woehl, E. U., & Dunn, M. F. (1999) *Biochemistry, 38,* 7118-7130.
9. Woehl, E. U., & Dunn, M. F. (1999) *Biochemistry, 38,* 7131-7141.
10. Rhee, S. Parris, K. D., Hyde, C. C., Ahmed, S. A., Miles, E. W., & Davies, D. R. (1997) *Biochemistry 36*, 7664-7680.
11. Brzovic, P. S., Sawa, Y., Hyde, C. C., Miles, E. W., & Dunn, M. F. (1992) *J. Biol. Chem. 267*, 13028-13038.
12. Drewe, W. F., Jr., & Dunn, M. F. (1985) *Biochemistry 24*, 3977-3987.
13. Drewe, W. F., Jr., & Dunn, M. F. (1986) *Biochemistry 25*, 2494-2501.
14. Pan, P. & Dunn, M. F. (1996) *Biochemistry 35*, 5002-5015.
15. Roy, M., Keblawi, S., & Dunn, M. F. (1988a) *Biochemistry 27*, 6698-6704.
16. Brzovic, P. S., Ngo, K., & Dunn, M. F. (1992) *Biochemistry 31*, 3831-3839.
17. Leja, C. A., Woehl, E. U., & Dunn, M. F. (1995) *Biochemistry 34*, 6552-6561.
18. Dunn, M. F., Aguilar, V., Brzovic, P. S., Drewe, W. F., Jr., Houben, D. F., Leja, C. A., & Roy, M. (1990) *Biochemistry 29*, 8598-8607.

Biochemistry and Molecular Biology of Vitamin B_6
and PQQ-dependent Proteins
ed. by A. Iriarte, H.M. Kagan and M. Martinez-Carrion

Equilibrium Isotope Effects: Evidence for Low-Barrier H-Bonding in Tryptophanase

Dimitri Niks*, Thomas Hellman Morton[§], and Michael F. Dunn*

Departments of Biochemistry* and Chemistry[§], University of California, Riverside, CA 92521

Summary

The PLP-requiring tryptophanase from *E. coli* reacts with L-Ala to give an equilibrating mixture of external aldimine E(Aex) and quinonoid E(Q) species. In agreement with the work of June et al. (1981), we find that the equilibrium for the interconversion of these two covalent intermediates is subject to a large equilibrium isotope effect, K_H/K_D, when the C-α hydrogen is replaced by deuterium. The magnitude of the isotope effect was found to be 3.87 ±0.4, which implies a low fractionation factor (0.29 ±0.03) for the base involved in abstracting the α-proton. The low fractionation factor qualifies as strong evidence for a low-barrier hydrogen bond (LBHB) in the mechanism of tryptophanase.

Introduction

Ab initio calculations have predicted that hydrogen bonds can be strengthened by external factors that impose compression. This effect becomes energetically even more important when the hydrogen-bonded heteroatoms meet the criteria for LBHB formation: namely, similar proton affinities, sufficiently short interatomic distances, and solvent-excluded environments. Frey and Cleland have noted that compression can be a general principle in enzymes, where enzyme-substrate complexes provide the necessary compression for H-bond strengthening (1).

The *E. coli* tryptophanase is a tetramer of 52kD subunits requiring pyridoxal 5'-phosphate (PLP) and monovalent cations for activity. It catalyzes the degradation of L-Trp to indole, pyruvate and ammonia. The step of interest in this pathway is the formation of an enzyme-bound quinonoid intermediate, E(Q), via abstraction of the proton from the C-α of the external aldimine structure, E(Aex). In order to quantify the magnitude of the EIE, our experiments employed a tryptophanase pseudo-substrate, L-Ala.

Materials and Methods

Tryptophanase from *E. coli* JM101/pMD6 was prepared as previously described (2). Protein concentrations were determined using a value of $A_{278} = 0.85$ for 1 mg/ml solutions (3). Stopped-flow kinetic studies were carried out as previously described (4,5). A Hewlett-Packard 8452A diode-array spectrophotometer was used for other static and kinetic experiments.

Reactions with α-^{1}H-L-Ala and α-^{2}H-L-Ala were carried out under identical conditions in 0.1 M K-PO_4, pH 8.0. The stock and dilutions were mixed with dialyzed tryptophanase. Maximum absorbance due to quinonoid formation was measured at $\lambda_{max} = 502$ nm. It was assumed that for the reaction conditions reported herein, the activity coefficients of α-^{1}H- and α-^{2}H-L-Ala are equal to unity. Reported concentrations and pH measurements describe conditions after mixing.

Results and Discussion

Upon reaction with L-Ala, the tryptophanase internal aldimine complex (λ_{max} = 337 and 420 nm) is converted to an equilibrating mixture of enzyme-bound PLP species dominated by a stable quinonoid with λ_{max} = 502 nm. Deuteration of L-Ala at C-α changes the rates of reaction as a consequence of a 1° KIE. The equilibrium E(Aex) $\leftrightarrows$ E(Q), is also strongly influenced by the substitution of ^{2}H for ^{1}H at C-α, with the accumulation of E(Aex) favored. The presence of a very large apparent equilibrium isotope effect (1° EIE) has been noted previously (6). The overlapping of E(Aex) spectral bands from two different tautomeric forms complicates the quantification of this 1° EIE. The reaction with L-Ala gives rise to a substantially more convoluted spectrum, thus, the value of the EIE has not previously been determined.

The magnitude of the EIE can be derived from the concentration-dependent formation of quinonoid, E(Q), as a function of the concentration (0-400 mM) of either α-^{1}H-L-Ala or α-^{2}H-L-Ala monitored at λ_{max} = 502 nm. The reaction of tryptophanase with α-L-Ala proceeds via four relaxation rates for α-^{1}H-L-Ala and five relaxations for α-^{2}H-L-Ala. The formation of E(Q) occurs during $1/\tau_1$, $1/\tau_2$, and $1/\tau_3$ while $1/\tau_4$ and $1/\tau_5$ constitute potentially complicating side reactions. Both the rate of transamination (3), $1/\tau_4$, and the rate of isotopic exchange, $1/\tau_5$, are significantly slower than either $1/\tau_1$, $1/\tau_2$, or $1/\tau_3$ ($1/\tau_4$ is at least 1/1000 times slower, while $1/\tau_5$ is at least 1/100 times slower than either relaxations $1/\tau_1$, $1/\tau_2$, or $1/\tau_3$). Thus, the amplitude of the spectral contribution from $1/\tau_4$ and $1/\tau_5$ to the total absorbance at λ_{max} = 502 nm is negligible on the timescale of quinonoid formation and can be ignored. The data points can be described precisely by the following hyperbolic equation: Abs = (Abs_{max} [L-Ala])/(K_{app} + [L-Ala]), where Abs_{max} = absorbance at λ_{max} = 502 nm at infinite [L-Ala], K_{app} = apparent hyperbolic constant, [L-Ala] = concentration of α-^{1}H- or α-^{2}H-L-Ala, and Abs = absorbance at λ_{max} = 502 nm for a given [L-Ala]. As a consequence of the EIE on K_3, the constant for interconversion between E(Aex) and E(Q), isotopic substitution affects the distribution of species in eq 1.

$$\text{E(Ain)} + \text{L-Ala} \underset{}{\overset{K_1}{\rightleftharpoons}} \text{E(GD)} \overset{K_2}{\rightleftharpoons} \text{E(Aex)} \overset{K_3}{\rightleftharpoons} \text{E(Q)} \qquad (1)$$

If the steps preceding said interconversion are unaffected by isotopic substitution (*i.e.*, $K_1^H = K_1^D$ and $K_2^H = K_2^D$), and the same equilibrating steps are involved for both α-^{1}H-L-Ala and α-^{2}H-L-Ala, then $K_{app} \propto 1/K_3$, and the 1° EIE can be expressed as $K_3^H/K_3^D = K_{app}^D/K_{app}^H$, where $K_{app}^H = K_1^H\ K_2^H\ K_3^H$ and $K_{app}^D = K_1^D\ K_2^D\ K_3^D$. Thus, the 1° EIE calculated from K_{app}^D/K_{app}^H (0.0670/0.0173) was found to have a magnitude of 3.87 ± 0.4. The additional relaxation rate ($1/\tau_5$), present only in the deuterated sample, is due to a slow exchange of ^{2}H for ^{1}H in the active site of the enzyme. Nevertheless, only a small percentage of ^{2}H actually exchanges on a time scale of hours. This lack of exchange is interpreted as evidence for a solvent-excluded active site environment in E(Q).

Cleland has shown that the 1° EIE assigned to the α-proton of an amino acid is typically 1.13. If this isotope effect also applies to the E(Aex) α-proton, a 1° EIE of 3.87 gives a calculated value for the fractionation factor (ϕ) of 0.29 ± 0.03 for the site residue to which the proton is bound after deprotonation of E(Aex) (7,8). Thus, the low fractionation factor determined herein along with the solvent-excluded active site provide evidence for LBHB involvement in the catalytic mechanism of tryptophanase. Since microscopic reversibility demands that abstraction of the α-proton be followed by reattachment to C-α, the slow rate of exchange requires that the same proton be transferred back and forth between the C-α and the protein. Thus, the data support a proposal that this proton become sequestered in a LBHB by conformationally–driven compression during the process of α-proton abstraction.

Although we have not been able to ascertain directly the basic residues involved in the proposed hydrogen bond, some hypotheses can be drawn from the published literature. The Lys that forms the Schiff-base linkage with the PLP (Lys-270 in tryptophanase) is thought to be the base responsible for the abstraction of the α-proton (9,10). There is also an abundance of proposed LBHB's with at least one partner consisting of a carboxylate moiety (11). The crystal structure of the *P. vulgaris* tryptophanase internal aldimine (12) has several conserved Glu and Asp residues, as well as a Tyr, a His, and adjacent Lys residues in the vicinity of the active site, any pair of which might act to trap the proton abstracted from the C-α in a strong H-bond. Initially we proposed that a Lys-269 can form a LBHB with the Lys-270 subsequent to proton abstraction. Our *ab initio* and DFT calculations indicate that two primary amines can in fact sequester a proton and exhibit large EIE's. We also compute a large barrier to rotation around the C-N bond that could account for the lack of scrambling seen in the experimental work. The K269R mutation yields an enzyme with highly impaired catalytic activity, where the formation and breakdown of the quinonoid intermediate are selectively affected (13). However, this mutation does not seem to abolish the large EIE. Therefore, we exclude the possibility of LBHB formation between two lysines in the active site of tryptophanase, although it seems probable that one atom for the LBHB involves the Lys-270.

References

1. Frey, P. A., and Cleland, W. W. (1998) *Biorg. Chem. 26*, 175-192.
2. Phillips, R S., & Gollnick, P.D. (1989) *J. Biol. Chem. 264*, 10627-10632.
3. Metzler, C.M., Viswanath, R., & Metzler, D.E. (1991) *J. Biol. Chem. 266*, 9374-9381.
4. Dunn, M.F., Bernhard, S.A., Anderson, D., Copeland, A., Morris, R.G., & Roque, I.-P. (1979) *Biochemistry 18*, 2346-2354.
5. Drewe, W.F., Jr., & Dunn, M.F. (1985) *Biochemistry 24*, 3977-3987.
6. June, D.S., Suelter, C.H., & Dye, J.L. (1981) *Biochemistry 20*, 2714-2719.
7. Cleland, W.W. (1980) *Methods Enzymol. 64*, 104-125.
8. Hur, O., Leja, C., & Dunn, M.F. (1996) *Biochemistry 35*, 7378-7386.
9. Nagata, S., Hyde, C.C., & Miles, E.W. (1989) *J. Biol. Chem. 264*, 6288-6296.
10. Hayashi, H. (1995) *J. Biochem. 118*, 463-473.
11. Mildvan, A.S., Harris, T.K., & Abeygunawardana, C. (1999) *Method. Enzymol. 308*, in press.
12. Isupov, M.N., Antson, A.A., Dodson, E.J., Dodson, G.G., Dementieva, I.S., Zakomirdina, L. N., Wilson, K.S., Dauter, Z., Lebedev, A.A., & Harutyunyan, E.H. (1998) *J. Mol. Biol. 276*, 603-623.
13. Phillips, R.S., Richter, I., Gollnick, P.D., Brzovic, P., & Dunn, M.F. (1991) *J. Biol. Chem. 266*, 18642-18648.

Biochemistry and Molecular Biology of Vitamin B_6
and PQQ-dependent Proteins
ed. by A. Iriarte, H.M. Kagan and M. Martinez-Carrion

3-DIMENSIONAL STRUCTURES OF RABBIT CYTOSOLIC AND *E. COLI* SERINE HYDROXYMETHYLTRANSFERASE

Verne Schirch, Neel Scarsdale, Martino di Salvo, Sonia Delle Fratte, and H. Tonie Wright

Department of Biochemistry and Molecular Biophysics and Institute for Structural Biology and Drug Discovery, Virginia Commonwealth University, Richmond, Virginia 23298

SUMMARY

The 3-dimensional structure of rabbit liver cytosolic serine hydroxymethyltransferase was determined in the absence of any ligands and with the external aldimine reduced to a secondary amine by sodium borohydride. The 3-dimensional structure of *E. coli* serine hydroxymethyltransferase was determined as a ternary complex with glycine and 5-formyltetrahydrofolate bound at the active site. The structures were used to elucidate the role of active site residues in the mechanism of the enzyme whose properties had previously been determined by kinetic studies of site mutants.

Introduction

Serine hydroxymethyltransferase (SHMT) catalyzes the reversible interconversion of serine and tetrahydrofolate to glycine and 5,10-methylenetetrahydrofolate (1). This reaction is the major source of one-carbon groups for the biosynthesis of purines, thymidylate, and choline, as well as numerous methylation reactions. All purified and characterized forms of SHMT contain pyridoxal phosphate (PLP) bound as an internal aldimine to a Lys residue (Lys229 in *E. coli* SHMT). The tetrahydrofolate behaves as a substrate carrying the one-carbon group to other enzymes involved in biosynthetic pathways. Considerable mechanistic work on the role of both pyridoxal phosphate and tetrahydrofolate have been published (2-9). Much of this work involved the use of site mutants of specific amino acid residues believed to be at the active site. These were selected because they are conserved in the sequence of all SHMTs, and correspond spatially to active site residues in aspartate aminotransferase, which has a similar fold as SHMT.

The mechanism of SHMT is considered to be a retroaldol cleavage of serine to a glycine anion and free formaldehyde. The glycine anion is stabilized as a quinonoid intermediate bound to pyridoxal phosphate, and the formaldehyde reacts with tetrahydrofolate. These two steps of the reaction mechanism are shown in Scheme I below.

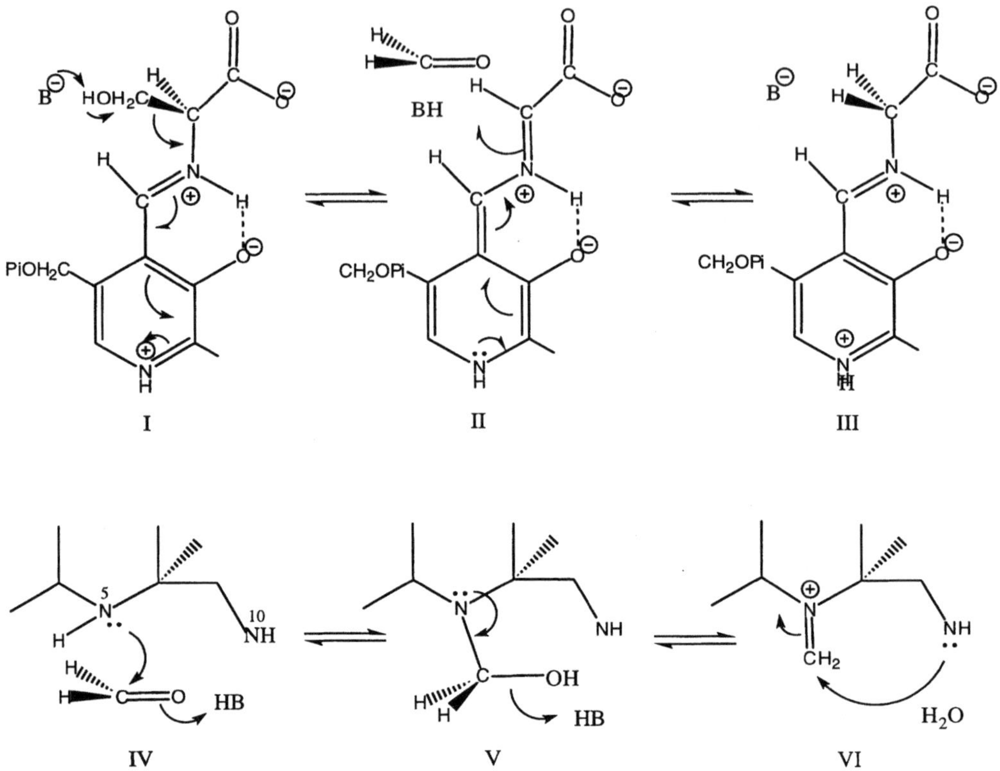

Scheme I

Materials and Methods

Materials. All chemicals were of the purest grade available. Both rabbit cytosolic and *E. coli* SHMTs were purified from *gly*A$^-$ *rec*A$^-$ *E. coli* cells containing plasmids that overproduce the respective enzyme (2, 10). Each enzyme was purified from cell extracts by procedures previously published (7, 10). The rabbit cytosolic enzyme is a mutant in which an Asn-Gly sequence at position 5-6 was changed to Gln-Gly. The Asn in this sequence has been shown to deamidate to both Asp and isoAsp residues, which introduces heterogeneity into the structure (11). Each enzyme was greater than 98 % pure as judged by gel electrophoresis.

Assays. Serine cleavage was monitored by coupling the oxidation of the product 5,10-methylenetetrahydrofolate to methylenetetrahydrofolate dehydrogenase and NADP and determining the rate of formation of NADPH (2). The rate of *allo*threonine conversion to glycine and acetaldehyde was measured by coupling to alcohol dehydrogenase and NADH (2). The rate of transamination of D- and L-alanine was followed by observing the decrease in absorbance of enzyme bound PLP at 425 nm during its conversion to PMP (3). The rate of exchange of the *pro* 2S proton of glycine with solvent protons was determined by measuring the release of tritium into solvent using 2-[^{3}H]glycine as substrate (4).

Affinity of ligands. Ligand affinities were determined from steady state kinetics, observing changes in the absorption spectrum of enzyme bound PLP upon titration with glycine and tetrahydrofolate, and changes in the circular dichroism spectra of the bound PLP (4, 5).

Results and Discussion

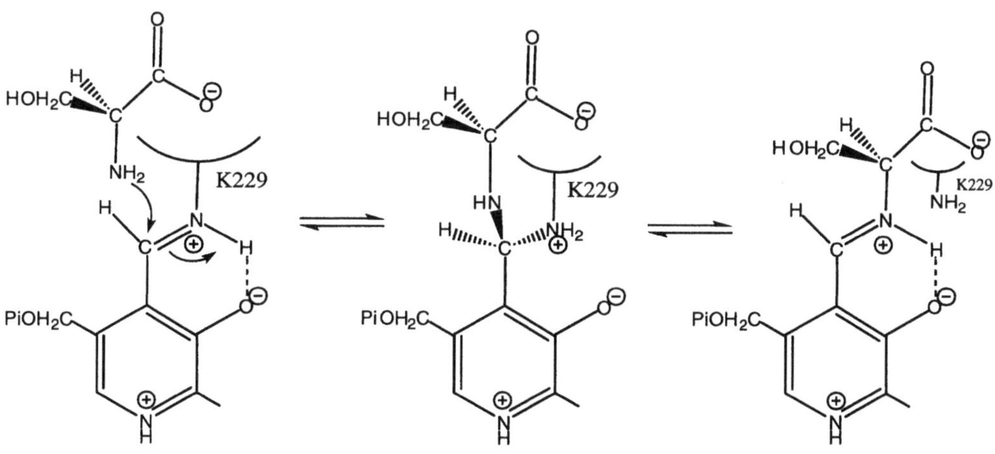

Scheme II

T226A mutant. The numbering system of the *E. coli* enzyme will be used for both rabbit cytosolic and *E. coli* SHMTs (12). A sequence of conserved residues in all SHMTs lies immediately upstream from the active site Lys229 that forms the internal aldimine with PLP. This series of residues is -Val-Val-Thr-Thr-Thr-Thr-His-Lys^{229}. Each of these Thr residues were changed to Ala and the enzymes expressed and purified. Only the T226A mutant enzyme shows unusual properties (5). It exhibited a 30-fold decrease in catalytic activity, but with almost unaltered affinity for substrate amino acids. The spectrum of the T226A enzyme was normal, but when saturated with amino acid ligands it exhibited a large absorption maximum at 343 nm with only a small absorption band at 425 nm. The wild-type enzyme exhibits only a small absorption band at 343 nm and a large band at 425 nm. This was interpreted as being the result of the accumulation of the *gem*-diamine complex (structure VIII, Scheme II) at the expense of the external aldimine, and it was inferred that Thr226 plays a role in catalyzing the conversion of the *gem*-diamine to the external aldimine (structure VIII to IX in Scheme II). In the *E. coli* ternary complex structure Thr226 forms a H-bond with the ε-amino group of Lys229. This H-bond rotates the Lys229 amino group away from C4' of PLP hindering its attack on C4' as shown in structure IX in Scheme II. Converting Thr226 to Ala removes this H-bond, resulting in a very slow conversion of VIII to IX because of the failure to remove the ε-amino group of Lys229 from proximity to C4', which it can attack in the reverse reaction. This results in the accumulation of intermediate VIII absorbing at 343 nm.

Arg363. Based on sequence homology it was proposed that Arg363 would form an ion pair with the oxygen atoms of the substrate amino acid carboxylate group (9). R363K and R363A mutants were made and the former shows a 3000-fold decrease in k_{cat} and the latter shows no detectable activity. Unlike the complete inactivity of wild-type SHMT in the cleavage of both serine ethyl ester and serinamide, the R363A mutant shows activity toward these substrates. The results were interpreted to mean that Arg363 was critical for substrate binding and activation of the bonds at the α-carbon. The ester or amide of the substrate amino acid could replace in part this activation by simulating the charge neutralization of the ion pair in the wild-type with uncharged substrate functional groups.

The crystal structure of *E. coli* SHMT in complex with glycine and 5-formyltetrahydrofolate clearly shows that the guanidinium group of Arg363 forms close contacts with the glycine carboxylate group. The two NH groups of the guanidinium group are 2.7 and 2.8 Å from the two carboxylate oxygens of glycine. In the rabbit cytosolic SHMT structure, which does not contain a bound amino acid, the guanidinium group is displaced from its site in the *E. coli* enzyme. Close examination of the two structures suggests that upon substrate binding the interaction of the substrate carboxylate group and Arg363 results in a small shift in position of Arg363, which results in a more substantial movement of segments of the small domain of the enzyme, which generates a more solvent inaccessible closed structure. This was predicted from the kinetic studies of the R363A mutant.

Lys229. Lys229 was known to be the residue which forms the external aldimine with PLP. The residue was changed to Gln, Arg, and His (7), none of which can form an external aldimine. The most interesting of these site mutants was K229Q. The purified mutant enzyme contains either tightly bound glycine or serine, and exhibits spectral properties of the external aldimine of the wild-type enzyme. Addition of tetrahydrofolate results in the rapid single turnover conversion of the bound serine to glycine and methylenetetrahydrofolate at a rate similar to the steady-state rate exhibited by the wild-type enzyme. It was concluded from these studies that Lys229 is required to expel amino acid ligands from the active site, which is in accordance with the mechanisms shown in Schemes I and II. Most important though, is the conclusion that the ε-amino group of the expelled Lys229 in the external aldimine is not the active site base that removes the proton for either the serine hydroxyl group or the *pro* 2S proton of glycine (structures I and III in Scheme I). The crystal structures of both the rabbit cytosolic and *E. coli* SHMTs confirm the role of Lys229 in forming the internal aldimine with PLP. The structures also confirm that the expelled ε-amino group of Lys229 in the external aldimine is on the wrong side of C4' to act as the active site base.

His228. As noted above, His228 is a residue that lies on the N-terminal side of Lys229 and is conserved in all known sequences of SHMT. Others have proposed that His228 is the active site base. This residue was changed to both Asp and Asn in the *E. coli* SHMT (2, 6) and these studies showed that His228 is not the active site base, since both mutant enzymes exhibit considerable enzymatic activity. The studies do suggest, however, that His228 plays a role in the open to closed conformational change and is involved in orienting the serine hydroxyl group for reactivity. Although we do not have a 3-dimenisonal structure with serine bound as an external aldimine, serine as a substrate can be modeled into the active site for both the rabbit and *E. coli* enzymes. These structures show that His228 is too far removed from the putative serine hydroxyl group to form any direct interaction that would explain its properties in the H228D and H228N mutants. Rather it appears from the crystal structures that the effect of His228 is mediated by Tyr55. To date this mutant has not been made and characterized. A double mutant of His228 and Y55 should confirm the roles of these residues in the mechanism of SHMT.

Asp200. In aspartate aminotransferase, an Asp residue forms a salt bridge with N1 of the PLP ring. Changing this Asp to an Ala results in loss of activity, but replacement with Glu results in enzyme with low activity. In sequence alignments, Asp200 of SHMT occupies an equivalent position as the critical Asp in aspartate aminotransferase. We made both D200A and D200E mutants of *E. coli* SHMT, which were expressed and purified by the same procedure used for the wild-type enzyme. The spectra of the mutant enzymes differ from the wild-type enzyme by having an absorption peak at 335 nm in addition to the one at 422 nm that is present in the wild-type enzyme. Addition of glycine and tetrahydrofolate does not give any evidence of a quinonoid complex absorbing at 494 nm. Neither mutant enzyme exhibits any detectable catalytic activity in any of the reactions catalyzed by SHMT. We conclude that Asp200 in SHMT is absolutely required to stabilize the quinonoid complex, which is an intermediate on the pathway for all reactions catalyzed by this enzyme. Attempts were made with the D200A enzyme to refold it in the presence of 1 M formate in the hope of trapping a formate in the position occupied by the β-carboxyl group of Asp, but this experiment gave only negative results.

Acknowledgements

This work was supported by NIH grants GM28143 and GM50209.

References

1. Schirch, V. (1998) Mechanism of Folate-Requiring Enzymes in One-Carbon Metabolism. In "Comprehensive Biological Catalysis" (ed. M. Sinnott) 211-252.

2. Hopkins, S. and Schirch, V. (1986) Properties of a Serine Hydroxymethyltransferase in Which an Active Site Sistidine Has Been Changed to an Asparagine by Site-Directed Mutagenesis. *J. Biol. Chem.* 261: 3363-3369.

3. Shostak, K. and Schirch, V. (1988) Serine Hydroxymethyltransferase: Mechanism of the Racemization and Transamination of D- and L-Alanine. *Biochemistry* 27: 8007-8014.

4. Schirch, V., Shostak, K., Zamora, M., and Gautam-Basak, M. (1991) The Origin of Reaction Specificity in Serine Hydroxymethyltransferase. *J. Biol. Chem.* 266: 759-764.

5. Angelaccio, S., Pascarella, S., Fattori, E., Bossa, F., Strong, W., and Schirch, V. (1992) Serine Hydroxymethyltransferase: Origin of Substrate Specificity. *Biochemistry* 31: 155-162.

6. Stover, P., Zamora, M., Shostak, K., Gautam-Basak, M., and Schirch, V. (1992) *Escherichia coli* Serine Hydroxymethyltransferase: The Role of Histidine 228 in Determining Reaction Specificity. *J. Biol. Chem.* 267: 17679-17687.

7. Schirch, D., Delle Fratte, S., Iurescia, S., Angelaccio, S., Contestabile, R., Bossa, F., and Schirch, V. (1993) Function of the Active-site Lysine in *Escherichia coli* Serine Hydroxymethyltransferase. *J. Biol. Chem.* 268: 23132-23138.

8. Pascarella, S., Schirch, V., and Bossa, F. (1993) Similarity Between Serine Hydroxymethyltransferase and Other Pyridoxal Phosphate-dependent Enzymes. *FEBS* 331: 145-149.

9. Delle Fratte, S., Iurescia, S., Angelaccio, S., Bossa, F., and Schirch, V. (1994) The Function of Arginine 363 as the Substrate Carboxyl-binding Site in *Escherichia coli* Serine Hydroxymethyltranserase. *Eur. J. Biochem.* 225: 395-401.

10. di Salvo, M. L., Delle Fratte, S., De Biase, D., Bossa, F., and Schirch, V. (1998) Purification and characterization of Recombinant Rabbit Cytosolic Serine Hydroxymethyltransferase. *Prot. Express. and Purif.* 13: 177-183.

11. Artigues, A., Birkett, A., and Schirch, V. (1990) Evidence for the In Vivo Deamidation and Isomerization of an Asparaginyl Residue in cytosolic Serine Hydroxymethyltransferase. *J. Biol. Chem.* 265: 4853-4858.

12. Scarsdale, J. N., Kazanina, G., Radaev, S., Schirch, V., and Wright, H. T. (1999) Crystal Structure of Rabbit Cyosolic Serine Hyroxymethyltransferase at 2.8 Å Resolution: Mechanistic Implications. *Biochemistry* 38: 8347-8358.

Biochemistry and Molecular Biology of Vitamin B_6
and PQQ-dependent Proteins
ed. by A. Iriarte, H.M. Kagan and M. Martinez-Carrion

Role of Y65 and E57 in *Escherichia coli* Serine Hydroxymethyltransferase.

S. Angelaccio, R. Contestabile, P. Di Giovine, F. Bossa and V. Schirch[§].

Dipartimento di Scienze Biochimiche "A. Rossi Fanelli" and Centro di Biologia Molecolare del CNR, Università di Roma "La Sapienza", Piazzale Aldo Moro 5, 00185 Roma, Italy and the [§]Department of Biochemistry, Virginia Commonwealth University, Richmond, Virginia 23298

Summary

The crystal structures of serine hydroxymethyltransferase from three different sources have recently been solved. Inspection of the active site reveals that, besides K229, whose role as catalytic base had been ruled out, either E57 or Y65 may be the putative active-site proton exchanger required for catalysis. Two mutant forms of the *E. coli* enzyme, E57Q and Y65F, were produced and their catalytic properties studied. The results show that neither E57 nor Y65 is the elusive catalytic base. The Y65F mutation substantially reduced k_{cat} and K_m for cleavage of β-hydroxyamino acids, although the catalytic efficiency was unaffected. A possible role of Y65 in the conformational change that is thought to take place upon binding of substrates is proposed.

Introduction

Serine hydroxymethyltransferase (SHMT) uses pyridoxal 5'-phosphate and tetrahydrofolate (H_4PteGlu) as coenzymes to catalyse the reversible interconversion of serine and glycine and the irreversible hydrolysis of 5, 10-methenyltetrahydrofolate to 5-formyltetrahydrofolate. In addition to these physiological reactions, SHMT also catalyses the retroaldol cleavage of several other 3-hydroxyamino acids in the absence of H_4PteGlu, such as *allo*threonine, and the slow transamination of D- and L-alanine. The crystallographic structure of SHMT from three different sources has recently been solved (1-3). The availability of the spatial data prompts the revision of the mechanistic studies based on mutagenesis experiments and gives new hints for their interpretation (2). Nevertheless, several important aspects of the catalytic mechanism remain unclear. The putative active-site base, required to remove the 3-hydroxyl group proton of substrates in the direction of retroaldol cleavage or the *pro* (2S) proton of glycine in the reverse direction, has so far eluded all investigations. It is known that both α-protons of glycine are exchanged with the solvent, even if the *pro* (2S) proton is exchanged 7400 times faster (4). Moreover, previous studies have shown that different bases at the active site are involved in removing the proton from the two alanine enantiomers (5). Inspection of the active site of the Gly•5-CHO-H_4PteGlu•*E. coli* SHMT (eSHMT) ternary complex (3) reveals that besides Lys 229, whose role has already been investigated (6) and that lies on the side of the *pro* (2R) proton of the glycine external aldimine, there are two other residues which may be involved in the acid-base catalysis and that are located on the *pro* (2S) proton side: Glu 57 and Tyr 65. Two eSHMT mutant forms, E57Q and Y65F, were produced and their catalytic properties were studied.

Materials and methods

Mutations in eSHMT were obtained by a PCR-based site-directed mutagenesis (7) and all enzymes were expressed, assayed for activity and purified as previously described (8). Enzyme expression was carried out in a *rec* A^- *E. coli* strain which has been shown neither to have any wild-type SHMT activity nor to undergo reversion of point mutations to form wild-type enzyme (7). The rate of exchange of the *pro* (2S) proton of glycine with solvent protons was determined as previously described (5).

Results and discussion

The two mutant eSHMTs catalysed the transamination of D-alanine with rate constants which are not vastly different from those observed with the wild-type enzyme (Table 1). Both mutant enzymes were also able to catalyse the solvent exchange of the *pro* (2S) proton of glycine (Table 1). With all eSHMT forms, the addition of H_4PteGlu increased the rate of exchange, however the rate constant of the reaction catalysed by the Y65F mutant was 0.27% the value of the wild-type. Apparent K_d values for glycine and serine binding to each enzyme form were determined by measuring the change of the cofactor optical activity in titration experiments (Table 1). The glycine proton exchange experiments and the unaltered capability of the mutant enzymes to transaminate D-alanine show that neither Y65 nor E57 are involved in the glycine α-proton abstraction. However, while the E57Q mutant retained activity in all of the other reactions experimented, the Y65F mutation had a considerable effect on the hydroxymethyltransferase activity of the enzyme. This mutant form displayed a 400-fold decrease in both K_m for serine and k_{cat} for serine cleavage. Thus, the catalytic efficiency (k_{cat}/K_m) remained unchanged. The same was substantially true for *allo*threonine cleavage, except the decrease of the kinetic constants was less dramatic. Apparently, in the cleavage of 3-hydroxyamino acids, the energy level of the transition state was not altered by the mutation, while the increased stability of the enzyme-substrate complex determined an equivalent increase of the activation energy. The affinities for glycine and for D-alanine were not vastly affected by the mutations, however, glycine bound about 330-fold tighter to the Y65F enzyme when this was saturated with H_4PteGlu.

Table 1

Ligand	Y65F	E57Q	**Reaction**	Y65F	E57Q
	K_m or K_d (mutant) / K_m or K_d (WT)			k (mutant) / k(WT)	
Serine					
- H_4PteGlu	0.027[a]	0.7[a]	Serine cleavage	0.0026	0.072
+ H_4PteGlu	0.0023[b]	0.5[b]			
Glycine			[2-^{3}H] glycine exch.[d]		
- H_4PteGlu	0.24[a]		- H_4PteGlu	0.52	0.3
+ H_4PteGlu	0.0033[a]		+ H_4PteGlu	0.0027	2
*allo*Threonine	0.053[b]	14[b]	*allo*Threonine cleav.	0.014	1.1
D-Alanine	0.27[c]	nd	D-Alanine transam.	4.1	0.3
H_4PteGlu	0.32[b]	0.24[b]			

[a] Determined from change in CD spectrum of cofactor
[b] Determined from catalytic activity
[c] Determined from ΔA_{492} nm
[d] Corrected for the difference in affinity for glycine between wild-type and mutants

The function of Tyr 65 in eSHMT is not clear. A possibility is that Y65 plays a role in the conformational change that is thought to take place upon binding of 3-hydroxyamino acids substrates or formation of the Gly•SHMT•H_4PteGlu ternary complex (9). Evidence in support of this hypothesis are the observed lack of effect of the Y65F mutation on the transition state energy level, the increased affinity for 3-hydroxyamino acid substrates and the greatly reduced dissociation constant of the ternary complex. Indeed, both the K_m and rate constant for transamination of D-alanine, whose binding does not appear to result in a conformational change (9), are largely unaffected by the mutation. The latter consideration may equally be applied to glycine binding and to the glycine proton exchange reaction, but only when H_4PteGlu is absent. Following this hypothesis, the replacement of Y65 with a Phe residue would affect the equilibrium position of the open-to-closed conformational change. The observed tighter binding of certain substrates would result from the shift of the equilibrium in favour of the closed form. The enzyme-substrate complex is then positioned into a deeper energy well, thereby increasing substantially the activation energy and decreasing the rate of the reactions.

Any attribution of a direct catalytic role to E57 proves difficult. It appears from the crystallographic data (3) that E57 may be hydrogen bonded to Y65. Therefore, the observed changes of the catalytic properties of SHMT caused by the E57Q mutation may result from the loss of this interaction with the more important Y65 residue.

Acknowledgements
This work was supported by funds from the Italian M.U.R.S.T. and by Grant GM 28143 (to V. S.) from the National Institutes of Health. Roberto Contestabile was the recipient of a fellowship from Istituto Pasteur-Fondazione Cenci Bolognetti.

References

1) Renwick, B., Snell, K. and Baumann U. (1998), *Structure* 6: 1105-1116.
2) Scarsdale, J.N., Kazanina, G., Radaev, S., Schirch, V. and Wright, H.T. (1999), *Biochemistry* 38: 8347-8358.
3) Scarsdale, J.N., Radaev, S., Kazanina, G., Schirch, V. and Wright, H.T (1999), *J. Mol. Biol.* in press.
4) Malthouse, J.P., Milne, J.J. and Gariani, L.S. (1991), *Biochem. J.* 274: 807-812.
5) Shostak, K. and Schirch, V. (1988), *Biochemistry* 27: 8007-8014.
6) Schirch, D., Delle Fratte, S., Iurescia, S., Angelaccio, S., Contestabile, R., Bossa, F. and Schirch, V. (1993), *J. Biol. Chem.* 268: 23132-38.
7) Iurescia, S., Condò, I., Angelaccio, S., Delle Fratte, S. and Bossa F. (1996), *Protein. Expr. Purif.* 7: 323-328.
8) Schirch, V., Hopkins, S., Villar, E. and Angelaccio, S. (1985), *J. Bacteriol.* 163: 1-7.
9) Schirch, V., Shostak, K., Zamora, M. and Gautam-Basak, M. (1991), *J. Biol. Chem.* 266: 759-764.

Biochemistry and Molecular Biology of Vitamin B_6
and PQQ-dependent Proteins
ed. by A. Iriarte, H.M. Kagan and M. Martinez-Carrion

Reaction and structure of 1-aminocyclopropane-1-carboxylate deaminase

M. Honma[1], T. Murakami[1], T. Ose[2], H. Matsui[1], I. Tanaka[2],

Graduate School of Agriculture[1] and Graduate School of science[2], Hokkaido University, Sapporo, 060-86589 Japan

Summary

1-Aminocyclopropane-1-carboxylate (ACC) deaminase was isolated from fungi *Hansenula saturnus* and *Penicillium citricum*, and their cDNA were analyzed. Amino acid sequences of the enzymes from *H. saturnus* and *P. citrinum* showed 61% and 52% of similarity respectively to that of a bacterial enzyme. Furthermore, the alignment of these sequences indicates their similarity to the tryptophan synthase β-subunit. This became clearly evident when crystallographical analysis of ACC deaminase from *H. saturnus* was conducted. From the three-dimensional figure, three residues found near the PLP were chosen for preparation of the mutant enzymes.

Introduction

1-Aminocyclopropane-1-carboxylate (ACC) is an immediate precursor of the plant hormone ethylene, which initiates the ripening of fruit and regulates various processes in plant growth and development (Adams and Yang, 1978). Various soil microorganisms grow on a medium containing ACC as a sole nitrogen source (Honma and Shimomura, 1978). Plant ACC oxidase cleaves two C-C bonds of a cyclopropane ring to release ethylene, while the microorgism ACC deaminase cleaves a C-C bond of the ring of ACC to form 2-oxobutyrate and ammonia. The ring-opening reaction by ACC deaminase is recognized as a sort of intramolecular γ elimination and followed by the same process as other similar enzymes. The cleavage of the ring opening occurs at the bond between C_α and *pro-S* methylene C, which corresponds to H atom at C_α in the case of D-serine. In fact, D-serine is degraded by ACC deaminase to form pyruvate and ammonia, but L-serine is not.

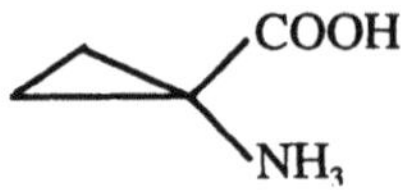

$\longrightarrow$ $CH_3CH_8COCOOH + NH_3$

D-Serine $\longrightarrow$ $CH_3COCOOH + NH_3$

A unique step in ACC deaminase catalysis is the cyclopropane-ring opening of the ACC-PLP aldimine. A possible mechanism for this ring-opening was proposed; that was the nucleophilic addition at *pro-S* methylene carbon followed by β-proton abstruction, which required a nucleophile and a base for the enzyme (Walsh et al., 1981). We report the structural similarity of ACC deaminase to tryptophan synthase β-subunit and amino acid residues found around PLP.

Results and Discussion

Several kinds of bacteria have been isolated from soil with the medium containing ACC as a sole nitrogen source and ACC deaminase genes have been cloned (Klee et al., 1991, Campbell and Thomson, 1996, Shah et al., 1998, Jia et al., 1998). The obtained sequences show an equal size of 338 residues and more than 80% of residues are identical to each other. Two sequences of ACC deaminase were given by analyses of the enzyme proteins and cDNAs which were isolated from fungi *H. saturnus* and *P. citrinum* (Fig. 1). The enzyme from *H. saturnus* had 341 amino acid residues and 61% of residues were identical to the *Pseudomonas* enzyme. The *P. citrinum* enzyme had 360 residues encoded in cDNA and was 52% identical to the *Pseudomonas* one. Several regions of these ACC deaminase sequences were similar to the corresponding regions of the tryptophan synthase β-subunit (Hyde et al., 1998, Minami et al., 1998).

The *Pseudomonas* enzyme was inhibited by dithiobis (2-nitrobenzoic acid) to a complete loss of enzyme activity, and the inhibition by iodoacetamide was stimulated by the presence of D-alanine but not L-alanine. This reactive cysteine residue was identified as Cys 162. However replacement of the cysteine residue maintained perfect enzyme activity (Murakami et al., 199). This same residue was not found in the sequence of the *Penicillium* enzyme. Only one cysteine residue was retained in three kinds of ACC deaminase in Fig. 1. This residue was also found in a corresponding site in the sequence of tryptophan synthase β-subunit, which was described to be close to pyridine N of PLP (Hyde et al., 1988).

```
PsACP   ---------------------------------------MNLQRFPRYPLTFGPTPIQPLARLSKHLGGKVH 33
H.sat   ---------------------------------------SGVAK AK       S  SN N   Q   S  N 33
P.cit   ------------------------MTDPNPVTLPTPFSTIPRTPLLLG SPIH L RTTADLAKNNPHPL T 48
TrpSB   MTTLLNPYFGEFGGMYVPQILMPALNQLEEAFVRAQKDPEFQAQ ADLLKNYAGR TALTKCQNITA TRTT 72
                                     H1            H2             S1

PsACP   LYAKREDCNSGLAFGGN-KTRKLEYLIPEALAQGCDTLVSIGGIQSNQ-TRQVAAVAAHLGMKCVLVQENWV 103
H.sat   V                - L     IV DIVEGDYTH      R    -  M   L  K  K    I  D   103
P.cit   I    D L   Y Y   -        LAD Q    T       V   H-         RS L AR    H   118
TrpSB     L  E ---- LH  AH  N--QV GQAL  KRMGKSEI AETGAG HGVAS LAS L  L  RIYMGAK- 137
          S2                   H3                S3        H4           S4

PsACP   NYSDA---VYDRVGNIQMSRILGADVRLVPDGFDIGFRRSWEDALES------------VRAAGGKPYAIPA 160
H.sat   PIPE EKD  N     EL   M     VIE      M K FAN  QE------------LED  H   P    163
P.cit   DWT P---G EST    L  LM      D S  G EHKNTVQRVV E------------A GD E   Y    175
TrpSB   -------D ERQSP VFRM LM  E IP HS SATLKDACN -  RDWSGSYETAHYMLGTA  PH  PTIV 201
                         H5      S5              H6             S6            H7

PsACP   GCSDHPLGGLGFVGFAEEVRAQEAE-LGFKFDYVVVCSVTGSTQAGMVVGFAADGRA------DRVIGVDAS 225
H.sat      E KY         DE IN  V -  I   KI   C     T  ILA M QY  Q------ D AI     228
P.cit    A          ARW FE  E  G -M VF  T L  A     F   IA  KLLE LYPADAKRR V I   246
TrpSB   REFQRMI --------  TK  ILDKE RLP A IA VGG NAI  FAD  IN -TS------VGL   EPG 258
                      H8                  S7         H9                  S9

PsACP   AKPAQTREQITRIARQTAEKV-GLERD-----------------------IMRADVVLDERFAGPEYGLPNE 273
H.sat   FTSEK K  TL   NN  KLI- V HE-------------------------FK FT T   Y C  V     274
P.cit    TVEA  A VL   KN  A I-  SE D---------------------- TED  M D  YHAGI V  DR 295
TrpSB   GHGIE G HGAPLKHGRVGIYF MKAPMMQTADGQIEESYSISAGLDFPSVGPQHAY NSIGRADYVSITDD 330
                                                              H10       S9

PsACP   GTLEAIRLCARTEGMLTDPVYEGKSMHGMIEMVRNGEFPEGSRVLYAHLGGVPALNGYSFIFRDG-- 338
H.sat     I    T  EQ  V           Q L ALIKEDY KP AN   V    A  SA  SF PTKTA 341
P.cit   Q  WD EYA KM AFI         FA  MD I R  IK- GNI      QL    A ELG TNE   360
TrpSB   EAL  FKTLC H  IIPALESSHALA ALKM REQP KEQLLV NLSGR DKDIFTVHDILKAR EI 397
           H11                   H12             S10            H13
```

Figure 1: ACC deaminases and tryptophan synthase β-subunit.

Sequences of ACC deaminases from *Pseudomonas* sp. ACP(PsACP), *H. saturnus* (H. sat), and *P. citrinum* (P. cit) and tryptophan synthase β-subunit (TrpSB)(Hyde et al., 1988) were alignel. The written amino-acid residues are different from residues of the *Pseudomonas* enzyme of a top line. Bars and letters H and S indicate the second structure of the tryptophan synthase β-subunit, H: α helix and S: β sheet (Hyde et al., 1988).

Crystallographical study gave a three-dimensional structure of the *H. saturnus* enzyme. The obtained structure of ACC deaminase monomer has a pair of two β sheets and its folding topology is very similar to that of tryptophan synthase β-subunit. The major differences of ACC deaminase from the other are deletion at the N terminal and a shortcut at the central part including a long loop of tryptophan synthase.

Amino acid residues around the PLP were sought by results from the crytallographical analysis. The aldehyde group of PLP froms a Schiff base with lysine 51 which is at the N-terminal end of a helix as shown in the structure of tryptophan synthase. This helix of ACC deaminase contains lysine 54, and structural analysis shows aspargine 79 being close to 3' hydroxyl group of the PLP.

Each mutation at K51, K54, or N79 caused an enzyme activity loss. The mutant enzyme K51A which had an absorption band at 405 nm, shifted the band to 422-426 nm by the addition of ACC, D-, L-alanine, D-, or L-serine. Dissociation constants of D-amino acids were higher than those of L-amino acids (Table 1). This specificity was also observed in the competitive inhibition constants to ACC deaminase activity (Table 2). A nucleophile and a base for the mechanism of ACC deaminase remain to be solved.

Table 1. Reaction of K51A enzyme with substrate or inhibitor.

Amino acid	λmax (nm)	K_d (μM)
None	405	
ACC	425	15
D-Alanine	422	246
L-Alanine	422	33
D-Serine	424	240
L-Serine	426	22

Table 2. Michaelis constant and competitive inhibition constant for ACC deaminase.

	K_m (mM)		K_i (mM)	
	PsACP	H. sat	PsACP	H. sat
L-Alanine			5.3	14.5
D-Alanine			660	280
L-Serine			0.90	2.9
D-Serine	18		16	13.5
L-Homoserine				59.0
L-α-Aminobutyrate			63	148
ACC	1.6	2.6		

PsACP: ACC deaminase from *Pseudomonas* sp. ACP
H. sat: ACC deaminase *H. saturnus*

References

Adams, D. O. and Yang, S. F. (1979) Ethylene Biosynthesis: Identification of 1-Aminocyclopropane-1-carboxylic Acid as an Intermediate in the Conversion of Methionine to Ethylene. *Proc. Natl. Acad. Sci. USA 76*: 170-174.

Campbell, B. G., and Thomson, J. A. (1996) 1-Aminocyclopropane-1-carboxylate Deaminase Genes from *Pseudomonas* Strains. *FEMS Microbiol. Let., 138*: 207-210.

Honma, M. and Shimomura, T. (1978) Metabolism of 1-Aminocyclopropane-1-carboxylic Acid. *Agric. Biol. Chem. 42*: 1825-1831.

Hyde, C. C., Ahmed, S. A., Padlan, E. A., Miles, E. W., and Davies, D. R. (1988) Three-dimensional Structure of Tryptophan Synthase α2β2 Multienzyme Complex from *Salmonella typhimurium*. *J. Biol. Chem. 263*: 17857-17871.

Jia, Y.-J., Kakuta, Y., Sugawara, M., Igarashi, T., Oki, N., Kisaki, M., Shoji, T., Kanetsuna, Y., Horita, T., Matsui, H., and Honma, M. (1999) Synthesis and Degradation of 1-Aminocyclopropanene-1-Carboxylic Acid by *Penicillium citrinum*. *Biosci. Biotechnol. Biochem. 63*: 542-549 .

Klee, H. J., Hayford, M. B., Kretzmer, K. A., Barry, G. F., and Kishore, G. M. (1994) Contol of Ethylene Synthesis by Expression of a Bacterial Enzyme in Transgenic Tomato Plants. *Plant Cell 3*: 1187-1193.

Minami, R., Uchiyama, K., Murakami, T., Kawai, J., Mikami, K., Yamada, T.,Yokoi, D., Ito, H., Matsui, H., and Honma, M. (1998) Properties, Sequence, and Synthesis in *Escherichia coli* of 1-Aminocyclopropane-1-carboxylate Deaminase from *Hansenula saturnus*. *J. Biochem. 123*: 1112-1118.

Murakami, T., Kiuchi, M., Ito, H., Matsui, H., and Honma, M. (1997) Substitution of Alanine for Cysteine at a Reactive Thiol Site and for Lysine at a Pyridoxal Phosphate Binding Site of 1-Aminocyclopropane-1-carboxylate Deaminase. *Biosci. Biotechnol. Biochem. 61*: 506-509.

Shah, S., Li, J., Moffat, B. A., and Glick, B. R. (1998) Isolation and Characterization of ACC Deaminase Genes from Different Plant Growth-promoting Rhizobacteria. *Can. J. Microbiol. 44*: 833-843.

Walsh, C., Pascal, R. A. Jr., Johnston, M., Raines, R., Dikshit, D., Kranz, A., and Honma, M. (1981) Mechanistic Studies on the Pyridoxal Phosphate Enzyme 1-Aminocyclopropane-1-carboxylate Deaminase from *Pseudomonas* sp. *Biochemistry 20*: 7509-7519.

TRANSAMINASES

Biochemistry and Molecular Biology of Vitamin B_6
and PQQ-dependent Proteins
ed. by A. Iriarte, H.M. Kagan and M. Martinez-Carrion

Refined Reaction Mechanism of Aspartate Aminotransferase

H. Hayashi, H. Mizuguchi, Y. Nakajima,* I. Miyahara,* K. Hirotsu,* and H. Kagamiyama

Department of Biochemistry, Osaka Medical College, Takatsuki 569-8686, and *Department of Chemistry, Osaka City University, Osaka 558-8585, Japan

Summary

The combined spectral, kinetic, and structural studies on the reaction of *E. coli* aspartate aminotransferase (AspAT) with aspartate and its analogue, 2-methylaspartate, provided a new, refined mechanism for this catalytic reaction. The association of the substrate aspartate and AspAT proceeds via dual routes (E + SH^+ and EH^+ + S). The Schiff base successively increases its pK_a from 6.8 in the unliganded state to 8.8 in the Michaelis complex, and finally to >10 in the external aldimine. These pK_a shifts are mainly brought about by relaxing the torsion around C4–C4' of the Schiff base through the interaction of AspAT with the substrate. The significance of the low pK_a value in the unliganded enzyme is interpreted to be the ground-state destabilization of the enzyme protein that increases the k_{cat}/K_m value.

Introduction

The reaction mechanism of aspartate aminotransferase has been extensively studied. In 1969, a mechanism of this enzyme was proposed by Karpeisky and Ivanov (*3*), and combined with the crystallographic analysis by Jansonius' group, Kirsch *et al.* described a detailed catalytic process, which has generally been accepted (*4*). In this mechanism, the Schiff base is almost unprotonated at neutral pH due to the positive charges of the two arginine residues, Arg292* and Arg386. The substrate binds to the Schiff-base-unprotonated form (E_L) and the two carboxylates neutralize the positive charges. The pK_a of the Schiff base is then increased, and the proton migrates from the substrate α-amino group to the Schiff base. The free α-amino group attacks the imine bond, resulting in the external aldimine. The liberated ε-amino group of Lys258 acts as a base catalyst in the subsequent reactions. Mutagenesis and kinetic analysis supported the validity of this mechanism. However, there remained the following unresolved problems. 1) Transient kinetics for the reaction of AspAT and aspartate shows anomalous behavior at low pH. 2) The origin of the low pK_a of the Schiff base is not clear. This must be decreased by 3 units from those of "normal" Schiff bases, and is increased by 2 upon ligand binding. However, mutation of the arginine residues to neutral amino acid residues gave only moderate increases in the pK_a (ΔpK_a = 0.5 for R386L and 0.2 for R292L). 3) The involvement of the enzyme protein is not fully understood. Then, we set out to study these problems. Most of the results in this proceeding has been recently published (*1*,*2*).

Materials and Methods

The wild type and mutant AspATs were purified from *E. coli* transformed with pUC19–*aspC* with appropriate mutations in the coding sequence. Spectroscopic and kinetic experiments were carried out in 50 mM buffer (MES, HEPES, or TAPS) in the presence of 0.1 M KCl. Absorption spectra were taken using Hitachi U-3300 spectrophotometer. Single- and multi-wavelength stopped-flow spectrophotometric measurements were done using Applied Photophysics SX.17MV spectrophotometer. Global kinetic analyses were performed using Glint® ver 1.1 (Applied Photophysics).

Results and Discussion

The rate for the conversion of the Schiff-base protonated form of the enzyme (E_LH^+) to the pyridoxamine form of the enzyme (E_M) did not show a Michaelis-Menten type dependency at lower pH values (Figure 1). From a pH jump study, we found that the interconversion of E_L and E_LH^+ is significantly slow, and based on Scheme 1, theoretical lines were drawn (Figure 1A). The line matched well with the experimental data at pH 6.1, but deviated with increasing pH and substrate concentration. We noticed that the rise in pH increases the fraction of the α-amino-group-unprotonated form of the substrate (S). If we take into account the association of E_LH^+ and S (Scheme 2), we could explain the data. These two routes have been pointed out to be kinetically equivalent in steady-state kinetics (*4*). However, we could separate them in transient kinetics.

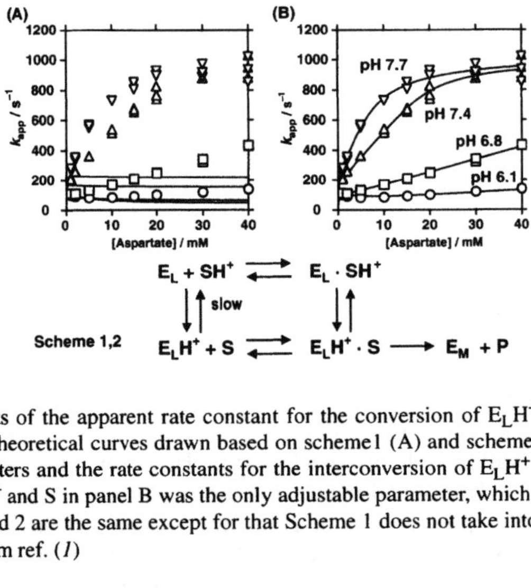

Figure 1: (A) and (B) Plots of the apparent rate constant for the conversion of E_LH^+ to E_L upon reaction with aspartate. Solid lines are theoretical curves drawn based on scheme1 (A) and scheme 2 (B) using the previously determined kinetic parameters and the rate constants for the interconversion of E_LH^+ and E_L. The rate constant for the association of E_LH^+ and S in panel B was the only adjustable parameter, which was determined to be $5.4 \times 10^6\ M^{-1}s^{-1}$. Schemes 1 and 2 are the same except for that Scheme 1 does not take into account the association of E_LH^+ and S. Modified from ref. (*1*)

Based on this result, we analyzed the reaction of AspAT with α-methylaspartate (MeAsp) (Figure 2). At pH 8.0, where the majority of the enzyme exists as E_L, the spectral change was fitted to a common two-step mechanism (Scheme 3). At pH 6.8, where both E_L and E_LH^+ exist in equal amounts, we should consider Scheme 4 (upper). With a sufficient concentration of MeAsp, E_L rapidly converts to the Michaelis complex, whereas E_LH^+ is slowly converted either to E_L or to the Michaelis complex. Scheme 4 (upper) can then be simplified to Scheme 4 (lower). The time-resolved spectra of Figure 2 were fitted to Schemes 3 and 4 (lower) using numerical integration, and the spectra of the two intermediates were obtained. At both pH values, the fraction of the protonated Schiff base is around 8% in the Michaelis complex and around 40% in the external aldimine. Thus, the Schiff base is partially protonated and this does not change with pH. Therefore, we should consider a mechanism in which a proton shuttles between the Schiff base and a base in the active site. In the Michaelis complex, the base is considered to be the substrate α-amino group. In the external aldimine, the ε-amino group of Lys258 is the candidate, as has been suggested by Fasella and Hammes (*5*). We then changed Lys258 to Ala and made a complex with MeAsp. MeAsp formed an external aldimine, and it was completely protonated at a pH up to 9.6. This supported that the Lys258 ε-amino group is the base and the intrinsic pK_a of the external aldimine is greater than 10.

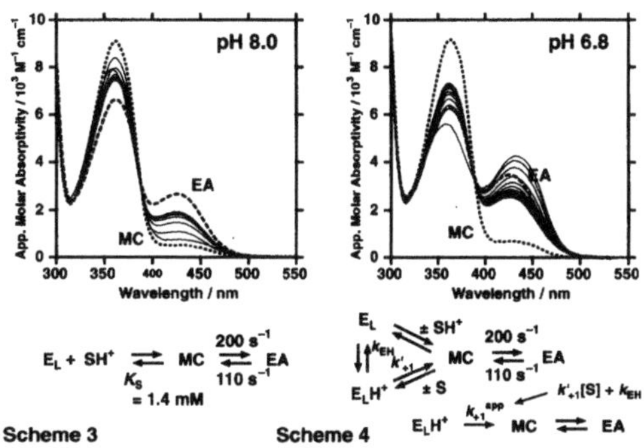

Figure 2: Time-resolved spectra (2.3 ms, 3.84 ms, and 2.56-ms intervals) of the reaction of AspAT with MeAsp at pH 8.0 and pH 6.8 at 298 K. The 430-nm absorption gradually increased (pH 8.0) or decreased (pH 6.8) after mixing. The spectra of the Michaelis complex (MC) and the external aldimine (EA) were obtained by fitting to Schemes 3 and 4 (lower).

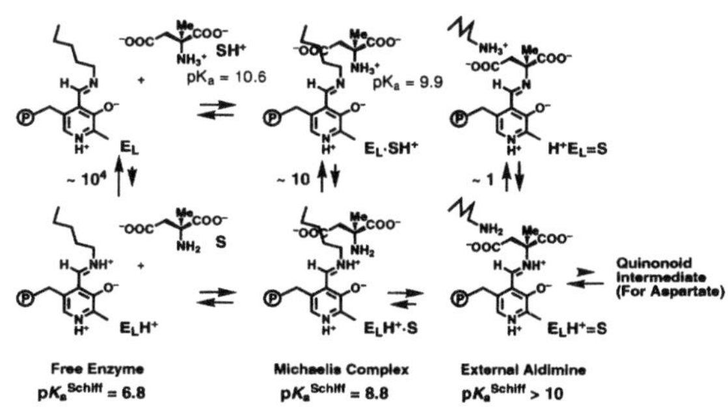

Scheme 5: Mechanism of reaction of AspAT and MeAsp leading to the external aldimine. If aspartate is used in place of MeAsp, E_LH^+=S leads to the quinonoid intermediate.

Combining these results altogether, we can draw the catalytic course of the reaction of AspAT with MeAsp leading to the external aldimine as shown in Scheme 5. There are two enzyme–substrate combinations (E_L–SH^+ and E_LH^+–S) that yields the Michaelis complex. The equilibrium between the two combinations is shifted upward by a factor of 10^4. However, this is decreased to ~10 in the Michaelis complex and to ~1 in the external aldimine. The pK_a of the Schiff base is shifted to 8.8 in the Michaelis complex, ~1 unit lower than the α-amino group of MeAsp, and to > 10 and matches that of the ε-amino group of lysine in the external aldimine. These shifts are important for efficient transaldimination and proton abstraction.

The next question is the mechanism that alters the pK_a of the Schiff base during the course of catalysis. Figure 3 shows the comparative structures of the active site of AspAT with the unprotonated Schiff base and that with the protonated Schiff base. The unprotonated imine is almost perpendicular to the pyridine plane. The protonated imine bond approaches the plane of the pyridine ring in order to form an intramolecular hydrogen bond. As this movement is restricted by the Lys258 side chain, there must be a compensatory rotation (counterclockwise in Figure 3) of the pyridine ring. However, this rotation is incomplete because of the hydrogen bond with Asn194, and van der Waals contact of the pyridine ring with Ala224.

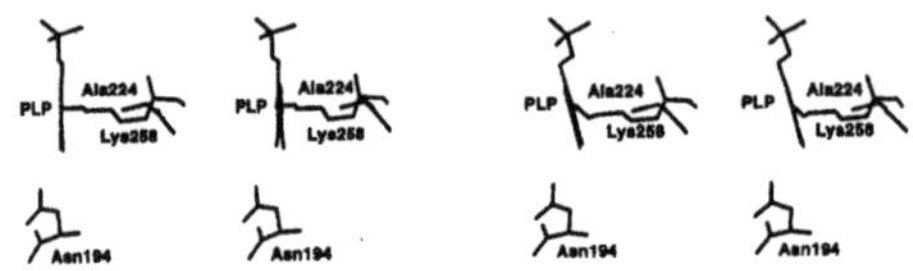

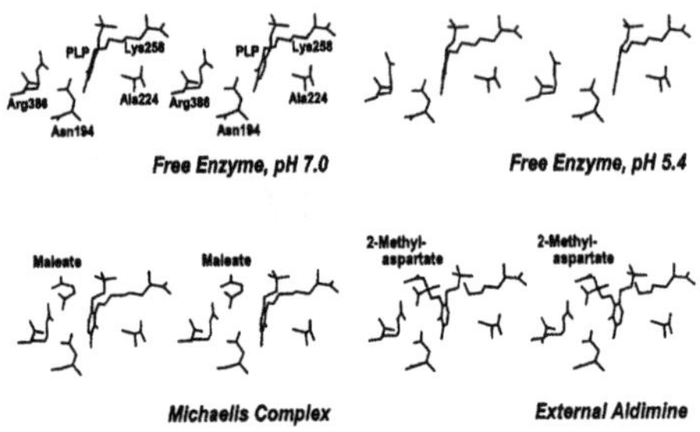

Figure 3 (left): Comparative structures of the active site of AspAT at pH 7 (*E. coli*, 1ARS) and at pH 5.4 (pig cytosolic, 1AJR).
Figure 4 (right): Comparative structures of the active site of AspAT at pH 7 (*E. coli*, 1ARS) and at pH 5.4 (pig cytosolic, 1AJR), and the complexes with maleate (*E. coli*, 1ASM) and α-methylaspartate (*E. coli*, 1ART). Note the position of the imine bond of the Schiff base relative to the pyridine ring and the torsion angle around C4–C4'.

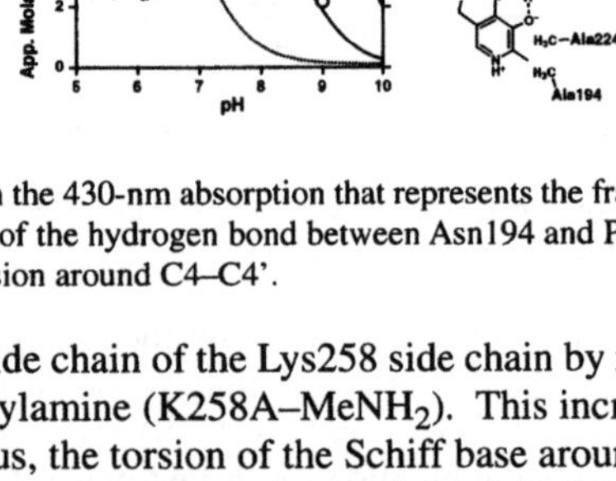

Figure 5: pH dependent change in the 430-nm absorption that represents the fraction of E_LH^+. Either cleavage of the Lys258 side chain or removal of the hydrogen bond between Asn194 and PLP increased the pK_a of the Schiff base through relaxation of the torsion around C4–C4'.

We have "cleaved" the side chain of the Lys258 side chain by reconstituting the Schiff base in K258A AspAT with methylamine (K258A–$MeNH_2$). This increased the Schiff base pK_a by 2.8 pH units (Figure 5). Thus, the torsion of the Schiff base around C4–C4' decreases the pK_a by 2.8. The Asn194 mutation also increased the pK_a by 1.7. The smaller increase in the pK_a by N194A mutation is considered to be due to the remaining van der Waals contact between the pyridine ring and the Ala224 side chain. Therefore, the torsion around C4–C4' is shown to be the principal factor that decreases the Schiff base pK_a value in the unliganded enzyme.

We next carried out a quantitative analysis using MOPAC calculations. With the increasing torsion angle of C4–C4', the protonated Schiff base is destabilized and the unprotonated Schiff base is stabilized. For direct comparison of the protonated and unprotonated species, the proton chemical potential was added to the unprotonated aldimine chemical potential. The protonated aldimine at $|\chi| = |\chi|_{min}$ and the unprotonated aldimine at $|\chi| = |\chi|_{max}$ should have the same chemical potential at a pH equal to the pK_a of the Schiff base. In K258A–MeNH$_2$ AspAT, there is no restriction on the torsion angle. Therefore, $0° \leq |\chi| \leq 90°$. The relative position of the two potential curves was determined in this way. In the wild type AspAT, $35° \leq |\chi| \leq 90°$. The chemical potential of the protonated aldimine at $|\chi|_{min} = 35°$ is equal to that of the unprotonated aldimine at $|\chi|_{max} = 90°$ and at pH 6.8. This pH value is just the pK_a value for the Schiff base of the wild-type enzyme in the unliganded state.

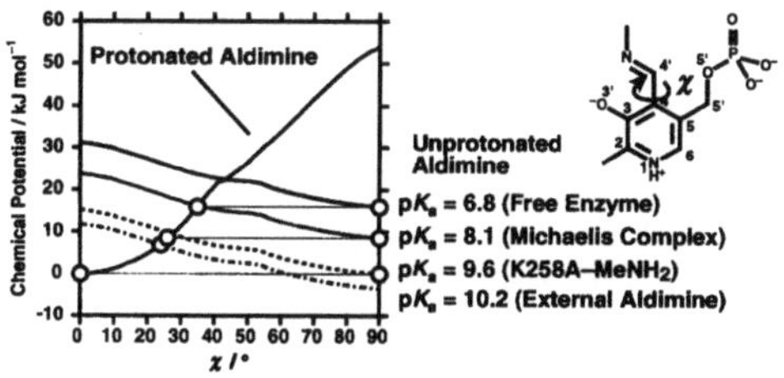

At pH = pK_a

$$\mu^\circ_{AldH^+}(|\chi|_{min}) = \mu^\circ_{Ald}(|\chi|_{max}) + \mu^\circ_{H^+} + RT\ln(10^{-pH})$$

Figure 6: Standard chemical potentials for the protonated form $\mu^\circ_{AldH^+}(\chi)$ and the unprotonated form plus proton $\mu^\circ_{Ald}(\chi) + \mu^\circ_{H^+} + RT\ln[H^+]$ as functions of the C4–C4' torsion angle χ. Horizontal bars indicate the range of the value allowed for $|\chi|$.

In the maleate complex that mimics the Michaelis complex, $|\chi|_{min}$ decreases to 27° (Figure 4). As a dicarboxylic ligand binds to the active site, AspAT undergoes a conformational change to the closed structure, and Arg386 approaches PLP, together with the side chain of Asn194. In addition, the newly formed hydrogen bond between the ligand carboxylate group and Asn194 will weaken the hydrogen bond between Asn194 and PLP. These are the reasons for the partial relaxation of the torsion of the Schiff base. In the external aldimine, the conformation is fixed, and χ is also fixed at –24° (Figure 4). Consequently, the pK_a values of these intermediates are determined to be 8.1 and 10.2. In addition, as the electrostatic effect of the substrate carboxylate groups is estimated to be 0.7 (the difference in pK_a of the Schiff base of R292L/R386L/N194A and N194A AspATs; unpublished results), the pK_a value is 8.8 in the Michaelis complex, and 10.9 in the external aldimine. These values are in good agreement with the experimental results. In this way, we can understand that the C4–C4' torsion angle of the Schiff bases, rather than the electrostatic effect of the two arginine residues, determines the pK_a of the Schiff base at each step in the AspAT catalysis.

The present results have important implications for general enzymology. One is the notion of pK_a in catalysis, and the other is the ground-state destabilization in enzyme catalysis. We can perform thermodynamic interpretations of the present results based on the free energy profile shown in Figure 7. If there is no strain in E_LH^+, the free energy profile is similar to panel A. Here, the combination $E_LH^+ + SH^+$ has the lowest free energy. The activation energy in the term of k_{cat}/K_m is the difference between $E_LH^+ + SH^+$ and the transition state. In the actual AspAT, the strain destabilizes the E_LH^+. Therefore, the activation energy is the difference between E_L + SH^+ and the transition state, and is decreased from that of the no-strain enzyme by 16 kJ mol^{-1}. Apparently, the destabilization energy of the protonated enzyme is used to decrease the activation energy (panel B). We can also decrease the pK_a by stabilizing E_L (panel C). The activation energy is again between E_L + SH^+ and the transition state. However, the activation free energy is the same as that with the no-strain enzyme (panel A). It is now clear that decreasing the Schiff base pK_a itself is not enough for efficient catalysis. It must be done by the destabilization of E_LH^+ and not by the stabilization of E_L. In other words, what is important for catalysis is not the "pK_a", but the free energy levels of the acid-base species.

The destabilization of E_LH^+ also destabilizes E_LH^+·S. As the reaction proceeds through this intermediate, it is better not to significantly increase the energy level. In AspAT, this energy level is lowered by the conformational change in the protein upon ligand binding, by the mechanism involving Asn194.

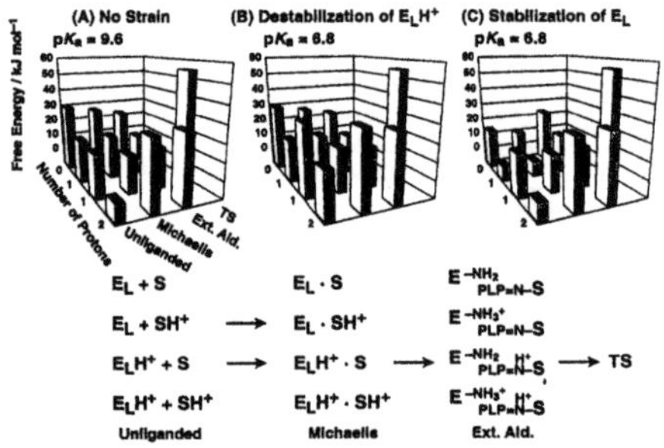

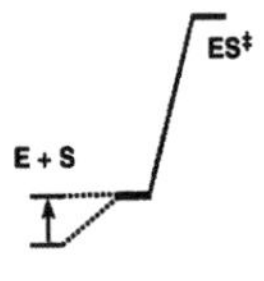

Figure 7: The free energy profiles of the catalytic reaction of AspAT with 1 mM aspartate at pH 7.0. The bottom comprises two coordinates, one of which represents the reaction step and the other, the total number of protons involved in the catalysis. Each bar corresponds to the schematic structure shown below. The energy levels at both ends in the Michaelis complex and the external aldimine are the minimum values estimated from the observation that the spectra do not change between pH 6.0 and 9.5.
Figure 8 (right): Ground-state destabilization of the free enzyme in order to increase k_{cat}/K_m.

Summary— The pK_a of the Schiff base is modulated by the conformation of the Schiff base during the course of catalysis. The torsion of the Schiff base destabilizes the predominant species in the unliganded enzyme. This torsion is released after the formation of the external aldimine. Therefore, this torsion energy can be considered to decease the activation energy, and hence increases the k_{cat}/K_m value (Figure 8). Until today, many studies have been devoted to clarify the mechanism that lowers the free energy level of the transition state. However, the present study show that the ground state destabilization of the enzyme, and not the substrate, can be another factor that enhances the catalytic reaction.

References

1. Hayashi, H., & Kagamiyama, H. (1997) *Biochemistry 36*, 13558–13569.
2. Hayashi, H., Mizuguchi, H., & Kagamiyama, H. (1998) *Biochemistry 37*, 15076–15085.
3. Ivanov, V.I., & Karpeisky, M.Y. (1969) *Adv. Enzymol. Relat. Areas Mol. Biol. 32*, 21–53.
4. Kirsch, J.F., Eichele, G., Ford, G.C., Vincent, M.G., Jansonius, J.N., Gehring, H., & Christen, P. (1984) *J. Mol. Biol. 174*, 497–525.
5. Fasella, P., & Hammes, G.G. (1967) *Biochemistry 6*, 1798–1804.

Biochemistry and Molecular Biology of Vitamin B_6
and PQQ-dependent Proteins
ed. by A. Iriarte, H.M. Kagan and M. Martinez-Carrion

^1H and ^{15}N NMR Spectroscopy of Aspartate Aminotransferase and Related Schiff Bases and Tautomerism in Enzyme Active Sites

David E. Metzler, Carol M. Metzler, Emilia T. Mollova, and Tsuyoshi Higaki

Iowa State University, Dept. of Biochemistry, Biophysics and Molecular Biology, Ames, Iowa, 50011, USA,

Summary

Downfield resonances from hydrogen bonded protons are delicate sensors of changes in charge distribution in active sites. They indicate that a tight electrostatic interaction of the phenolate oxygen with the charged amino group of substrate, coenzyme, or catalytic lysine is maintained in the various enzyme-substrate complexes which can be viewed as a set of interconverting tautomers. The resonance of the Schiff base N-H proton was observed in nonenzymatic Schiff bases of PLP with amino acids in $90\%H_2O/10\%D_2O$ at ~15 ppm.

Introduction

Although aspartate aminotransferase (AspAT) is a large protein of M_r 93,000 significant results have been obtained by many investigators using ^{31}P, ^{19}F, ^{13}C, ^{15}N and ^1H NMR spectroscopy. We will discuss observations of ^1H and ^{15}N nuclei from previously reported collaborative research [1–3] and will relate these to tautomerization in the 3-hydroxypyridine ring system. Figure 1 shows a sketch of the active site of pig cytosolic aspartate aminotransferase (cAspAT) with many H and N nuclei indicated. Most protons present in proteins resonate with chemical shift values in the 0–10 ppm range. A few lie further downfield in the 10–20 ppm range. Many of the latter are imidazole, indole, or amide NH protons which are shifted downfield as a result of hydrogen-bonding to negatively charged oxygen atoms of carboxylate or phosphate groups.[4] Because these protons exchange rapidly with solvent their resonances may be difficult to observe with commonly used pulse sequences and may sometimes be obliterated by phosphate-buffer catalyzed exchange.[2, 4]

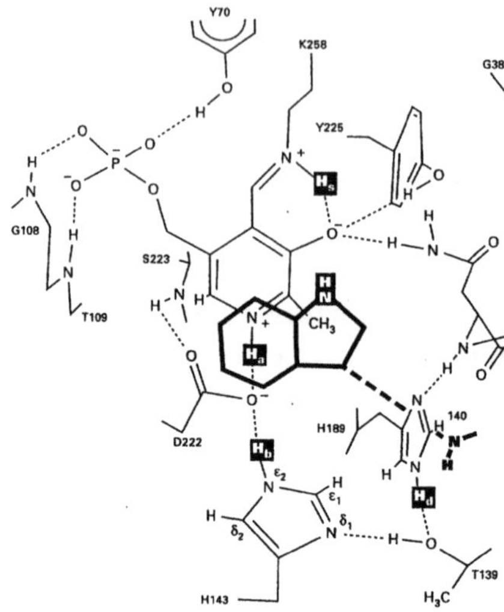

Figure 1. Sketch of active site region of cytosolic aspartate aminotransferase (cAspAT). In the *E. coli* enzyme T139 is replaced by serine and S223 by phenylalanine.

Results and Discussion

We will focus on the hydrogen atoms labeled H_a, H_b, H_d, and H_s in Fig. 1. Figure 2 shows NMR spectra of homogeneous cAspAT and *Escherichia coli* AspAT under several conditions. Peaks A, B, C, D, and D' all represent single proton resonances.[1] Peak A represents H_a, the proton on the ring nitrogen of the coenzyme pyridoxal phosphate (PLP). It is absent in the apoenzyme, and shifts in the pyridoxamine phosphate (PMP) form. A strong nuclear Overhauser effect (NOE) from peak A identified peak B as representing the imidazole proton H_b ($N^{\varepsilon 2}$). Other NOEs identified peak D as arising from $N^{\varepsilon 2}$ of His 189. Irradiation of peak D of cAspAT also elicited NOEs to ring CH protons at 8.25 and 7.57 ppm, indicating that the imidazole ring of His 189 is not cationic, a conclusion verified by HMQC spectroscopy on *E.coli* AspAT.[3] However, the positions of the CH resonances for the His 143 ring are anomolous suggesting that the ring may be partially dissociated to an anionic form.[2]

The position of peak A varies with pH in a striking way. It moves from a low pH chemical shift of 17.4 ppm to a high pH value of 15.4 ppm around a pK_a value of 6.15, a change of 2.0 ppm. At the same time peak B moves in the opposite direction ~ 1.0 ppm. This is consistent with the existence of a strong hydrogen bond from the Asp 222 carboxylate oxygen to H_a at low pH and of a weak hydrogen bond at high pH. Proton H_b forms a weak hydrogen bond at

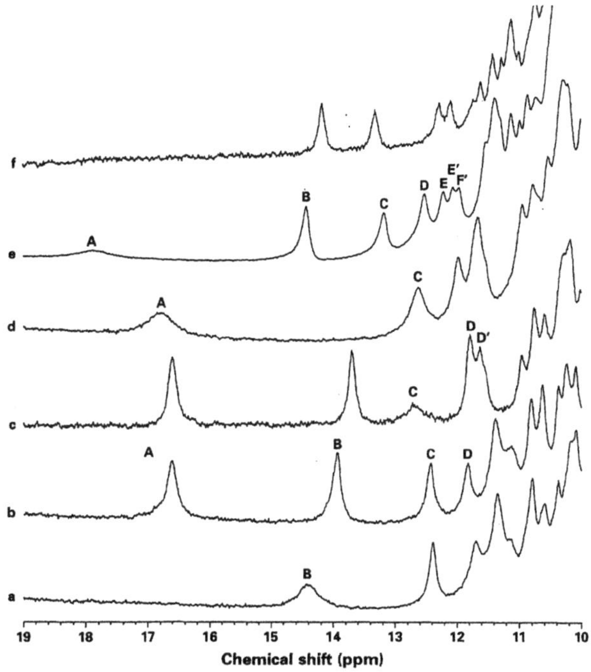

Figure 2. ^{1}H NMR spectra of porcine cAspAT and *E. coli* AspAT measured under the conditions described previously at 35° C. Chemical shifts are relative to sodium 2,2-dimethyl-2-silapentane-5 sulfonate (DSS) *a*, Apoenzyme form of cAspAT, pH 6.7; *b*, PMP form of cAspAT, pH 5.7; *c*, PMP form of *E. coli* AspAT, pH 6.7; *d*, H143A mutant of *E. coli* AspAT, pH 8.3; *e*, glutarate complex of cAspAT, pH 6.8; *f*, glutarate complex of the H189Q mutant of cAspAT produced in bacteria, pH 6.0.[1]

low pH and a stronger one at high pH. The pK_a of 6.15 is that determined spectrophotometrically for H_S, the Schiff base proton. This behavior is related to the chemistry of the 3-hydroxypyridine ring. As indicated in the following scheme two protons dissociate from the hydroxypyridine cation with pK_a values of 4.9 and 8.6 at 25°C. However, the monoprotonated form is a mixture of dipolar ionic and neutral uncharged species in a constant ratio $R \approx 1$. The first dissociation constant is the sum of two microscopic dissociation constants; $K_1 = K_A + K_B$, while $1/K_2 = 1/K_C = 1/K_D$. Notice that pK_A = 5.2 for the pyridinium cation but pK_C is 3.4 units higher because of the much increased donation of electron density from the phenolate oxygen atom into the ring in the dipolar ion. The two monoprotonated species are in a constant ratio *R*. The chemical shift of peak A in AspAT is explained in a similar way. The Schiff base proton H_S is hydrogen bonded to the phenolate oxygen, inhibiting donation of electrons into the ring and allowing formation of a strong hydrogen bond from H_a to the Asp 222 carboxylate. When the Schiff base proton dissociates the hydrogen bond is weakened and the microscopic pK_a (not observed) of the ring NH^+ is raised. It was somewhat surprising to find that peak A

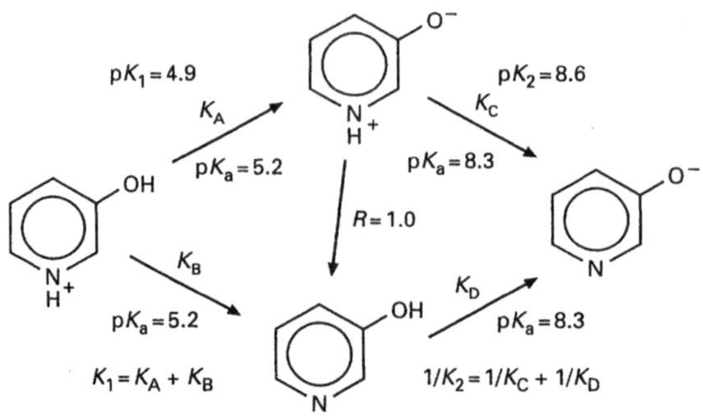

of the PMP form has a pH-independent position of 16.5 ppm, closer to the low pH position of the peak in the PLP form than to the high pH form. We interpret this as indicating a strong electrostatic interaction of the phenolate oxygen with the charged amino group of the PMP in this form of the enzyme. The active site of AspAT contains a constellation of charged groups appropriate for binding of an amino acid substrate. Two arginines neutralize the two positive charges of the substrate while the substrate – NH_3^+ group again interacts with the 3-hydroxypyridine group (Fig. 3). This interaction was shown in a clear way by titration of cAspAT with α-methylaspartate at pH 8.3. As this quasisubstrate was added peak A shifted from its high pH value down to ~16.6 ppm, a position we assign to the Michaelis complex with a charged amino group.[5] Again a tight electrostatic interaction with the phenolate oxygen is indicated. At the same time a well-resolved peak representing the external aldimine appeared at 17.2 ppm. Its position indicates an even stronger hydrogen bond from H_s to the phenolate oxygen.

Another comparison with 3-hydroxypyridine can be made. The absorption spectrum of a mixture of interconverting glutamate and 2-oxoglutarate saturating the active site of cAspAT is nearly constant over the pH range 5–11. The various components are present in constant ratios. The maximum velocity is also almost constant. This implies one very low pK_a and one very high pK_a. Let us assume values of 5.5 and 10.0. We can write, as we did for 3-hydroxypyridine that $K_1 = K_{1a} + K_{1b} + K_{1c} + \cdots$ and $1/K_2 = 1/K_{2a} + 1/K_{2b} + 1/K_{2c} + \cdots$. If we can estimate the fraction f_i of each component present, we can also estimate the microscopic dissociation constants for formation of each intermediate from the low pH form and for dissociation of each intermediate to the high pH form: $K_{1i} = f_i K_1$; $K_{2i} = K_2/f_i$. The fractions f_i in Table 1 were estimated from absorption spectra.[6] Since species 3, 6, 7, and 8 have heavily overlapping bands their total of 70% was somewhat arbitrarily divided, making use of some data from the α-methylaspartate complex. The estimation of fractions is crude but the pK_a

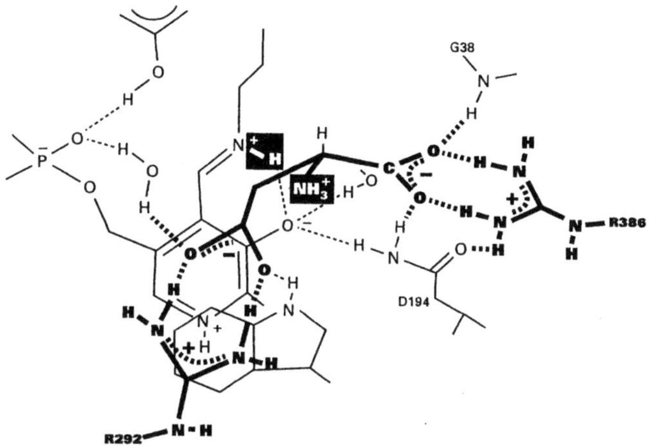

Fig. 3. Sketch of the active site with a bound substrate (or dicarboxylate) in front of the coenzyme. The latter will carry the Schiff base proton H_s for dicarboxylate binding but this will be replaced by the substrate $-NH_3^+$ for amino acid binding. In both cases a close interaction with the phenolate anion of the coenzyme is expected and observed.

values are precise enough to be meaningful. In each of the eight species listed in Table 1 the proton is tightly bound to a group that interacts with the phenolate ion of the coenzyme and has a high pK_a. When a proton moves to another site the pK_a of the latter increases sharply while that of the previous proton acceptor drops. For example, the pK_a of 9.3 in the first Michaelis complex can be attributed largely to the $-NH_3^+$ of the bound substrate, but after the proton jumps to the Schiff base the pK_a falls to 7.0. We can view the system as a set of interacting tautomers in which the proton donated by the substrate provides the mobility needed in catalysis but also acts as a glue to keep the intermediates bound into the active site.

The role of the Schiff base proton H_s is clear but it has been hard to observe in either the free enzyme or in the external aldimine, probably because of rapid exchange. It may have been seen in the H143N mutant of *E.coli* AspAT at 6°C as a broad resonance at 14.8 ppm and in a glutarate complex of the H143Q mutant of cAspAT at 14.14 ppm.[2] We have now observed the corresponding proton in Schiff bases of PLP with [^{15}N]L-aspartate a-methyl [^{15}]DL-aspartate and [^{15}N]L-alanine in 90%H_2O/10%D_2O at ~15 ppm confirming the expected strong hydrogen bonding. The ^{15}N chemical shift is ~200 ppm changing to ~45 ppm in the reduced Schiff base. Details will be published elsewhere.

Table I. Pig cytosolic aspartate aminotransferase nearly saturated with 0.1 M glutamate + 1.0 mM 2-oxoglutarate, an equilibrium mixture

Component	λ_{max} (nm)	Assumed fraction	pK_{1i}	pK_{2i}[a]
1. Michaelis complex	356	.20		
Substrate$-N^{+}H_3$ • PLP$-C=N-$			6.2	9.3
2. 2nd Michaelis complex	430	.03		
Substrate$-NH_2$ • PLP$-C=N^{+}H-$			7.02	8.48
3. Geminal diamines	330	.10		
Substrate$-NH-CH(PLP)-N^{+}H_2-$Lys			6.5	9
4. External aldimine	420	.06		
Lys$-NH_2$ • PLP$-C=N^{+}H-$			6.72	8.78
5. Quinonoid-carbanionic	490	.01		
Lys$-NH_3^{+}$ • Quinonoid			7.5	8
6. Ketimine	327	.30		
Lys$-NH_2$ • PMP$-N^{+}H=C-$			6.02	9.48
7. Carbinolamine	320	.10		
Lys$-NH_3^{+}$ • PMP$-NH-CH(OH)-$			6.5	9
8. PMP$-$oxoglutarate	320	.20		
Lys$-NH_2$ • $H_3^{+}N-$PMP			6.2	9.3

[a] Apparent microscopic p*K*a values if $pK_1 = 5.5$ and $pK_2 = 10$. Some of these values are complex constants involving adduct formation etc.

References

[1] Metzler, D. E., Metzler, C. M., Mollova, E. T., Scott, R. D., Kintanar, A., Tanase, S., Kogo, K., Higaki, T., Morino, Y., Kagamiyama, H., Yano, T., Kuramitsu, S., Hayashi, H., Hirotsu, K., and Miyahara, I. (1994) Using ^{1}H- and ^{15}N-NMR spectroscopy to observe active sites of porcine cytosolic and *E. coli AspAT.* in *Biochemistry of vitamin B6 and PQQ* (Marino, G., Sannia, G., and Bossa, F., eds), pp. 55–59, Birkhäuser, Basel

[2] Metzler, D. E., Metzler, C. M., Scott, R. D., Mollova, E. T., Kagamiyama, H., Yano, T., Kuramitsu, S., Hayashi, H., Hirotsu, K., and Miyahara, I. (1994) NMR Studies of ^{1}H Resonances in the 10-18 ppm Range for Aspartate Aminotransferase from *Escherichia coli*. *J. Biol. Chem.* **269**, 28027–28033

[3] Mollova, E. T., Metzler, D. E., Kintanar, A., Kagamiyama, H., Hayashi, H., Hirotsu, K., and Miyahara, I. (1997) Use of ^{1}H - ^{15}N Heteronuclear Multiple-Quantum Coherence NMR Spectroscopy to Study the Active Site of Aspartate Aminotransferase. *Biochemistry* **36**, 615–625

[4] Metzler, D. E. (1997) Nuclear Magnetic Resonance in Study of Active Sites of Pyridoxal Dependent Enzymes. *Meth. Enzymol.* **280**, 30–40

[5] Scott, R. D., Jin, P., Miura, R., Chu, Wen-C., Kintanar, A., Metzler, C. M., and Metzler, D. E. (1991) Fluorine-19, Phosphorus-31, and Proton NMR Spectroscopy of AspAT containing 6-Fluoropyridoxal Phosphate or 6-Fluoro-Pyridoxamine Phosphate. in *Enzymes dependent on pyridoxal phosphate and other carbonyl compounds as cofactors* (Fukui, T., Kagamiyama, H., Soda, K., and Wada, H., eds), pp. 129–143, Pergamon, Oxford

[6] Metzler, C. M., and Metzler, D. E. (1987) Quantitative Description of Absorption Spectra of a Pyridoxal Phosphate-containing Enzyme Using Lognormal Distribution Curves. *Anal. Biochem.* **166**, 313–327

Biochemistry and Molecular Biology of Vitamin B_6
and PQQ-dependent Proteins
ed. by A. Iriarte, H.M. Kagan and M. Martinez-Carrion

Structure, Induced Fit and Substrate Recognition of *E. coli* Branched-Chain Amino Acid Aminotransferase

K. Hirotsu, M. Goto, I. Miyahara, H. Hayashi[1] and H. Kagamiyama[1], and K. Okada[2]

Department of Chemistry, Graduate School of Science, Osaka City University, Sumiyoshi-ku, Osaka 558-8585, Japan; [1]Department of Biochemistry, Osaka Medical College, Takatsuki 569-8686, Japan; 2Department of Molecular Biology, Nara Institute of Science and Technology, Ikoma, Nara, 630-0101, Japan

Summary

Structures of the branched-chain amino acid aminotransferase from *E. coli* and its complexes with substrate analogues have been solved by X-ray crystallographic method. BCAT is in a hexameric form, and catalyzes the transamination of branched-chain amino acids. The hexamer can be regarded as the assembly of three dimer units around a 3-fold axis. On binding of the substrate analogue, the flexible loop, which is disordered in the unliganded enzyme, moves toward the active-site entrance and shields the substrate from the solvent region. α-Carboxylate of the substrate directly interacts with an OH group of Tyr and two NH groups of β-turn and indirectly with Arg. The side chain of the branched-chain amino acid is in the cavity formed by aromatic rings and an isopropyl group of the active site residues.

Introduction

A number of X-ray crystallographic studies on aspartate aminotransferases have been performed to elucidate the structure, function, and catalytic mechanism of the enzymes (Jansonius, 1998). Branched-chain amino acid aminotransferase (BCAT) exhibits no primary sequence homology with aspartate aminotransferase, and belongs to an evolutionary class different from that of aspartate aminotransferase (Kuramitsu et al., 1985; Inoue et al., 1988). Based on the classification of the PLP dependent enzymes by Grishin et al., (1995) aspartate aminotransferase and many of the aminotranferases are placed in the fold type I, while BCAT is placed in the fold type IV together with D-amino acid aminotransferase. The structure of the D-amino acid aminotransferase has been determined as the first structure of the fold type IV by Sugio et al (1995). The structure of the unliganded BCAT was then determined as the first structure of L-amino acid aminotransferase of the fold type IV (Okada et al., 1997).

BCAT reversibly catalyzes the transfer of α-amino group of branched-chain amino acids to 2-oxoglutarate to form glutamate. BCAT is a homo hexamer with a subunit molecular weight of 34,000, and each subunit is composed of 308 amino acids residues, and has a covalently bound pyridoxal 5'-phophate (PLP). BCAT has two absorption maxima at 410 and 330nm (Inoue et al.,

1988). In contrast to aspartate aminotransferase, BCAT shows no detectable pH dependent spectral change in the pH range of 5.5 to 8.5. Here we present the structure, induced fit and substrate recognition of BCAT on the basis of X-ray crystallographic studies of *E. coli* BCAT and its complexes with substrate analogues.

Materials and Methods

The overproduced wild type BCAT and its complexes with 4-methylvalerate and 2-methylleucine were crystallized from PEG400 at pH 7.5. All crystals are isomorphous and belong to the orthorhombic space group $C222_1$. All data sets were collected to 2.1 to 2.4 Å resolution on a Rigaku Raxis IIc imaging plate detector. All the structures were refined by the program XPLOR (Brünger et al., 1987) to give R-factors in the range of 0.17 to 0.20. A detailed account for the structure solutions and refinements is in preparation (K. Okada et al., in preparation).

Results and Discussion

The overall structure of the unliganded BCAT is shown in Figure 1 (a). Although the enzyme molecule consists of six subunits, the hexamer can be regarded as the assembly of three dimer units around the three-fold axis. The dimeric unit along a two-fold axis can be seen in the center of the molecule in Figure 1 (a). As was expected, the folding of this dimer unit is quite different from that of the aspartate aminotransferase (Jansonius, 1998), and is quite similar to that of the

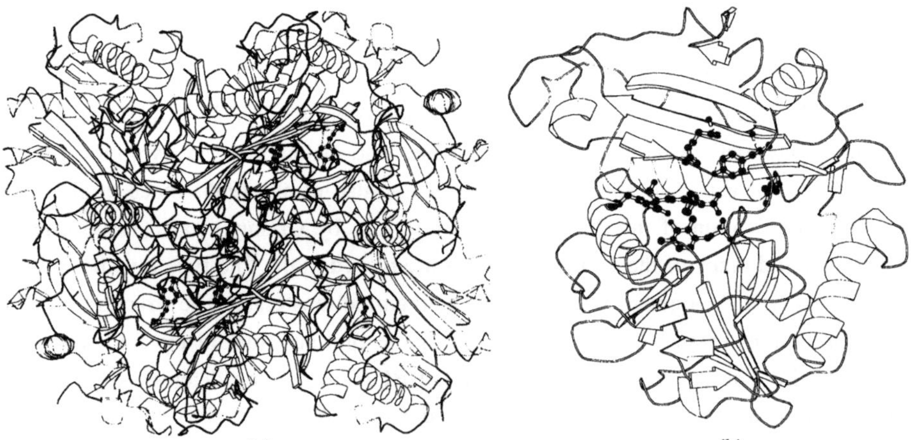

(a) (b)

Figure 1. (**a**) View of the molecule along the 2-fold axis. The 2-fold symmetric dimer drawn in dark gray is shown in the center, with another two dimer units on two sides. (**b**) The overall subunit structure of BCAT in complex with 4-methylvarelate. The small and large domains comprise the upper and lower parts of this figure.

D-amino acid aminotransferase (Sugio et al., 1995, Peisach et al., 1998), which has a 28% sequence identity with BCAT.

The subunit structure of BCAT in complex with 4-methylvarelate is given in Figure 1 (b). The subunit is divided into two domains. The upper half is the small domain characterized by the α/β structure and the lower half is the large domain characterized by the pseudo barrel structure. The coenzyme PLP is bound to the central cavity of the subunit, and two loops drawn in black approach the PLP. These loops are from the small domain of the other subunit of the dimer. Thus, the active site is comprised of the residues from both domains of one subunit and the small domain of the other subunit.

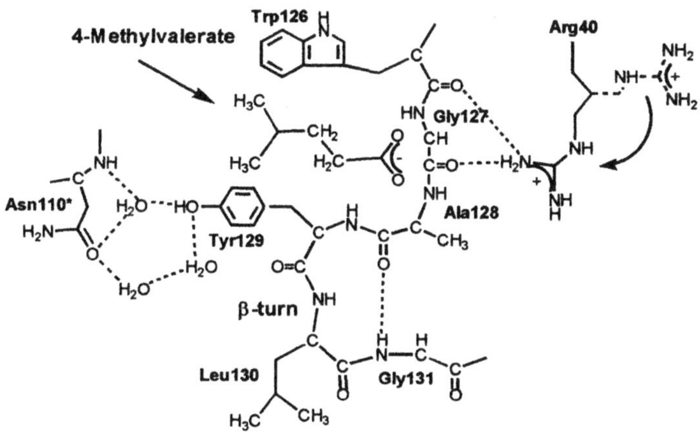

Figure 2. Interactions of the interdomain loop with active-site residues with 4-methylvalerate.

It is well known that aspartate aminotransferase shows a conformational change from the open form to the closed form (McPhalen et al., 1992, Jäger et al., 1994, Okamoto et al., 1994, Miyahara et al., 1994, Nakai et al., 1999). The small domain moves to close the active site depending on the binding of the substrates. BCAT does not exhibit any domain movement, but shows the active site closure by the induced fit of the interdomain loop from Pro125 to Glu133. Thus, there exist open and closed forms in BCAT. In the unliganded form, the interdomain loop is not visible except for N-terminal two residues because this loop is disordered. On the binding of the substrate analogue, the interdomain loop, drawn in black, shows its ordered structure to close the active site. The interactions of the interdomain loop with the active site residues and 4-methylvalerate as the substrate analogue are given schematically in Figure 2. The loop starts with the extended form, then makes a β-turn and then makes a further turn by a 3_{10} helix followed by a short α-helix. The carboxylate of the bound substrate analogue induce the side chain movement of Arg40 from the solvent side to the protein side. The guanidino group of Arg40 interacts with two carbonyl groups of the interdomain loop. The hydroxyl group of Tyr129 is linked to Asn110* through water molecules. The side chains of Trp126 and Tyr129 approach the substrate analogue and make a van der Waals contact with the hydrophobic moiety of the substrate analogue.

When the active site structure of the open form is compared with that of the closed form, Arg40 is the only residue that significantly changes its side chain conformation and directly interacts with the interdomain loop. In this respect, the directional change of the side chain in Arg40 may play an important role in the structural change to the closed form. In addition, two glycine residues of the interdomain loop are important for the induced fit, since the residues make the main chain of the loop flexible. The Pro125, Trp or Val126, Gly127, and Tyr129 in the loop are conserved in many BCAT enzymes, suggesting that the interdomain loops have similar interactions with the active site residues and substrates in these BCAT enzymes. Interestingly, D-amino acid aminotranferase does not have the consensus sequence of the interdomain loop, and possibly the loop behaves differently (Peisach et al., 1998).

The active site structures of the unliganded BCAT and its complex with 2-methylleucine are drawn schematically in Figure 3 (a) and (b), respectively. The most important feature of the active site, which distinguishes between the fold type I and IV, is the orientation of PLP. The orientation of the PLP-Schiff base (Figure 3 (a)) is viewed from the solvent side. In the fold type IV enzyme, the phosphate group of PLP is on the right side of the pyridine ring, while in the fold type I enzyme, the phosphate group is on the left side (Yoshimura et al., 1993, Sugio et al., 1995,

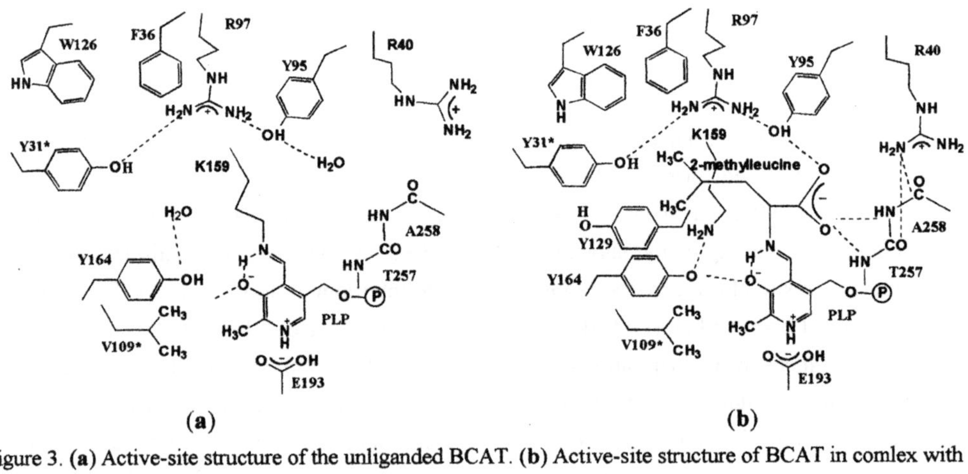

Figure 3. (**a**) Active-site structure of the unliganded BCAT. (**b**) Active-site structure of BCAT in comlex with 2-methylleucine.

Okada et al., 1997). Although PLP of the fold type IV enzyme rotates 180° compared with that of the fold type I enzyme, the substrate L-amino acid takes the same orientation for both fold types.

In the active site of the unliganded form (Figure 3 (a)), PLP is bound to the bottom of the active site cavity and makes a Schiff base with the catalytic residue Lys159. The Schiff base linkage is coplanar with the pyridine ring of PLP. The pyridine ring of PLP is sandwiched by Leu217 and the loop at Gly196. Glu193 forms an ion pair with the protonated nitrogen atom of the pyridine ring of PLP, and Tyr164 is hydrogen bonded to O3' of PLP. On the left side of PLP, there exists a hydrophobic cluster formed by Phe36, Tyr31*, Tyr164, and Val109*. Tyr31* is hydrogen bonded to Arg97, and Arg97 is hydrogen bonded to Tyr95. Arg40 directs its side chain toward the solvent side.

In the active site of BCAT complex with 2-methylleucine (Figure 3 (b)), the Schiff bond linkage between PLP and Lys159 is cleaved and a new Schiff linkage is formed between PLP and 2-methyleucine. The released Lys159 is hydrogen bonded to Tyr164, and PLP rotates about 30° toward the solvent side. Arg40 change its side chain direction to make hydrogen bonds with the main chain carbonyl groups of the β-turn. The isopropyl group of 2-methylleucine is bound to the hydrophobic pocket constructed by the hydrophobic cluster (Phe36, Tyr31*, Tyr164, and Val109*).

Arg40 was expected to make a salt bridge with α-carboxylate of the substrate. Contrary to our expectation, the α-carboxylate of the substrate interacts with two NH groups of Thr257 and Ala258 at the β-turn and with the hydroxyl group of Tyr95. Interestingly, Arg40 indirectly recognizes the α-carboxylate from the solvent side of the β-turn. Arg97, indirectly recognizes the α-carboxylate through Tyr97. Consequently, Arg97, Tyr95, Arg40 and the β-turn form the recognition site for α-carboxylate of the substrates.

Acknowledgment

This work was supported by Grant-in Aid for Scientific Research (No. 11133254) from the Ministry of Education, Science, and Culture of Japan, and by the Research Grant from Japan Society for the Promotion of Science (Research for the Future).

References

Brünger, A.T., Kuriyan, J. and Karplus, M. (1987) Crystallographic R factor refinement by molecular dynamics. *Science* **235**, 458-460

Grishin, N.V., Phillips, M.A. and Goldsmith E.J. (1995) Modeling of the spatial structure of eukaryotic ornithine decarboxylases. *Protein Sci.* **4**, 1291-1304.

Inoue, K., Kuramitsu, S., Aki, K., Watanabe, Y., Takagi, T., Nishigai, M., Ikai, A. and Kagamiyama, H. (1988) Branched-chain amino acid aminotransferase of Escherichia coli: overproduction and properties. *J. Biochem.* **104**, 777-784

Jäger, J., Moser, M., Sauder, U. and Jansonius, J.N. (1994) Crystal structures of *Escherichia coli* aspartate aminotransferase in two conformations. Comparison of an unliganded open and two liganded closed forms. *J. Mol. Biol.* **239**, 285-305

Jansonius, J.N. (1998) Structure, evolution and action of vitamin B6-dependent enzymes. Curr. Opin. Struct. Biol. **8**, 759-769.

Kuramitsu, S., Ogawa, T., Ogawa, H. and Kagamiyama, H. (1985) Branched-chain amino acid aminotransferase of Escherichia coli: nucleotide sequence of the ilvE gene and the deduced amino acid sequence. *J. Biochem.* **97**, 993-999

McPhalen, C.A., Vincent, M.G., Picot, D. and Jansonius, N.J. (1992) Domain Closure in Mitochondrial Aspartate Aminotransferase. *J. Mol. Biol.* **227**, 197-213.

Miyahara, I., Hirotsu, K., Hayashi, H. and Kagamiyama, H. (1994) X-Ray crystallographic study of pyridoxamine 5'-phosphate-type aspartate aminotransferase from *Escherichia coli* in three forms. *J. Biochem.* **116**, 1001-1012

Nakai, T., Okada, K., Akutsu, S., Miyahara, I., Kawaguchi, S., Kato, R., Kuramitsu, S. and Hirotsu, K. (1999) Structure of *Thermus thermophilus* HB8 Aspartate Aminotransferase and its Complex with Maleate. *Biochemistry* **38**, 2413-1414.

Okada, K., Hirotsu, K., Sato, M., Hayashi, H. and Kagamiyama, H. (1997) Three-Dimensional Structure of *Escherichia coli* Branched-Chain Amino Acid Aminotransferase at 2.5Å Resolution.*J. Biochem.* **121**, 637-641.

Okamoto, A., Higuchi, T., Hirotsu, K., Kuramitsu, S. and Kagamiyama, H. (1994) X-Ray crystallographic study of pyridoxal 5'-phosphate-type aspartate aminotransferase from *Escherichia coli* in open and closed form. *J. Biochem.* **116**, 95-107

Peisach, D., Chipman, D.M., Van Ophem, P.W., Manning, J.M. and Ringe, D. (1998) Crystallographic Study of Steps along the Reaction Pathway of D-Amino Acid Aminotransferase. *Biochemistry*, **37**, 4958-4967.

Sugio, S., Petsko, G.A., Manning, J.M., Soda, K. and Ringe, D. (1995) Crystal structure of a D-amino acid aminotransferase: How the protein controls stereoselectivity. *Biochemistry* **34**, 9661-9669

Yoshimura, T., Nishimura, K., Ito, J., Esaki, N., Kagamiyama, H., Manning, J.M. and Soda, K. (1993) Unique stereospecificity of D-amino acid aminotransferase and branched-chain L-amino acid aminotransferase for C-4'-hydrogen transfer of the coenzyme. *J. Am. Chem. Soc.* **115**, 3897-3900

STRUCTURE AND MECHANISMS OF QUINOENZYMES

Biochemistry and Molecular Biology of Vitamin B_6
and PQQ-dependent Proteins
ed. by A. Iriarte, H.M. Kagan and M. Martinez-Carrion

TRYPTOPHAN TRYPTOPHYLQUINONE ENZYMES: STRUCTURE AND FUNCTION

Victor L. Davidson, Department of Biochemistry, University of Mississippi Medical Center, Jackson, MS 39216-4505, USA

Summary

Tryptophan tryptophylquinone (TTQ), the prosthetic group of methylamine dehydrogenase (MADH), is formed by covalent cross-linking of two tryptophan residues and incorporation of two oxygen atoms into one of the indole rings. MADH converts primary amines to their corresponding aldehyde plus ammonia. Catalysis occurs via a transaminase-type mechanism involving a Schiff base intermediate between the substrate amine and a TTQ carbonyl. In addition to participating in chemical reactions at the enzyme active site, TTQ forms a bridge from the active site to the protein surface and mediates electron transfer from substrate to a copper protein, amicyanin.

Introduction

Methylamine dehydrogenase (MADH) and aromatic amine dehydrogenase (AADH) catalyze the oxidative deamination of primary amines to their corresponding aldehydes plus ammonia. For each enzyme a type I copper protein, amicyanin for MADH and azurin for AADH, is the electron acceptor which reoxidizes the enzyme to complete the reaction cycle. The prosthetic group which participates in catalysis and electron transfer is tryptophan tryptophylquinone (TTQ) [1]. This paper will focus on the properties of MADH from *Paracoccus denitrificans*.

Structure and biosynthesis

TTQ is formed via the posttranslational modification of two tryptophan residues of the polypeptide chain (Figure 1). Two atoms of oxygen are incorporated into the indole ring of Trp^{57} and a covalent bond between the indole rings of Trp^{57} and Trp^{108} is formed. The mechanism by which this occurs has not yet been elucidated, but it requires the action of other proteins which are subject to the same genetic regulation as the structural genes for the enzyme. The methylamine utilization (*mau*) gene cluster contains several genes (Table 1), at least four of

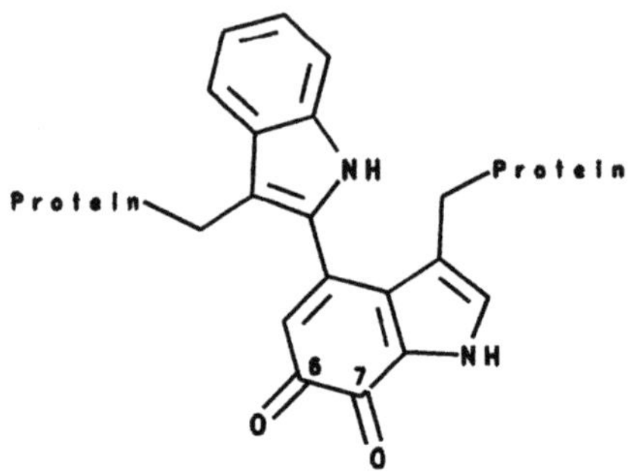

Figure 1. Tryptophan tryptophylquinone

GENE	FEATURES INFERRED FROM SEQUENCE
mauR	LysR-type transcription activator
* *mauF*	-
* *mauB*	MADH large subunit
* *mauE*	Membrane protein
* *mauD*	Motif for disulfide bond formation
* *mauA*	MADH small subunit
mauC	Amicyanin
mauJ	-
* *mauG*	Similarity to peroxidases
mauM	Similarity to ferredoxins
mauN	Similarity to ferredoxins

Table 1. The *mau* gene cluster of *P. denitrificans*

which encode proteins that are required for biosynthesis of MADH (indicated by * in Table 1), in addition to the two MADH structural genes [reviewed in 2]. None of these gene products have been isolated, but from their sequences; the *mauE* product appears to be a membrane protein, the *mauD* product may be involved in disulfide bond processing, the *mauG* product may be a heme-bearing peroxidase, and *mauF* shows no homology to any other proteins. These gene products must play essential roles in translocation of MADH subunits across the cell membrane to the periplasm, formation of the six disulfide bonds which are present in the TTQ-bearing β-subunit, and TTQ biosynthesis.

Mechanism of catalysis

A mechanism for the overall oxidation-reduction reaction of MADH has been proposed based on the results of several kinetic and biochemical studies. The amine substrate forms a Schiff base adduct with the C6 carbonyl of TTQ. An active site base then abstracts a proton from the methyl carbon which leads to reduction of the TTQ cofactor [3]. Nucleophilic attack of the imine by water leads to cleavage of C-N bond and release of the formaldehyde product to yield a reduced aminoquinol [4]. This intermediate is reoxidized by two one-electron transfers to amicyanins.

The first requires the presence of a monovalent cation at the active site [5]. The second product, ammonia, is not released until after TTQ has been fully reoxidized. In the absence of more substrate, water hydrolyzes the iminoquinone to regenerate the quinone. Alternatively, in the steady-state the amino nitrogen of another molecule of substrate, rather than water, may react directly with the iminoquinone to form the next enzyme-substrate adduct with concomitant release of the ammonia product [6]. This mechanism requires that multiple active site residues interact with reaction intermediates and TTQ to catalyze the overall reaction. The crystal structure of MADH from *P. denitrificans* [7] reveals that the side chains of Asp^{76}, Thr^{122}, Asp^{32} and Tyr^{119} are in close enough proximity to C6 of TTQ to participate in these reaction steps (Figure 2). Site-directed mutagenesis studies are in progress to elucidate the specific roles of each residue in the overall reaction mechanism.

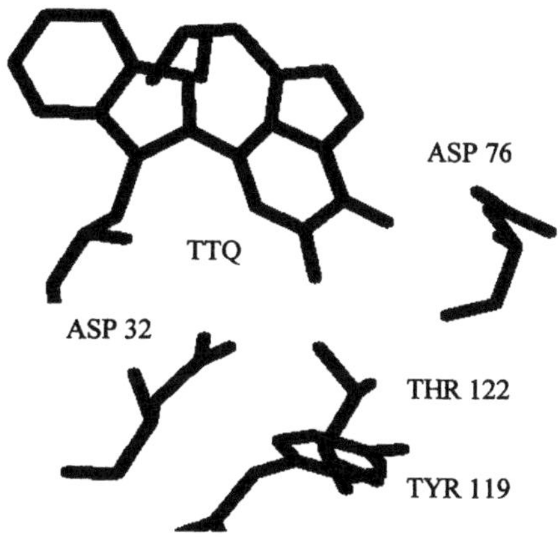

Figure 2. Active site of MADH

Mechanisms of electron transfer (ET)

The edge of the non-quinolated indole ring of TTQ is exposed at the MADH surface. Crystal structures of MADH in complex with amicyanin reveal that TTQ is able to act as a bridge that physically connects the enzyme active site to the point of surface-mediated ET from MADH to copper [8]. It is possible to study ET from several different redox forms of MADH to amicyanin. Since TTQ is a two-electron carrier and the type I copper is a one-electron carrier, two sequential oxidations of fully reduced TTQ by amicyanins are required to completely reoxidize MADH. O-quinol and O-semiquinone forms of MADH may be generated by reduction by dithionite (Figure 3A) [9]. The product of the reduction of MADH by the substrate amine is an N-quinol which retains the covalently bound substrate-derived amino group after release of the aldehyde product [4]. An N-semiquinone is the product of the first one-electron oxidation of the N-quinol (Figure

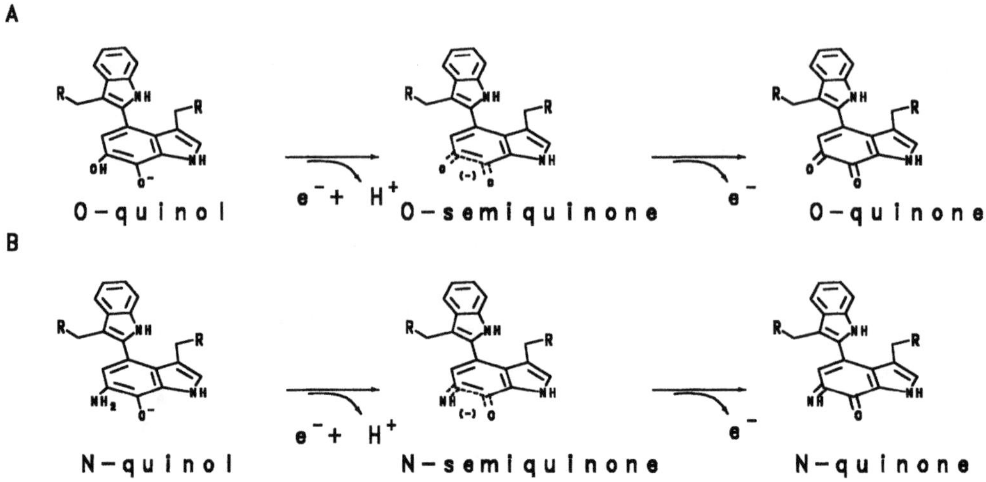

Figure 3. Electron transfer from different redox forms of TTQ in MADH

3B) [9]. The N-semiquinone may also be generated in vitro [6]. The rates of the ET reactions from each of these four redox forms of MADH to amicyanin are different (Table 2).

	O-quinol	O-semiquinone	N-quinol	N-semiquinone
k_{ET} (10 mM KP_i, pH 7.5)				
+ 0 mM KCl	14 s^{-1}	> 500 s^{-1}	12 s^{-1}	28 s^{-1}
+ 200 mM KCl	12 s^{-1}	> 500 s^{-1}	130 s^{-1}	51 s^{-1}
+ 400 mM KCl	12 s^{-1}	> 500 s^{-1}	> 500 s^{-1}	84 s^{-1}
k_{ET} 10 mM KP_i + 0.2 M KCl				
pH 6.5	12 s^{-1}	> 500 s^{-1}	42 s^{-1}	21 s^{-1}
pH 7.5	12 s^{-1}	> 500 s^{-1}	130 s^{-1}	51 s^{-1}
pH 9.0	11 s^{-1}	> 500 s^{-1}	> 500 s^{-1}	110 s^{-1}
KSIE ($^{H2O}k/^{D2O}k$) at pH 7.5	1.5	-	6.5	1.8
Marcus Parameters				
λ (eV)	2.3 ± 0.1	2.3 ± 0.1	3.4 ± 0.1	2.4 ± 0.1
H_{AB} (cm^{-1})	12 ± 7	12 ± 7	>20,000	13 ± 4
r (Å, for β=1)	9.6 ± 0.7	9.6 ± 0.7	< 0	9.4 ± 1.2
Rate-limiting Step	Electron Transfer	Electron Transfer	Proton Transfer	Electron Transfer

Table 2. Electron transfer from different redox forms of MADH to amicyanin

ET reactions from the O-quinol, O-semiquinone and N-semiquinone, are rate-limited by the ET event. The rates of these reactions vary predictably with driving force (i.e., the redox potential difference between the reactants). Analysis by ET theory of the temperature dependence of the

ET rate of each of these reactions [10,11] yielded identical values for the reorganizational energy (λ) and electronic coupling (H_{AB}) associated with the reaction, and predicted the electron transfer distance which is seen in the crystal structure of the complex [8]. Analysis of the $\Delta G°$-dependence of these three reactions [11,12] by ET theory yielded values for λ and H_{AB} that were identical to those obtained from the temperature dependencies. Furthermore, the rates of these reactions were relatively insensitive to pH and buffer composition (Table 2), as would be expected for an ET reaction occurring within a protein complex.

Covalent incorporation of substrate-derived N into the TTQ semiquinone alters the ET rate simply by altering the redox potential of TTQ relative to the O-semiquinone [11]. However, the effects of covalent incorporation of substrate-derived N into the fully reduced TTQ were much more profound [5]. The driving force for the reaction of the N-quinol has apparently been changed such that the reaction has become both extremely slow and thermodynamically unfavorable. In order for this physiologic reaction to occur, a prerequisite step is now required to activate TTQ for ET. It is proposed that deprotonation of the N-quinol amino group is required to generate the "activated" intermediate from which ET occurs [5]. ET from this intermediate is so rapid that the redox reaction becomes rate-limited (i.e., gated) by the deprotonation event. This hypothesis is supported by the observation that only the redox reaction of the N-quinol exhibits a significant deuterium kinetic solvent isotope effect, consistent with rate-limitation by the transfer of an exchangeable proton. The rate of this redox reaction also varies markedly with pH and salt (Table 2). The salt dependence is due specifically to monovalent cations [5]. To account for these observations, a detailed model has been proposed in which the pH-dependent binding of a monovalent cation to the MADH active site facilitates the rate-limiting deprotonation of the N-quinol amino group [5]. A similar pattern of gated and ungated ET has been reported for the ET reactions from TTQ in AADH to copper in azurin [13].

Acknowledgements

This work was supported by NIH grant GM41574.

References

1. McIntire, W. S., Wemmer, D. E., Christoserdov, A. Y. and Lindstrom, M .E., (1991) A new cofactor in a prokaryotic enzyme: Tryptophan tryptophylquinone as the redox prosthetic group in methylamine dehydrogenase. *Science 252*: 817-824.

2. Graichen, M. E., Jones, L. H., Sharma, B., van Spanning, R. J. M., Hosler, J. P. and Davidson, V. L. (1999) Heterologous expression of correctly assembled methylamine dehydrogenase in *Rhodobacter sphaeroides*. *J. Bacteriol. 181*: 4216-4222.
3. Brooks, H. B., Jones, L. H. and Davidson, V. L. (1993) Stopped-flow kinetic and deuterium kinetic isotope effect studies of the quinoprotein methylamine dehydrogenase from *Paracoccus denitrificans*. *Biochemistry 32*: 2725-2729.
4. Bishop, G. R., Valente, E J., Whitehead, T. L., Brown, K. L., Hicks, R. T., and Davidson, V. L. (1996) Direct detection by ^{15}N-NMR of the tryptophan tryptophylquinone aminoquinol reaction intermediate of methylamine dehydrogenase. *J. Am. Chem. Soc. 118*:12868-12869.
5. Bishop, G. R., and Davidson, V. L. (1997) Catalytic role of monovalent cations in the mechanism of proton transfer which gates an interprotein electron transfer reaction. *Biochemistry 36*: 3586-13592.
6. Zhu, Z., and Davidson, V. L. (1999) Identification of a new reaction intermediate in the oxidation of methylamine dehydrogenase by amicyanin. *Biochemistry 38*: 4862-4867.
7. Chen, L., Dol, M., Durley, R. C. E., Chistoserdov, A. Y., Lidstrom, M. E., Davidson, V. L., and Mathews, F. S., 1998, Refined crystal structure of methylamine dehydrogenase from *Paracoccus denitrificans* at 1.75 Å resolution. *J. Mol. Biol. 276*: 131-149.
8. Chen, L., Durley, R., Mathews, F. S. and Davidson, V. L. (1994) Structure of an electron transfer complex: Methylamine dehydrogenase, amicyanin and cytochrome *c*-551i. *Science 264*: 86-90.
9. Davidson, V. L., Brooks, H. B, Graichen, M. E., Jones, L. H. and Hyun, Y-L. (1995) Detection of intermediates in TTQ enzymes. *Methods Enzymol. 258*: 176-190.
10. Brooks, H. B., and Davidson, V. L. (1994) Kinetic and thermodynamic analysis of a physiologic intermolecular electron transfer reaction between methylamine dehydrogenase and amicyanin. *Biochemistry 33*: 5696-5701.
11. Bishop, G. R., and Davidson, V. L. (1998) Electron transfer from the aminosemiquinone reaction intermediate of methylamine dehydrogenase to amicyanin. *Biochemistry 37*: 11026-11032.
12. Brooks, H. B., and Davidson, V. L. (1994) Free energy dependence of the electron transfer reaction between methylamine dehydrogenase and amicyanin. *J. Am. Chem. Soc. 116*: 11201-11202.
13. Hyun, Y-L., Zhu, Z., and Davidson, V. L. (1999) Gated and ungated electron transfer reactions from aromatic amine dehydrogenase to azurin. *J. Biol. Chem. 274*: 29081-29086.

Biochemistry and Molecular Biology of Vitamin B_6
and PQQ-dependent Proteins
ed. by A. Iriarte, H.M. Kagan and M. Martinez-Carrion

Continuous-Flow Column Electrolytic Spectroelectrochemical Method for Determination of Protein Redox Potentials – Application to Quinoproteins

Tokuji Ikeda, Akihiro Sato, Kazuyoshi Takagi, Masaki Torimura, and Kenji Kano
Division of Applied Life Sciences, School of Agriculture, Kyoto University, Sakyo-ku, Kyoto 606-8502, Japan

A new spectroelectrochemical method based on flow column electrolysis has proved to be useful for the determination of the redox potentials of (1) alcohol dehydrogenase form *Gluconobacter suboxydans*, (2) soluble glucose dehydrogenase from *Acinetobacter calcoaceticus*, (3) methylamine dehydrogenase from *Paracoccus denitrificans* and (4) a new quinohemoprotein amine dehydrogenase from *Paracoccus denitrificans*. They are (1) 401 (heme 1), 370 (heme 2), 216 (heme 3), 101 (heme 4), and −167 (PQQ) mV vs. NHE at pH 7.0, (2) 33 ($PQQ_{ox/sem}$) and −12 ($PQQ_{sem/red}$) mV at pH 7.0, (3) 129 (TTQ) mV at pH 7.5, and (4) 192 (heme), 190 ($Q_{ox/rem}$), and 100 ($Q_{sem/red}$) mV at pH 7.0, Q being a quinone of unknown structure.

Introduction

Redox potential ($E°'$) is one of the most important physicochemical parameters characterizing the catalytic properties of oxidoreducatses. The redox potential measurement of proteins is, however, not an easy task primarily because of sluggish electrochemistry of many proteins. We have developed a new method of measuring protein redox potentials, which relies upon rapid and quantitative bulk electrolysis by flow column electrolysis coupled with spectroscopy [1,2]. Here, we demonstrate that our method has successfully been applied to the determination of redox potentials of four kinds of quino(hemo)proteins.

Materials and Methods

Materials: alcohol dehydrogenase (ADH) form *Gluconobacter suboxydans* (IFO12528), a new quinohemoprotein amine dehydrogenase (QH-AmDH) from *Paracoccus denitrificans* (IFO 12442), and methylamine dehydrogenase (MADH) from *Paracoccus denitrificans* (IFO 12442) were isolated from the respective bacteria as described elsewhere [1, 3, 4]. Soluble glucose dehydrogenase (sGDH) from *Acinetobacter calcoaceticus* was donated from Prof. Duine. All chemicals were of analytical reagent grade and used as received except for phenazine methosulfate-2-sulfonate (PMS-S), which was synthesized from phenazine ethosulfate (PES^+) [4].

Methods: the method of $E^{\circ\prime}$ measurements is based on the spectroscopic detection of the redox state of proteins in a continuous-flow buffer regulated with the potential applied to the column electrode. Details of the principle, instruments, and methods are described in the literature [2]. All the potentials in this paper are referred to the normal hydrogen electrode (NHE).

Results and Discussion

ADH

This enzyme is a membrane-bound quinohemoprotein functioning as the primary dehydrogenase of the ethanol respiratory chain of acetic acid bacteria to transfer electrons from ethanol to ubiquinone. It consists of three subunits I (78 kDa), II (48 kDa), and III (14 kDa). Subunit I contains one mole each of pyrroloquinoline quinone (PQQ) and heme *c* and subunit II is a triheme cytochrome *c*, while subunit III has no prosthetic group [5]. ADH is an enzyme capable of direct electron transfer communication with a carbon electrode [6], thus ADH exhibits reproducible potential (E)-dependent spectral change by the direct electrolysis at the carbon fiber column electrode with the potential changing in a staircase mode. Because of a large molar absorption co-efficient of ADH and high sensitivity of the photo-diode array detector, ADH concentrations as low as ca. 1×10^{-7} M suffice for the spectroscopic detection. Figure 1 shows the absorbance (A)-E plot, which is composed of three sigmoidal parts with four plateaus. During the potential change from state A to C, increase of the α and β bands and a red shift of the Solet band were observed. The spectral changes are reasonably assigned to the reduction of the heme *c* moieties. On the potential change from state C to D, the Solet band as well as α band increased slightly, although the maximum wavelength of the Solet band remained unchanged. Then we conclude that the four hemes *c* are completely reduced at the sate C and that the spectral change from state C to state D is due to the two-electron reduction of PQQ, which could affect the electronic circumstance of a heme(s) *c*, most probably that in subunit I. Regression curve obtained on the basis of a five-step electron transfer mechanism reproduces the data well. Evaluated $E^{\circ\prime}$ values are 401 (heme 1), 370 (heme 2), 216 (heme 3), 101 (heme 4), and –167 (PQQ) mV at pH 7.0.

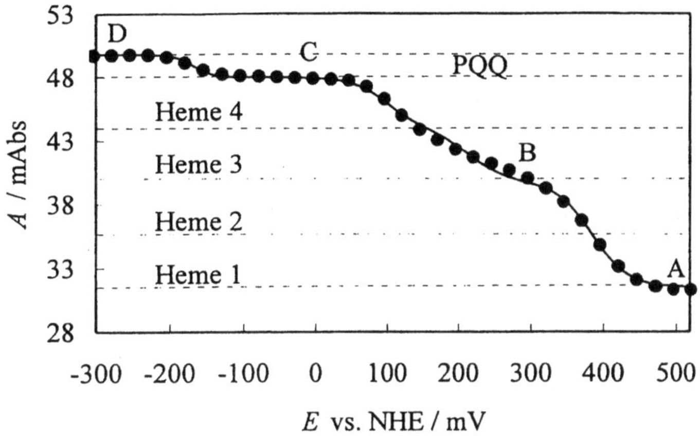

Fig. 1 A-E curve of ADH in pH7.0 phosphate buffer at 422 nm

sGDH

sGDH exists in the periplasm of the bacterium *Acinetobacter calcoaceticus*. It is a homodimeric enzyme (subunits of 50 kDa) containing PQQ and Ca^{2+} and catalyzes the oxidation of D-glucose and other aldoses to their corresponding aldono-δ-lactones [7]. $E^{\circ\prime}$ measurement of sGDH is based on mediator-assisted continuous–flow column electrolytic spectroelectrochemical technique as detailed in the literature [2]. The redox state of sGDH in solution is indirectly controlled by the column electrode potential through the redox mediation of PES^+, and a reproducible absorption spectrum of sGDH is obtained by the correction for a large but stable background spectrum. The *A-E* curves at 338 nm and 485 nm are given in Fig. 2, which clearly show the existence of a stable semiquinone form of PQQ in sGDH. In accord with this, sGDH generated very stable ESR spectrum on the partial reduction with D-glucose. The $E^{\circ\prime}$ values are calculated to be 33 ($PQQ_{ox/sem}$) and –12 mV ($PQQ_{sem/red}$) vs. NHE at pH 7.0. Ca^{2+} seems to stabilize the semiquinone intermediate in sGDH. Similar stabilization of semiquinone form was observed with free PQQ in the presence of Ca^{2+}; $E^{\circ\prime}$ of free PQQ was changed from $E^{\circ\prime}$ = 74 mV (one-step two electron transfer) to 115 for $PQQ_{ox/sem}$ and 71 mV for $PQQ_{sem/red}$ (two-step one-electron transfer) in the presence of Ca^{2+}.

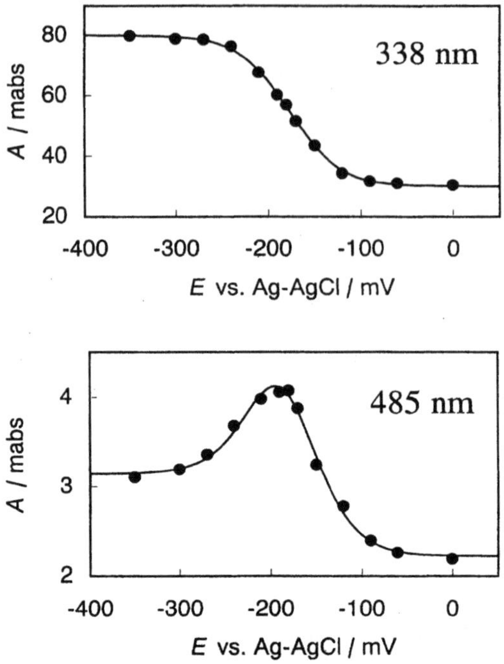

Fig. 2 *A-E* curves of sGDH at pH 7.0

MADH

Tryptophan tryptophylquinone (TTQ) is the cofactor of MADH, which exists in each of the β subunits of the enzyme having an $\alpha_2\beta_2$ structure [8]. In a similar manner to sGDH, potential-dependent spectral change of MADH is obtained in the presence of both PES^+ and PMS-S as mediators, where the protein redox state reaches in equilibrium with the electrode potential within 150 s. Thus, $E^{\circ\prime}$ of MADH (one-step two electron transfer) is determined as 129 mV vs. NHE at pH 7.5, which is slightly positive than the reported one (100 mV at pH 7.5 [9,10]). When we use only one kind of mediator compounds, PES^+ or PMS-S, however, the equilibrium state is not attained in the time scale of the conventional column electrolysis, ca. 20 to 200 s. This is because the redox reaction rate of the protein with the mediator is not sufficiently high, which can be confirmed as the time-dependent change of A at a given wave length. We have analyzed the absorbance change by a simple kinetic theory and shown that the kinetic analysis also gives the $E^{\circ\prime}$ value agreeing with that obtained above [4].

QH-AmDH

We have isolated a new quinohemoprotein, QH-AmDH, from periplasm of *P. denitrificans*. It is a heterodimer constituted of α (59.5 kDa) and β subunits (36.5 kDa); the former contains one unknown quinonoid cofactor and one heme c per molecule, while the latter has no prosthetic group. QH-AmDH catalyzes the oxidative deamination of primary aliphatic and aromatic amines [3]. The present method of indirect column electrolysis with 2,6-dimethyl

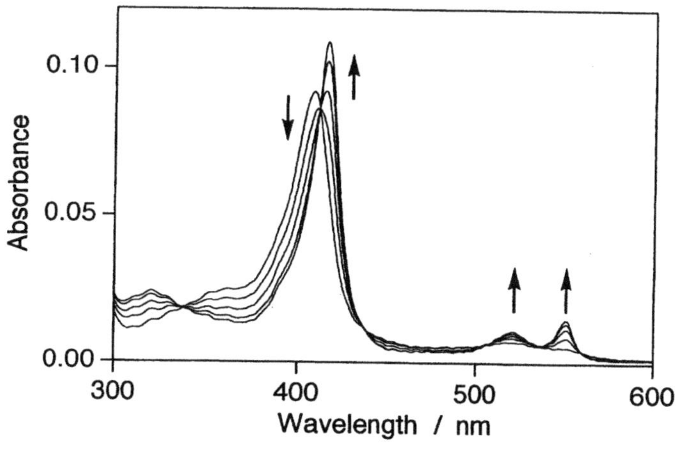

Fig. 3 UV-visible absorption spectra of QH-AmDH

benzoquinone as a mediator allowed us to determine $E^{\circ\prime}$ of the heme *c* moiety as 192 mV from the potential-dependent spectral change (Fig. 3). Although the spectrum attributable to the quinone moiety was not observed, the $E^{\circ\prime}$ values of the quinone/semiquinone and semiquinone/quinol redox couples could be evaluated as 190 and 110 mV, respectively, from the analysis of the substrate titration curve [3].

Acknowledgements

We thank Prof. O. Adachi, Department of Biological Chemistry, Yamaguchi University, Japan, for his kind advice in isolation of MADH and Prof. K. Matsushita, the same University, for his helpful discussion on the redox property of ADH, and Prof. J. A. Duine, Delft University of Technology, The Netherlands, for his kind donation of sGDH and helpful information on the properties of sGDH.

References

1. Torimura, M., Kano, K., Ikeda, T., and Ueda, T., *Chem. Lett.* **1997**, 525.
2. Torimura, M., Mochizuki, M., Kano, K., Ikeda, T., and Ueda, T. (1998) *Anal. Chem.*, **70**, 4690.
3. Takagi, K., Torimura, M., Kawaguchi, K., Kano, K., and Ikeda, T., (1999) *Biochemistry*, **38**, 6935.
4. Sato, A., Torimura, M., Takagi, K., Kano, K., and Ikeda, T., (1999) *Anal. Chem.*, in press.
5. Matsushita, K., Yakushi, T., Toyama, H., Shinagawa, E., and Adachi, O., (1996) *J. Biol. Chem.*, **271**, 4850.
6. Ikeda, T., Kobayashi, D., Matsushita, F., Sagara, T., and Niki, K., (1993) *J. Electroanal. Chem.*, **361**, 221.
7. Olsthoorn, A. J. J. and Duine, J. A., (1996) *Achiv. Biochem. Biophys.* **336**, 42.
8. McIntire, W. S., Wemmer, D. E., Chistoserdov, A., and Lindstrom, M. E., (1991) *Science*, **252**, 817.
9. Husain, M., Davidson, V. L., Gray, K. A., and Knaff, D. B., (1987) *Biochemistry*, **26**, 4139.
10. Zhu, Z. and Davidson, V. L., (1998) *J. Biol. Chem.*, **273**, 14254.

Biochemistry and Molecular Biology of Vitamin B_6
and PQQ-dependent Proteins
ed. by A. Iriarte, H.M. Kagan and M. Martinez-Carrion

Enantioselectivity of PQQ-containing Alcohol Dehydrogenases: Kinetic, Thermodynamic, and Molecular Modeling

A. Jongejan and J.A. Jongejan

Biotechnology Department, Delft University of Technology, Julianalaan 67, 2628 BC Delft, The Netherlands

Summary

Quino(hemo)protein alcohol dehydrogenases, containing pyrroloquinoline quinone, PQQ as an organic cofactor, show appreciable enantioselectivity in the oxidation of chiral alcohols. Fundamental insight into the mechanistic and structural factors of importance for the observed enantio-selectivity has been obtained from the evaluation of the thermodynamic fingerprint of the kinetic features. It has been found that entropic contributions play a major role in the enantiopreference. Molecular modeling strategies, i.e. molecular mechanics and quantum mechanics calculations aimed at the evaluation of the Gibbs free energy difference between the enantioselectivity-controlling transition states have been developed. Preliminary results of deuterium kinetic isotope experiments appear relevant for the location of the transition states along the reaction coordinate.

Introduction

Dye-linked alcohol dehydrogenases with pyrroloquinoline quinone (2,7,9-tricarboxy-1*H*-pyrrolo[2,3-*f*]quinoline-4,5-dione, PQQ) and Ca^{2+} as cofactors include methanol dehydrogenases, MDHs, from methylotrophic bacteria (Anthony & Ghosh, 1998) and ethanol dehydrogenase from *Pseudomonas* sp. (Schrover et al., 1993). Quino*hemo*protein alcohol dehydrogenases, QH-ADHs, contain heme *c* as an additional organic cofactor. Monomeric (70 to 80 kDa) QH-ADHs, with one molecule of PQQ, one Ca^{2+}, and a single *c*-type heme, have been isolated from ethanol-grown *Comamonas testosteroni* (de Jong et al., 1995), vanillyl alcohol grown *Rhodopseudomonas acidophila* (Yasuda et al., 1996) and *P. putida* (Toyama et al., 1995). Membrane-associated QH-ADHs composed of different subunits, of which a large subunit (72 to 80 kDa) containing one molecule of non-covalently-bound PQQ, one Ca^{2+}, and one heme *c*, is thought to possess the dehydrogenase activity (Matsushita et al., 1994) and a medium-sized subunit (43 to 53 kDa) contains three heme *c* moieties, have been isolated from acetic acid bacteria.

X-ray crystallographic investigations of PQQ-containing MDHs (Ghosh et al., 1995, Xia et al., 1992, Xia et al., 1999), have provided insight into the structural aspects of quinoprotein alcohol dehydrogenases. A structural model of the QH-ADH from *C. testosteroni* based on homology modeling has been proposed (Jongejan et al., 1998). The X-ray structure of the related QH-ADH from *P. putida* has been announced (Mathews and Matsushita, These Proceedings). However, important questions concerning the catalytic mechanism of the PQQ-containing alcohol dehydrogenases remain to be solved. This is also the case for the quinohemoprotein alcohol dehydrogenases that were shown to possess appreciable enantioselectivity (*E*-values 10 to 100) in the oxidation of low-molecular chiral alcohols of which the pure enantiomers are of potential use in industrial applications (Machado et al., 1999). Since their performance compares favorably to that currently realized with non-biological catalysts (i.e. (Noyori & Hashiguchi, 1997)), where $E = 40$ is judged satisfactory and $E = 200$ is a remarkable exception (Corey et al., 1992), the molecular origin of their enantioselectivity is of interest.

Kinetic Modeling

Despite their ready availability, MDHs are not particularly suited for the study of the catalytic mechanism. Several complications exist (see (Frank et al., 1989a)), however, by combining spectroscopic and kinetic techniques a convincing kinetic scheme could be devised (Frank et al., 1989b). The situation is less complicated for QH-ADHs. These enzymes have been shown to obey regular bi-bi ping-pong kinetics for a variety of (chiral) substrates (Geerlof et al., 1994, Machado et al., 1998). Contrary to MDHs, high pH, ammonia, or amines, are not required for activity.

Thermodynamic Modeling

From the definition of the enantiomeric ratio and application of Eyring TST the following relation can be obtained:

$$E = \frac{\left(k_{cat} / K_M\right)^R}{\left(k_{cat} / K_M\right)^S} = \frac{\exp\left(-\Delta G^{\#}_{R,spec} / RT\right)}{\exp\left(-\Delta G^{\#}_{S,spec} / RT\right)} = \exp\left(-\Delta\Delta G^{\#} / RT\right)$$

From the temperature dependence of *E* thermodynamic information of the enthalpic and entropic contributions to the enantioselectivity can be obtained according to:

$$R \ln E = -\Delta\Delta G^{\#} / T = \Delta\Delta S^{\#} - \Delta\Delta H^{\#} / T$$

This relation has been shown to give a satisfactory representation of the temperature dependence of various enantioselective enzyme-catalyzed reactions (See (Phillips, 1996) for a review). The difference in activation enthalpy and entropy for enantiomeric substrates, $\Delta\Delta H^{\#}$ and $\Delta\Delta S^{\#}$, respectively, can be estimated from Arrhenius-type plots of $R\ln E$ versus $1/T$. From the results obtained in the enantioselective oxidation of 2-butanol (QH-ADH from *A. pasteurianus*) and solketal (QH-ADH from *C. testosteroni*) (Machado et al., 1999), it can be concluded that at room temperature the enantioselectivity of these enzymes contains a large *entropic* contribution. This finding is in contrast with the data reported for the enantioselective oxidation of 2-butanol using NAD-dependent secondary alcohol dehydrogenase from *Thermoanaerobacter ethanolicus* (Tripp et al., 1998) and the enantioselective esterification of 2-butanol using lipase B from *Candida antarctica.* (Overbeeke et al., 1998). Clearly, models designed to summarize the enantioselective properties of QH-ADHs will have to take the entropic contribution into account.

Molecular Modeling

Molecular modeling experiments aimed at the validation of transition state structures relevant to the enantioselective features of quinoprotein alcohol dehydrogenases have been carried out. From the results of molecular mechanics simulations we concluded that a hydride transfer mechanism (Anthony, 1996, Zheng & Bruice, 1997) would explain the observed enantiopreference, whereas an addition-elimination mechanism (Itoh et al., 1998, Itoh et al., 1993) would give rise to the opposite enantiopreference (Jongejan et al., 2000). Recently, we calculated the potential energy surface for hydride transfer to C(5) of PQQ using quantum mechanical ab initio methods at different levels of theory. From the results obtained, it appears that this step would be feasible indeed. As the eventual formation of $PQQH_2$ would require a subsequent tautomerization step to take place, we investigated the rate- and enantioselectivity-determining character of the hydride transfer step using deuterated alcohols. Preliminary results show that the primary deuterium KIE for non-chiral alcohols is of the order 7-9, while the effect of deuteration in the kinetic resolution of solketal appears to be more modest. Current explanations of these findings (see (Jongejan et al., 2000) for a theoretical model) support the partial rate-determining character of a non-enantiodiscriminatory step, for which the tautomerization of the initially formed C(5)-H structure to $PQQH_2$ is a likely candidate. A similar step has been proposed for the oxidation of glucose by quinoprotein GDH (Duine, These Proceedings).

Acknowledgement
We gratefully acknowledge financial support from the Organization for the Advancement of Pure Research/NWO, Foundation for Chemical Research/SON, The Netherlands.

References

Anthony, C. (1996) Quinoprotein-catalysed reactions, *Biochem. J. 320*, 697-711.

Anthony, C., & Ghosh, M. (1998) The structure and function of the PQQ-containing quinoprotein dehydrogenases, *Prog. Biophys. Mol. Biol. 69*, 1-21.

Corey, E. J., Loh, T.-P., Roper, T. D., Azimioara, M. D., & Noe, M. C. (1992) The origin of greater than 200:1 enantioselectivity in a catalytic Diels-Alder reaction as revealed by physical and chemical studies, *J. Am. Chem. Soc. 114*, 8290-8292.

de Jong, G. A. H., Geerlof, A., Stoorvogel, J., Jongejan, J. A., de Vries, S., & Duine, J. A. (1995) Quinohaemoprotein ethanol dehydrogenase from *Comamonas testosteroni.*, *Eur. J. Biochem. 230*, 899-905.

Frank, J., Dijkstra, M., Balny, C., Verwiel, P. E. J., & Duine, J. A. (1989a) Methanol dehydrogenase: Mechanism of action, in *PQQ and Quinoproteins* (Jongejan, J. A., & Duine, J. A., Eds.) pp 13-22, Kluwer Academic Publishers, Dordrecht.
Frank, J. J., van Krimpen, S. H., Verwiel, P. E. J., Jongejan, J. A., Mulder, A. C., & Duine, J. A. (1989b) On the mechanism of inhibition of methanol dehydrogenase by cyclopropane-derived inhibitors, *Eur. J. Biochem. 184*, 187-195.
Geerlof, A., Rakels, J. J. L., Straathof, A. J. J., Heijnen, J. J., Jongejan, J. A., & Duine, J. A. (1994) Description of the kinetic mechanism and the enantioselectivity of quinohaemoprotein ethanol dehydrogenase from Comamonas testosteroni in the oxidation of alcohols and aldehydes, *Eur. J. Biochem. 226*, 537-546.
Ghosh, M., Anthony, C., Harlos, K., Goodwin, M. G., & Blake, C. (1995) The refined structure of the quinoprotein methanol dehydrogenase from *Methylobacterium extorquens* at 1.94 Å, *Structure 3*, 177-187.
Itoh, S., Kawakami, H., & Fukuzumi, S. (1998) Model studies on Calcium-containing quinoprotein alcohol dehydrogenases. Catalytic role of Ca2+ for the oxidation of alcohols by coenzyme PQQ (4,5-dihydro-4,5-dioxo-1*H*-pyrrolo[2,3-*f*]quinoline-2,7,9-tricarboxylic acid), *Biochemistry 37*, 6562-6571.
Itoh, S., Ogino, M., Fukui, Y., Murao, H., Komatsu, M., Ohshiro, Y., Inoue, T., Kai, Y., & Kasai, N. (1993) C-4 and C-5 adducts of cofactor PQQ (Pyrroloquinoline quinone). Model studies directed towards the action of quinoprotein methanol dehydrogenase, *J. Am. Chem. Soc. 115*, 9960-9967.
Jongejan, A., Jongejan, J. A., & Duine, J. A. (1998) Homology model of the quinohaemoprotein alcohol dehydrogenase from *Comamonas testosteroni, Protein Engineering 11*, 185-198.
Jongejan, A., Machado, S. S., & Jongejan, J. A. (2000) The Enantioselectivity of Quinohaemoprotein Alcohol Dehydrogenases: Mechanistic and Structural Aspects, *J. Mol. Catal. B: Enzymatic*, in press.
Machado, S. S., Jongejan, A., Geerlof, A., Jongejan, J. A., & Duine, J. A. (1999) Entropic and enthalpic contributions to the enantioselectivity of quinohaemoprotein alcohol dehydrogenases from *Acinetobacter pasteurianus* and *Comamonas testosteroni* in the oxidation of primary and secondary alcohols, *Biocatalysis and Biotransformations 16*, 1-29.
Machado, S. S., Wandel, U., Jongejan, J. A., Straathof, A. J. J., & Duine, J. A. (1998) Characterization of the enantioselective properties of the quinohaemoprotein alcohol dehydrogenase of *Acetobacter pasteurianus* LMG 1635, *Biosci. Biotechnol. Biochem. 63*, 10-20.
Matsushita, K., Toyama, H., & Adachi, O. (1994) Respiratory chains and bioenergetics of acetic acid bacteria, *Adv. Microb. Physiol. 36*, 247-301.
Noyori, R., & Hashiguchi, S. (1997) Asymmetric transfer hydrogenation catalyzed by chiral ruthenium complexes, *Acc. Chem. Res. 30*, 97-102.
Overbeeke, P. L. A., Ottosson, J., Hult, K., Jongejan, J. A., & Duine, J. A. (1998) The contribution of enthalpy and entropy to the chiral discrimination of enzymes, *Biocatalysis and Biotransformations 17*, 61-79.
Phillips, R. S. (1996) Temperature modulation of the stereochemistry of enzymatic catalysis: prospects for exploitation, *TIBTECH 14*, 13-16.
Schrover, J. M. J., Frank, J., van Wielink, J. E., & Duine, J. A. (1993) Quaternary structure of quinoprotein ethanol dehydrogenase from *Pseudomonas aeruginosa* and its reoxidation with a novel cytochrome *c* from this organism, *Biochem. J. 290*, 123-127.
Toyama, H., Fujii, A., Matsushita, K., Shinagawa, E., Ameyama, M., & Adachi, O. (1995) Three distinct quinoprotein alcohol dehydrogenases are expressed when *Pseudomonas putida* is grown on different alcohols, *J. Bacteriol. 177*, 2442-2450.
Tripp, A. E., Burdette, D. S., Zeikus, J. G., & Phillips, R. S. (1998) Mutation of serine-39 to threonine in thermostable secondary alcohol dehydrogenase from *Thermoanaerobacter ethanolicus* changes enantio-specificity, *J. Am. Chem. Soc. 21*, 5137-5141.
Xia, Z.-x., Dai, W.-w., Xiong, J.-p., Hao, Z.-p., Davidson, V. L., White, S., & Mathews, F. S. (1992) The three-dimensional structures of methanol dehydrogenase from two methylotrophic bacteria at 2.6 Å-resolution, *J. Biol. Chem. 267*, 22289-22297.
Xia, Z.-x., He, Y.-n., Dai, W.-w., White, S. A., Boyd, G. D., & Mathews, F. S. (1999) Detailed active site configuration of a new crystal form of methanol dehydrogenase from *Methylophilus* W3A1 at 1.9 Å resolution, *Biochemistry 38*, 1214-1220.
Yasuda, M., Cherepanov, A., & Duine, J. A. (1996) Polyethylene glycol dehydrogenase activity of *Rhodopseudomonas acidophila* derives from a type I quinohaemoprotein alcohol dehydrogenase, *FEMS Microbiol. Lett. 138*, 23-28.
Zheng, Y.-J., & Bruice, T. C. (1997) Conformation of coenzyme pyrroloquinoline quinone and role of Ca^{2+} in the catalytic mechanism of quinoprotein methanol dehydrogenase, *Proc. Natl. Acad. Sci. USA 94*, 11881-11886.

Biochemistry and Molecular Biology of Vitamin B_6
and PQQ-dependent Proteins
ed. by A. Iriarte, H.M. Kagan and M. Martinez-Carrion

Structural Studies of a soluble monomeric Quinohemoprotein Alcohol Dehydrogenase from *Pseudomonas putida* HK5

F. Scott Mathews, Zhi-wei Chen, Department of Biochemistry and Molecular Biophysics, Washington University, St. Louis MO, USA, 63110; Kazunobu Matsushita, Tetsuo Yamashita, Nahoko Aoki, Hirohide Toyama and Osao Adachi, Department of Biological Chemistry, Yamaguchi University, Yamaguchi, Japan, 753-8515 and Henry D. Bellamy, Stanford Synchrotron Radiation Laboratory, Stanford, CA, USA, 94309.

Summary

We have traced the polypeptide chain at 3 Å resolution of the soluble, monomeric alcohol dehydrogenase (ADH-IIB) isolated from *Pseudomonas putida*. The enzyme contains a large N-terminal 8-stranded β-propeller domain (~60 kDa) similar to methanol dehydrogenase (MDH) and a small C-terminal *c*-type cytochrome domain (~10 kDa) similar to the cytochrome subunit of *p*-cresol methylhydoxylase (PCMH), also from *P. putida*. The heme is inclined by about 70° to the PQQ and their planes are about 14 Å apart at the closest point. The propionic acids of the heme group are exposed to solvent suggesting them to be the site for interaction with its azurin electron transfer partners.

Introduction

A family of alcohol dehydrogenases (ADHs) that are independent of pyridine nucleotides and contain non-covalently bound pyrroloquinoline quinone (PQQ) as a redox cofactor is found in many aerobic bacteria [1]. One class of ADH, called type II is composed of monomeric, soluble quinohemoproteins of ~70-75 kDa that contain PQQ and covalently-bound heme, and are found in several strains of *P. putida* [2] and in *Comamonas testosteroni* [3]. Sequence analysis of the *C. testosteroni* enzyme indicates that the first 60 kDa are homologous to methanol dehydrogenase and the remaining 15 kDa appear to form a *c*-type cytochrome domain. One form of type II ADH, ADH-IIB, can be isolated from *P. putida* strain HK5 cells grown on 1-butanol. It is estimated to have a molecular weight of 69 kDa on the basis of SDS-PAGE and has been shown to be monomeric [2]. Its amino acid sequence is unknown. *In vitro*, potassium ferricyanide can serve as an artificial electron acceptor for ADH-IIB, and is used in a dye-linked assay. An azurin isolated from the same organism has been shown to be an efficient electron acceptor for ADH-IIB [4] suggesting that it acts *in vivo* as an electron transfer mediator in a PQQ-dependent alcohol respiratory chain.

Materials and Methods

ADH-IIB was prepared as described previously [2]. Crystals are obtained with 22% PEG8000, 200mM sodium citrate, 100mM HEPES-NaOH, pH 7.5 and 6% 2-propanol [5]. The crystals are triclinic with cell parameters a=54.8 Å, b=57.4 Å, c=67.5 Å, α=89.6°, β=69.4°, γ=68.4° and contain 1 molecule in the unit cell. X-ray data were collected at 100K from a cryoprotected crystal to 1.9 Å resolution (Rmerge=6.4%, 87% completeness).

Figure 1: Stereo view of the ADH-IIB molecule. The PQQ and calcium ion are in the large β-propeller domain at the bottom and the heme group is in the small cytochrome domain at the top.

The data were analyzed by molecular replacement (MR) with AMORE [6] using MDH from *Methylophilus methylotrophus* W3A1 as the search molecule. An anomalous difference electron density map based on the MR phases contained two prominent peaks. One peak (14σ) is ~20 Å from the PQQ position in the MDH search molecule and was attributed to iron and the other (9σ) is close to the position of the calcium ion in MDH that bridges PQQ to protein side chains and was attributed to calcium.

Two compounds, $K_3UO_2F_5$ and CH_3HgCl showed 12%-15% isomorphous structure factor changes. MIR data were collected in St. Louis and MAD data from the mercury and UO_2 derivatives at the SSRL and APS synchrotrons, respectively. Protein phases were calculated with SHARP [7]. Using maps based on combinations of MIR and MAD phases, the quinoprotein and the cytochrome domains could be traced. Since the ADH-IIB sequence was unknown, we fit the sequence of the *C. testosteroni* ADH quinoprotein domain wherever possible and polyalanine elsewhere. The structure was partially refined using CNS [8] (R=31%, Rfree=39%, 40-3 Å resolution).

Results and Discussion

The current model of ADH IIB consists of 664 amino acid residues. Residues 1-556 comprise the PQQ-containing 8-bladed β-propeller domain, residues 582-664 comprise the cytochrome domain and residues 557-581 form a 55-residue linker segment connecting the two domains (Fig. 1). The current electron density provides a reasonably accurate representation of the path of the polypeptide chain through the structure although the model is limited by the lack of sequence information; however, the *C. testosteroni* ADH is likely to be reasonably homologous.

The PQQ is located in the funnel of the β-propeller that forms a depression in the top of the domain. As in MDH, the PQQ is sandwiched between a tryptophan side chain on the bottom and a vicinal disulfide group on the top. The calcium ion is in the plane of the PQQ and is liganded by atoms O5, N6 and O7A of PQQ plus a glutamic acid and a gultamine side chain.

The cytochrome domain is very small, containing 3 α-helices in 83 residues. The heme group is covalently attached to Cys596 and Cys599 via thioether linkage to the two heme vinyl side chains while His600 and Met636 are coordinated to the iron atom. The propionic acid side chains of the heme are located on the surface of the protein and project into solution. Among bacterial cytochromes, the cytochrome domain of ADH-IIB most closely resembled the cytochrome subunit of *p*-cresol methylhydroxylase [9]. Both of these proteins contain 3 α-helices rather than the usual 4 and have solvent- exposed propionate groups. The heme iron atom and C5 atom of PQQ are separated by about 20.5 Å and the distance between the two rings is about 16 Å. The heme and PQQ planes are tilted by about 70° to each other.

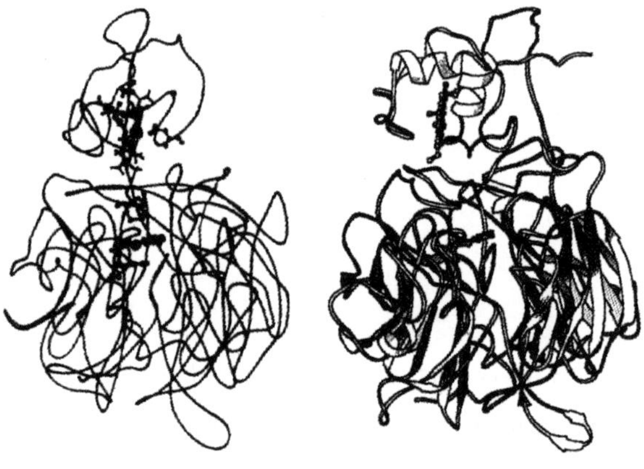

Figure 2. Comparison of the hypothetical model of ADH from *C. testosteroni* (left) and ADH-IIB (right). The left panel was repoduced by permission from [10].

A hypothetical molecular model of ADH from *C. testosteroni* has been constructed by homology modelling based on sequence similarities with the known structures of MDH from *M.* methylophilus W3A1 and *Methylobacterium extorquens* AM1[10]. The structure of ADH-IIB is compared with the published model of the *C. testosteroni* in Fig. 2. The large β-propeller domains of the two structures appear remarkably similar. The orientation of the cytochrome domains are also quite similar. However, the heme group in ADH-IIB is asymmetrically displaced with respect to the funnel axis of the large domain and to the PQQ ring. Detailed comparison of the *C. testosteroni* ADH model and the ADH-IIB structure awaits completion of the sequence and full refinement of the latter.

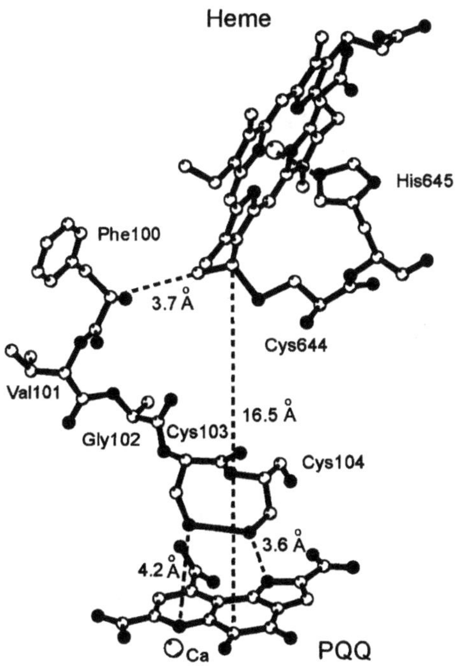

Figure 3. Relative positions and orientations of the PQQ and heme groups showing potential sigma-bond tunneling pathway for electron transfer between them. The separation of the PQQ and heme planes is also indicated. Carbon, calcium and iron atoms are shown as white spheres while nitrogen and oxygen atoms are shown as black spheres.

The PQQ and heme group are linked by a covalent chain of 4 or 5 amino acids, residues 99-102 or 103, that may provide a pathway for electron flow between the two prosthetic groups (Fig. 3). The path involves a through-space jump of about 4 Å from the plane of PQQ to the Cys102-Cys103 vicinal disulfide, travel along the backbone chain from residue Cys102 to Phe99 and a through space jump of 3.7 Å to Cys599 which forms one of the covalent attachment sites of the heme.

Acknowledgement.

This work is supported by USPHS Grant No. GM-20530 (FSM) and by Grants-in-Aid for scientific research from the Ministry of Education, Science and Culture, Japan (09044228 and 10660091 to KM). SSRL and the APS are funded by the Department of Energy, Office of Basic Energy Sciences and Office of Health and Environmental Research.

References

1. Matsushita, K., Toyama, H. and Adachi, O. (1994) Respirtory chains and bioenergetics of acetic acid bacteria. *Adv. Microb. Physiol.* **636**, 247-301.
2. H. Toyama, H., Fujii, A., Matsushita, K., Shinagawa, E., Ameyama, O. & Adachi, O. (1995) Three distinct quinoprotein alcohol dehydrogenases are expressed when *Pseudomonas putida* is grown on different alcohols; *J. Bacteriol.* **177,** 2442-2450.
3. Groen, B., van Kleef, M. A. G. and Duine, J. A. (1986) Quinohemoprotein alcohol dehydrogenase apoenzyme from *Pseudomonas testosteroni. Biochem. J.* **234**, 611-615.
4. Matsushita, K., Yamashita, T., Aoki, N., Toyama, H. & Adachi, O. (1999) Electron transfer from quinohemoprotein alcohol dehydrogenase to blue copper protein azurin in the alcohol oxidase respiratory chain of *Pseudomonas Putida* HK5. *Biochemistry*, **38**, 6111-6118.
5. Chen, Z.-w., Baruch, P., Mathews, F. S., Matsushita, K., Yamashita, T, Toyama, H. and Adachi, O. (1999) Crystallization and Preliminary Diffraction Studies of Two Quinoprotein Alcohol Dehydrogenases (ADH): a soluble monomeric ADH from *Pseudomonas putida* HK5 (ADH-IIB) and a heterotrimeric membrane-bound ADH from *Gluconobacter suboxydans* (ADH-GS). *Acta Cryst.* **D55**, 1933-1936.
6. Navaza, J. (1994) AMoRe: an automated package for molecular replacement. *Acta Cryst.* **A50**:157-163.
7. La Fortelle, E. de and Bricogne, G. (1997) Maximum-Likelihood Heavy-Atom Parameter Refinement in the MIR and MAD Methods. *Methods Enzymol.* **276**, 472-494.
8. Brünger, A.T., Adams, P.D., Clore, G.M., DeLano, W.L., Gros, P., Grosse-Kunstleve, R.W., Jiang, J.S., Kuszewski, J., Nilges, M., Pannu, N.S., Read, R.J., Rice, L.M., Simonson, T. & Warren, G.L. (1998) Crystallography & NMR system: A new software suite for macromolecular structure determination. *Acta Cryst.* **D 54**, 905-921.
9. Cunane, L. M., Chen, Z.-w., Shamala, N., Mathews, F. S., Cronin, C. N. and McIntire, W. S., (2000) Structures of the flavocytochrome p-cresol methylhydroxylase and its enzyme-substrate complex: gated substrate entry and proton relays support the proposed catalytic mechanism. *J. Mol. Biol.* **295**, 357-374.
10. Jongejan, A., Jongejan, J. A. and Duine, J. A. (1998) Homology model of the quinohaemoprotein alcohol dehydrogenase from Comamonas testosteroni. *Protein Eng.* **11,** 185-198.

Biochemistry and Molecular Biology of Vitamin B_6
and PQQ-dependent Proteins
ed. by A. Iriarte, H.M. Kagan and M. Martinez-Carrion

Electron Transport Systems for Quinohemoprotein Type II Alcohol Dehydrogenase of *Pseudomonas putida* HK5

Kazunobu Matsushita, Tetsuo Yamashita, Nahoko Aoki, Hirohide Toyama, and Osao Adachi.
Department of Biological Chemistry, Faculty of Agriculture, Yamaguchi University, Yamaguchi, Yamaguchi 753-8515, Japan

Summary

To understand the electron transport system with quinohemoprotein type II alcohol dehydrogenase (ADH; ADH IIB) in *Pseudomonas putida* HK5 grown on butanol, an ADH-dependent electron transfer reaction was examined in vivo and in vitro. Azurin purified from the periplasm of the same organism was shown to react with ADH IIB, where the electron transfer seems to occur by a freely reversible on and off binding process. The azurin-dependent alcohol oxidation activity could be reproduced by reconstituting ADH IIB on the right-side out membrane vesicles and also with purified cytochrome oxidase, which is completely CN-sensitive. However, alcohol oxidation activity was shown to be CN-insensitive in the intact cells, and also in azurin-independent, reconstituted system with the right-side out vesicles. Thus, there seems to be two different alcohol oxidase electron transport systems in *P. putida* HK5 that utilize ADH IIB, one of which transfers electrons via azurin to a CN-sensitive terminal oxidase, and the other presumably via ubiquinone to both a CN-sensitive and -insensitive terminal oxidases.

Introduction

Quinoprotein dehydrogenases can be found only in the periplasm of the most highly evolved bacteria, purple bacteria, α, β and γ subdivisions, but not in any other Bacteria, Archaea, and Eukarya so far. Alcohol dehydrogenase (ADH) is the largest group of quinoproteins, and can be divided into 3 subgroups (type I to III) according to their molecular and catalytic properties, and cellular localization [1]. Type I ADH, containing only pyrroloquinoline quinone (PQQ) as the prosthetic group, is structurally very similar to methanol dehydrogenase (MDH) of methylotorophic bacteria, but is highly specific for ethanol but not methanol. It has been found in *Pseudomonas aeruginosa* [2, 3] and *Pseudomonas putida* [4]. Type II ADH has an additional heme *c* domain in its C-terminal portion, and thus is called a quinohemoprotein, and has been found in *Comamonas testosteroni* [5] and *P. putida* [4]. Type III ADH is a membrane-bound enzyme found in the cytoplasmic membrane of acetic acid bacteria [1]. In contrast to type I and II ADHs, type III ADH consists of three subunits, in which the subunit I is a quinohemoprotein similar to type II ADH, subunit II is a tri-heme cytochrome *c* containing membrane-binding and ubiquinone-reacting domains and subunit III has unknown function [6].

The electron transport systems of these quino(hemo)protein ADHs have been elucidated for type I and III ADHs as well as for MDH. The MDH of methylotorophs has been shown in most cases to transfer electrons to the terminal oxidase via two cytochromes *c*, c_L and c_H [7]. Similar to MDH, type I ADH has been shown in *P. aeruginosa* to react with a soluble cytochrome c_{550} and then with azurin (DeVeries et al., personal communication). In the case of type III ADH of acetic acid bacteria, the situation is somewhat different because the enzyme donates electrons to ubiquinone which then go to the terminal ubiquinol oxidase [1, 6]. However, no physiological electron acceptor for type II ADH has been reported to date.

Pseudomonas putida HK5 produces three different quinoprotein ADHs, one type I ADH and two type II ADHs (ADH IIB and ADH IIG), which are induced when the organism is grown on, respectively, ethanol, 1-butanol and 1, 2-propanediol [4]. As described here, ADH IIB and a blue copper protein, azurin, were found in the periplasm of *P. putida* HK5 grown on 1-butanol and purified from the soluble fraction; the interaction of AHD IIB with azurin was then examined. Furthermore, the electron transport system including ADH IIB and azurin was reconstituted with a right-side out membrane vesicles and also with cytochrome cbb_3 oxidase obtained from the same organism, and examined.

Materials and Methods

Bacterial growth and membrane preparations. Pseudomonas putida HK5 was grown at 30•C up to late logarithmic phase in a minimal medium with 0.3% 1-butanol as the sole carbon source [4]. Cells were collected by centrifugation, and then used to prepare soluble and membrane fractions [8]. Spheroplasts and periplasm were prepared based on the method developed for *Pseudomonas aeruginosa* with some modifications and right-side out (RSO) membrane vesicles were prepared from the spheroplasts by osmotic burst (Matsushita et al., manuscript in preparation).

Purification of ADH IIB, azurin, and cytochrome oxidase. ADH IIB was purified from the soluble fraction by column chromatography utilizing DEAE-cellulose, Butyl-Toyopearl, DEAE-Toyopearl, and Superdex S-200, as described [8]. Azurin was separated from ADH IIB during the DEAE-cellulose step and purified by chromatography using Butyl-Toyopearl, DEAE-Toyopearl, and Superdex S-200 [8]. Cytochrome oxidase (cbb_3-type oxidase) from *P. putida* HK5 was solubilized with 1 % dodecyl maltoside (DM) from the membrane fraction after washing them with 0.5% Triton X-100, and then purified by Q-Sepharose and Superdex S-200 column chromatography in the presence of 0.01% of DM (Matsushita et al., manuscript in preparation).

Enzyme Assays. The azurin reductase activity of ADH IIB reduced with 1-butanol was measured spectrophotometrically by monitoring the decrease in absorbance at 620 nm [8]. Oxidase activity for several substrates was measured polarographically with a Clark-type oxygen electrode.

Results and Discussion

Interaction of ADH IIB with azurin. When the butanol-grown cells were sufficiently well aerated to get reach a completely oxidized state before cell disruption, and provided that contaminating alcohol was completely depleted from the buffer system used for the purification, a blue copper protein-containing fraction could be seen in the Butyl-Toyopearl chromatography step eluting between two cytochrome *c* fractions having high and low ADH IIB activities. The blue copper protein was then purified and characterized. The N-terminal amino acid sequence of the purified protein [8] and PCR cloning (Toyama et al., unpublished) clearly showed that the blue copper protein is highly homologous to azurins form several bacterial species, especially to those of *Pseudomonas putida* NCIB 9869 [9] and of *Pseudomonas fluorescens* B [10].
The purified azurin was shown to be reduced by ADH IIB and the direct electron transfer from ADH IIB to azurin was shown to occur at a rate of 48-70 s^{-1}. The apparent Km value of ADH IIB for azurin, determined by steady-state kinetics, was lowered several fold by increasing ionic strength, suggesting that hydrophobic interaction is involved in this reaction. Furthermore, fluorescence quenching of ADH IIB due to the interaction with azurin increased with increasing ionic strength, although the binding constant between ADH IIB and azurin was unchanged. To obtain more data on the interaction between ADH IIB and azurin, redox titrations were carried out. The redox potential of the heme *c* of ADH and the copper center of azurin were 188 and 280 mV (pH 8.0), respectively, when measured independently. The potential of azurin increased by 12 mV after incubation with ADH but not vice versa, suggesting azurin but not ADH IIB may have undergone some conformational change due to the interaction. Furthermore, when the ionic strength was increased, the redox potential of ADH decreased from 189 to 172 mV, but that of azurin increased 6 mV. Thus, the redox potential gap between ADH and azurin was increased from 102 mV to 126 mV by increasing ionic strength, which is consistent with the increased electron transfer rate from ADH to azurin at high ionic strength.

From these experimental results obtained by steady state kinetics, fluorescence quenching and redox titration, the following speculations can also be made (Fig. 1). i) Hydrophobic interaction is involved in the binding of the two proteins, which may stimulate the electron transfer activity from ADH to azurin. ii) Binding of ADH, or the electron transfer step, may induce a conformational change in azurin, which may stimulate their dissociation. iii) The association and dissociation may both be stimulated by higher ionic strength, which would explain why the Kd-values are not altered by ionic strength. Thus, it is conceivable that electron transfer between

ADH IIB and azurin may occur by a freely reversible on and off binding process. The present results strongly suggest that azurin functions as an electron transfer mediator in a PQQ-dependent alcohol oxidase respiratory chain in *P. putida* HK5.

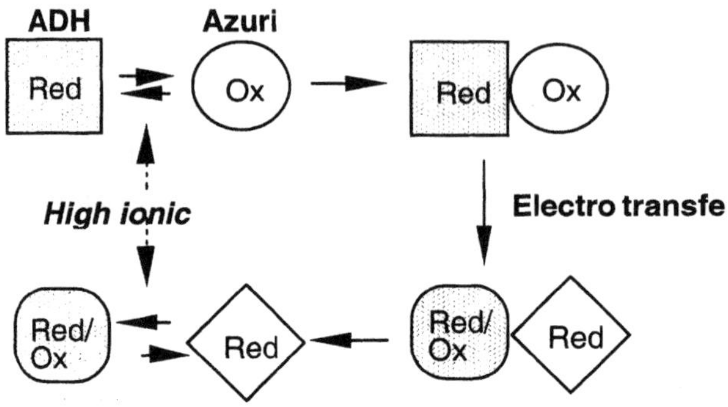

Fig. 1. Schematic model for the mode of interaction between ADH IIB and azurin.

Red: reduced form, Ox : oxidized form, Red/Ox : semi-reduced or oxidized form.

Azurin-dependent ADH IIB alcohol oxidase electron transport. In order to understand the role of azurin in electron transport with ADH IIB, alcohol oxidase respiratory activity was reconstituted on RSO membrane vesicles. As shown in Table 1, some alcohol oxidase activity was reproduced by only the addition of ADH to the RSO membrane vesicles, which was also shown to be CN-resistant. Upon the addition of azurin to this system, the oxidase activity was greatly increased, and turned out to be CN-sensitive. Thus, in the azurin-dependent alcohol oxidase activity reconstituted on RSO membrane vesicles, the turnover number of ADH for butanol oxidation was estimated to be around 1.8 if saturating amounts of azurin is present in the reconstituted system. The electron transfer activity of the reconstituted system and its CN-sensitivity, comparable to that of the TMPD oxidase, suggest that azurin donates electrons to the terminal cytochrome oxidase. This was directly demonstrated by a reconstitution performed with the purified cytochrome oxidase, where alcohol oxidase activity was observed only after the addition of azurin and completely inhibited by cyanide (Table 1). Although the cytochrome oxidase was purified as a cytochrome bc_1-cbb_3 super complex (Matsushita et al., manuscript in preparation), only the cytochrome cbb_3 is likely to be involved in the electron transfer from azurin to oxygen (data not shown). The reconstituted system with the purified cytochrome oxidase exhibited a reasonably high turnover number (40/sec) with ADH, which is comparable to the expected number for intact butanol oxidase respiratory activity (Table 1). This suggests that azurin has the ability to transfer electrons from ADH IIB to the terminal cytochrome oxidase at a rate that can account for that of the intact system.

However, since some azurin-independent alcohol oxidase activity was observed in the reconstituted system with RSO membrane vesicles but not with the cytochrome oxidase, and also the difference in CN-sensitivity was observed in the reconstituted azurin-dependent and -independent alcohol oxidase activities, the reconstituted azurin-dependent alcohol oxidase system may not reflect the complete electron transfer activity of the intact cell.

Table 1. Reconstituted alcohol oxidase respiratory activity with RSO membrane vesicles and with the purified cytochrome oxidase.

Systems	Respiratory activity		Turnover number	CN-sensitivity
	Km μM	Vmax U/mg	per ADH IIB e/s/mol	Remaining activity at 1 mM KCN
Intact cells	-	0.563	9~40	48%
Reconstituted systems				
RSO + ADH	11.4 (ADH)	0.064	0.035	42%
RSO/4 μM ADH + azurin	14.8 (azurin)	0.313	1.8	7.5%
Purified oxidase/20 μM ADH + azurin	3.6 (azurin)	28.5	39.6	0%

Azurin-independent electron transport system with ADH IIB. In order to examine whether both azurin-dependent and azurin-independent systems really function in the respiratory chain of *P. putida* HK5, the CN-sensitivity of the alcohol oxidase respiratory activity was also examined with the intact cells. The alcohol oxidase activity of the intact cells was found to be CN-insensitive, which is similar to other respiratory activities such as lactate, malate and succinate oxidases (data not shown). The data clearly show that an alternative, azurin-independent electron transfer pathway must be present in addition to the azurin-dependent one, in the alcohol oxidase electron transport system of *P. putida* HK5.

Although there is no direct evidence on how such an alternative pathway is working, some preliminary data could be obtained. In the purification of ADH IIB, besides the normal active enzyme, an additional ADH IIB fraction was found having high cytochrome *c* level and low ADH activity (Matsushita et al., unpublished). The ADH fraction was further separated into three cytochrome *c* fractions; normal ADH IIB, a newly found cytochrome *c* of 35 kDa, and a high molecular weight complex of ADH and the cytochrome. Since the 35 kDa band was heavily stained in a heme-staining SDS-PAGE, and also cross-reacted with antibody for type III ADH, the cytochrome *c* seems to be a multi-heme c cytochrome similar to the subunit II of type III ADH [1, 6]. Furthermore, the ADH-cytochrome *c* complex and the free 35 kDa-cytochrome *c* were also detected directly in the periplasmic fraction. Thus, the ADH complex may have an important functions in an azurin-independent alternative electron transport system. Although, at this moment, we do not have any evidence to show this ADH complex has such an electron transfer activity, the alternative oxidase system seems to function via ubiquinone in a manner similar to type III ADH of acetic acid bacteria, as shown in Fig 2.

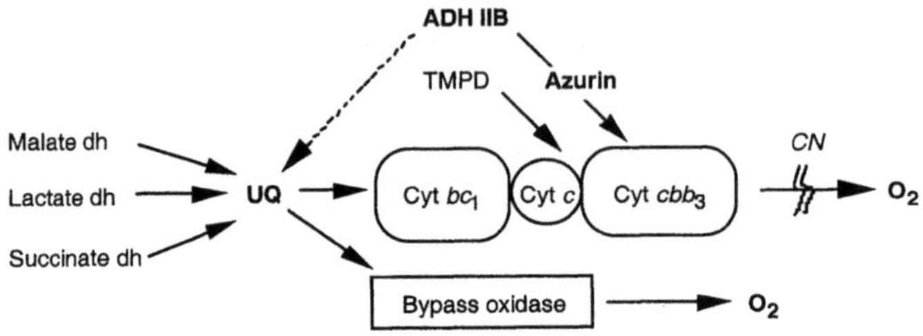

Fig. 2. Schematic model for alcohol oxidase respiratory chain in *Pseudomonas putida* HK5

Thus, there seems to be two different alcohol oxidase electron transport systems in *P. putida* HK5, one in which ADH IIB transfers electrons via azurin to a CN-sensitive terminal oxidase, and the other in which it transfers them, presumably, to ubiquinone which then transfers electrons to both CN-sensitive (via cytochrome bc_1-cytochrome cbb_3) and CN-insensitive (bypass oxidase) terminal oxidase systems.

Acknowledgment

This work is supported by Grants-in-Aid for scientific research from Ministry of Education, Science and Culture, Japan (09044228 and 10660091 to KM). We thank Prof. Scott Mathews, Washington University, St. Louis for critical reading and discussion.

References

[1] Matsushita, K., Toyama, H. & Adachi, O. (1994) *Adv. Microbial. Physiol.* 36, 247-301.
[2] Rupp, M. & Gorisch, H. (1993) *Biol. Chem. Hoppe Seyler* 369, 431-439.
[3] Schrover, J. M., Frank, J., van Wielink, J. E. & Duine, J. A. (1993) *Biochem. J.* 290, 123-127.
[4] Toyama, H., Fujii, A., Matsushita, K., Shinagawa, E., Ameyama, M. & Adachi, O. (1995) *J. Bacteriol.* 177, 2442-2450.
[5] Groen, B. W., van Kleef, M. A. & Duine, J. A. (1986) *Biochem. J.* 234, 611-615.
[6] Matsushita, K., Yakushi, T., Toyama, H., Shinagawa, E. & Adachi, O. (1996) *J. Biol. Chem.* 271, 4850-4857.
[7] Anthony, C. (1993) Methanol dehydrogenase in Gram-negative bacteria, p. 17-45. In V. L. Devidson (ed.), Principles and applications of quinoproteins. Marcel Dekker, Inc., New York.
[8] Matsushita, K., Yamashita, T., Aoki, N., Toyama, H. & Adachi, O. (1999) *Biochemistry* 38, 6111-6118.
[9] Barber, M. J., Trimboli, A. J. & McIntire, W. S. (1993) *Arch. Biochem. Biophys.* 303, 22-26.
[10] Ambler, R. P. & Brown, L. H. (1967) *Biochem. J.* 104, 784-825.

CELLULAR AND MOLECULAR BIOLOGY: MODELS WITH PLP-DEPENDENT PROTEINS

Biochemistry and Molecular Biology of Vitamin B_6
and PQQ-dependent Proteins
ed. by A. Iriarte, H.M. Kagan and M. Martinez-Carrion

Activation and transformation of cells induce translocation of ornithine decarboxylase (ODC) to the surface membrane

Marja Heiskala,[1,2] Jian Zhang,[1] Shin-ichi Hayashi,[3] Erkki Hölttä,[1] and Leif C. Andersson[1,4]
[1]Department of Pathology, Haartman Institute, and Helsinki University Hospital, University of Helsinki Haartmaninkatu 3, FIN-00014 Helsinki, Finland, [2]The R.W. Johnson Pharmaceutical Research Institute, 3210 Merryfield Row, San Diego, California 92121, USA, [3]Department of Biochemistry II, The Jikei University School of Medicine, Minato-ku Tokyo 105, Japan, [4]Institute for Oncology and Pathology, Karolinska Institute S-171 76 Stockholm, Sweden

Summary
Ornithine decarboxylase (ODC) is up regulated in cells in response to activating stimuli. We now show that the initial cytosolic increase of ODC is followed by the translocation of a fraction of the enzyme to the surface membrane. The ODC membrane translocation is mediated by a p47phox membrane-targeting motif-related sequence that has the protein kinase C phosphorylation consensus motif. The motif-mediated translocation of ODC is of fundamental importance for its biologic function.

Introduction

Ornithine decarboxylase (ODC) regulates the rate of polyamine synthesis by catalyzing the generation of putrescine and CO_2 from ornithine. ODC activity and/or polyamines are crucial to cell proliferation and transformation (1). Transformed cells display constitutively high ODC activities (1, 2) that are involved in the mechanisms leading to the transformed phenotype (2, 3). ODC regulates the level of tyrosine phosphorylation of the Src substrate p130^{cas} (4), which is involved in the formation of signaling complexes at the cytoplasmic side of the plasma membrane in focal adhesion sites (5). ODC is also associated with the transmembrane calcium and/or calmodulin signaling (6), and with activation-mediated membrane translocation of protein kinase C (PKC) (7). Given the above associations with transmembrane signaling we hypothesized that ODC might exert its activity in the vicinity of membranes.

Materials and methods

The Rat-1 tsRSVLA29 and NIH 3T3 cell line cells were used in the experiments. NIH 3T3 cells were transfected with empty vectors pCHA and pNHA, and with pCHA-ODC, pNHA-ODC-K-ras-2 adhesion tag, pCHA-ODC.C360A (mutated enzymatically inactive ODC), and pNHA-ODC.C360A-K-ras-2 adhesion tag. Rat-1 tsRSVLA29 cells were transfected with these plasmids and with the pCHA-ODC.S167A (mutated ODC lacking the phosphorylation site Ser167 that is central in the ODC membrane targeting motif) plasmid.

The sequence surrounding serine 167 of human ODC, and a fusion peptide consisting of the ODC S167 surrounding octamere linked via a proline residue to an antennapedia peptide that promotes an active cellular intake, were synthesized.

To monitor the subcellular localization of ODC by measuring the enzyme activity, cells were harvested in hypotonic buffer, intact cells and nuclei were removed, and cytosol and membrane fractions were isolated by a Beckman airfuge, 15 minutes at 30 psi. The ODC activity of the samples was measured according the method of Djurhuus (8).

For immunostaining, a combination of two monoclonal anti-ODC antibodies (9, 10) and the monoclonal antibody to the HA epitope were used as first antibodies, and biotinylated anti-mouse, FITC/avidin, biotinylated anti-avidin, and FITC/avidin were applied. The cells were inspected with a confocal microscope.

Results and Discussion

Rat-1 tsRSVLA29 cells were induced to up regulate ODC activity by introducing the following stimuli: (1) TPA; (2) hypotonic medium; (3) dialyzed human AB serum (15%) after 13 serum starvation; and (4) transfer to permissive temperature (35°C) to induce ts-v-src-driven transformation. Cells subjected to the first three stimuli displayed a transient membrane staining for ODC, whereas cells transformed at the permissive temperature showed a prolonged membrane decoration (Fig 1, 2).

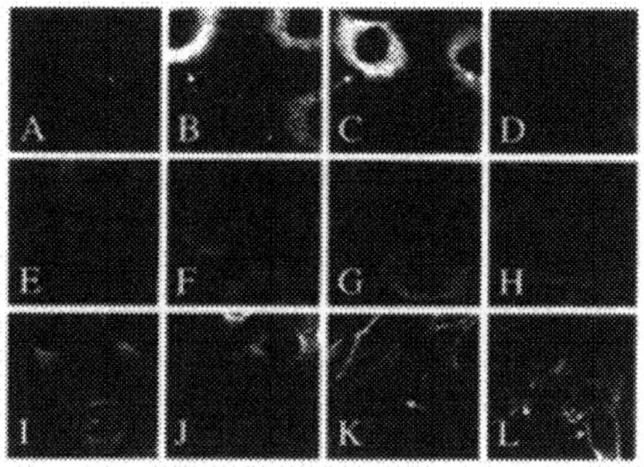

Fig. 1
Fluorescent immunostaining of ODC in quiescent Rat-1 tsRSVLA29 cells (A) and in cells activated by 100 nM TPA for 2, 4, and 24 hours (B, C, and D) is shown. As a control, similarly treated cells were stained using monoclonal anti-CD3 (OKT3) as the primary antibody (E, F, G, and H). ODC immunostaining in quiescent cells is shown (I) as well as in cells exposed to the permissive temperature for 1, 4, and 24 hours (J, K, and L), respectively. Arrowheads indicate membrane staining.

The ODC activity in the membrane and soluble fractions of activated and control Rat-1 tsRSVLA29 cells was measured, and the result was consistent with the immunostaining pattern (Fig. 2).

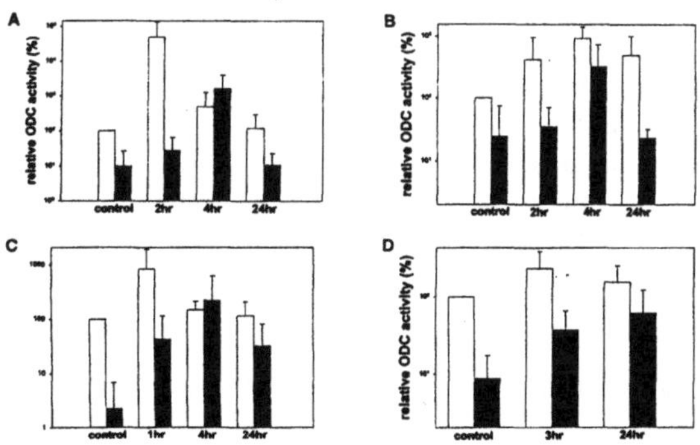

Fig. 2
The relative ODC activity/mg of protein in Rat-1 tsRSVLA29 cell soluble (open bars) and membrane fractions (filled bars), expressed as mean +/- SD of the percentage of the activity of the soluble fraction of quiescent control cells indicated by 100% (3-7 experiments/group). The mode of activation was TPA (A), hypotonia (B), 15% serum (C), and permissive temperature (D).

Serine 167 (S167) of ODC is surrounded by the sequence **AVCRLSVKFG**, which is homologous to the membrane-targeting motif of p47phox, a component of the NADPH-dependent leukocyte respiratory burst oxidase, and also corresponds to the protein kinase C (PKC) phosphorylation consensus site. By treating Rat-1 tsRSVLA29 cells with 0.2 to 2 μg/ml of the synthetic peptide RLSVKFGA linked via a proline to an antennapedia (RQIKIWFQNRRMKWKK) peptide to facilitate transmembrane transport, the temperature-induced transformation was delayed (Fig3).

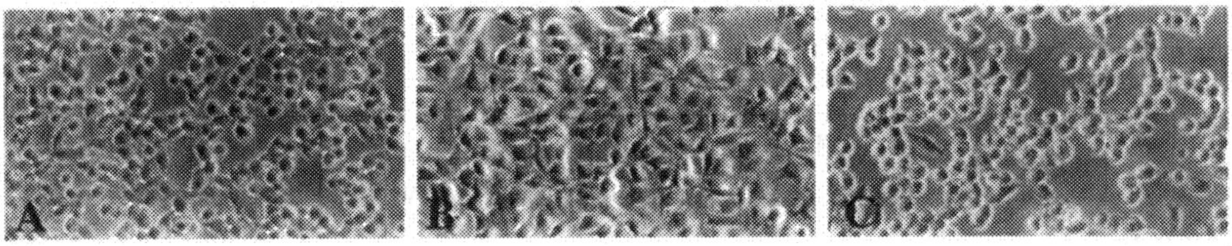

Fig. 3
Rat-1 tsRSVLA29 cells were treated for 3 hours either with the solvent (PBS, A), with 1 μg/ml of the RLSVKFGA (ODC S167 surrounding octapeptide, homologous to phox47 membrane-targeting motif)-antennapedia peptide chimera (B), or with 1 μg/ml of the control peptide-antennapedia chimera (C), whereafter, they were transferred to the permissive temperature (35°C) to induce v-src-driven transformation. The cells were photographed 5 hours later.

Measuring ODC activities from the peptide-treated Rat-1 tsRSVLA29 cells consistently revealed a dose dependent suppression of the permissive temperature-induced increase in the membrane-associated ODC activity.

To investigate the impact of constitutive ODC membrane adhesion, NIH 3T3 cells were transfected to over express wt ODC, wt ODC-K-ras-2 adhesion sequence, ODC.C360A, and ODC.C360A-K-ras-2 adhesion sequence. Morphological changes, transformation and defective cytokinesis leading to polykaryosis in the early passages, respectively, only occurred when a construct encoding for an enzymatically active wt ODC was used, whereas positive immunofluorescence for ODC was evident in the membranes of all the cells transfected with a plasmid encoding for an ODC-K-ras-adhesion chimera (Fig. 4).

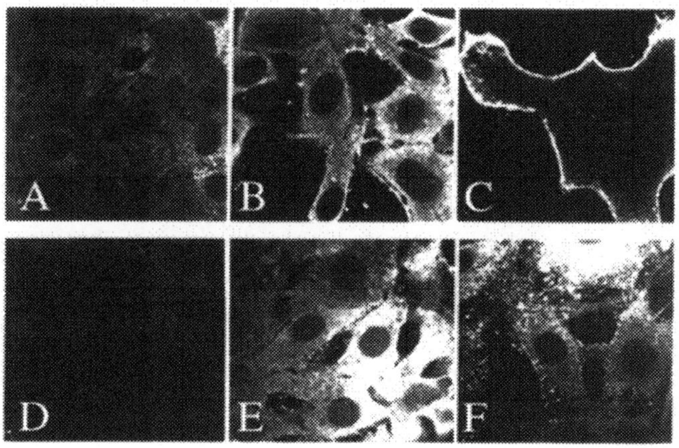

Fig. 4
NIH 3T3 cells, transfected with empty pCHA (A) and pNHA vectors (D), pCHA-ODC (B) and pNHA-ODC-K-ras-2 adhesion tag (C), pCHA-ODC.C360A (E), and pNHA-ODC.C360A-Ki-ras-2 adhesion tag (F) were immunostained using monoclonal anti-HA as the first antibody.

Cells transfected to over express the wt ODC showed elevated ODC activity in both the membrane and in the cytosolic fraction (5-fold), whereas those transfected to over express wt ODC-K-ras-2 adhesion tag displayed an extremely high activity in the membrane fraction (40-fold) indicating that the chimeric construct induced an accumulation of functional ODC in the membrane. In contrast cells expressing the enzymatically inactive ODC (ODC.C360A) showed a subnormal ODC activity in both fractions suggesting that the construct has a dominant negative impact.

The findings demonstrate that ODC-related alterations in cell morphology and growth behavior are dependent on intact ODC enzyme activity, and on ODC membrane targeting via the phosphorylation-regulated, p47phox-related membrane targeting motif surrounding the serine 167 in human ODC. Constitutive (irrelevant) membrane targeting of the active enzyme leads to defective cytokinesis. Given the importance of the enzyme activity and the phosphorylation-induced membrane translocation, the ultimate significance of the function of ODC is probably controlling the local subcellular concentration of polyamines, to support transmembrane signal transduction leading eventually to reorganization of actin cytoskeleton.

Acknowledgments

We thank Ms. Hannele Laaksonen for her technical help. This work has been supported by the Finnish Academy of Sciences, the Sigrid Juselius Foundation, the Finnish Life and Pension Insurance Companies, the Finnish Cancer Society, the Stockholm Cancer Society, and the University of Helsinki. The pictures were reproduced with the permission of the EMBO Journal.

References

1. Pegg,A.E. (1988) Polyamine metabolism and its importance in neoplastic growth and as a target for chemotherapy. *Cancer Res.,* **48**, 759-774.
2. Auvinen,M., Paasinen,A., Andersson,L.C. and Hölttä,E. (1992) Ornithine decarboxylase activity is critical for cell transformation. *Nature,* **360**, 355-358.
3. Auvinen,M., Laine,A., Paasinen-Sohns,A., Kangas,A., Kangas,L., Saksela,O., Andersson,L.C. and Hölttä,E. (1997) Human ornithine decarboxylase-overproducing NIH3T3 cells induce rapidly growing, highly vascularized tumors in nude mice. *Cancer Res.,* **57**, 3016-3025.
4. Auvinen,M., Paasinen-Sohns,A., Hirai,H., Andersson,L.C. and Hölttä,E. (1995) Ornithine decarboxylase- and ras-induced cell transformations: reversal by protein tyrosine kinase inhibitors and role of $pp130^{cas}$. *Mol. Cell. Biol.,* **15**, 6513-6525.
6. Parsons,J.T. (1996) Integrin-mediated signalling: regulation by protein tyrosine kinases and small GTP-binding proteins. *Curr. Opin. Cell Biol.,* **8**, 146-152.
7. Ginty,D.D. and Seidel,E.R. (1989) Polyamine-dependent growth and calmodulin-regulated induction of ornithine decarboxylase. *Am. J. Physiol.,* **256**, G342-G348.
8. Groblewski,G.E., Ways,D.K. and Seidel,E.R. (1992) Protein kinase C regulation of IEC-6 cell ornithine decarboxylase. *Am. J. Physiol.* **263**, G742-G749.
9. Djurhuus, R. (1981) Ornithine decarboxylase (EC 4.1.1.17) assay based upon regulation of IEC-6 cell ornithine decarboxylase. *Am. J. Physiol.,* **263**,G742-G749.
10. Matsufuji,S., Fujita,K., Kameji,T., Kanamoto,R., Murakami,Y. and Hayashi,S. (1984) A monoclonal antibody to rat liver ornithine decarboxylase. *J. Biochem.,* **96**, 1525-1530.
11. Nishiyama,M., Matsufuji,S., Kanamoto,R., Takano,M., Murakami,Y. and Shin-ichi,H. (1988) Two-step purification of mouse kidney ornithine decarboxylase. *Preparative Biochem.*, **18**, 227-238.

Biochemistry and Molecular Biology of Vitamin B_6
and PQQ-dependent Proteins
ed. by A. Iriarte, H.M. Kagan and M. Martinez-Carrion

A partially folded conformation is not the only requirement for import of mitochondrial aspartate aminotransferase.

J. R. Mattingly, Jr., C. Torella, and A. J. Yañez

School of Biological Sciences, University of Missouri–Kansas City, Kansas City, MO USA 64110-2499

Summary

Limited proteolysis and binding to GroEL were used to probe the conformations of the nascent mitochondrial aspartate aminotransferase synthesized in reticulocyte lysate and the denatured protein refolding *in vitro*. Results suggest that the conformations of the partially folded proteins are similar. The GroEL–bound, refolding precursor can be imported into mitochondria in the presence of reticulocyte lysate but not in its absence. This suggests that a partially folded conformation alone is insufficient for import and that reticulocyte lysate factors are also needed.

Introduction

Most mitochondrial proteins are synthesized in the cytosol and translocated into mitochondria [reviewed in 1]. This process can be simulated *in vitro* by synthesizing mitochondrial precursors in a cell–free extract then adding this mixture to isolated mitochondria. (See [2] for an example.) Some chemically denatured mitochondrial precursor proteins can be directly imported into mitochondria [3,4]. These precursors, however, often rapidly lose their ability to be translocated when transferred to the native conditions of the import reaction. Various cytosolic factors have been ascribed a role in maintaining mitochondrial precursors in an import competent conformation [5-10]. Nevertheless, there is no comprehensive model to explain the factors required for *in vitro* import of a particular protein, let alone the factors involved *in vivo*.

In this report, we describe efforts to develop a model system for identifying cytosolic factors necessary for the import of the mitochondrial aspartate aminotransferase precursor (pmAspAT).

Materials and Methods

^{35}S-labeled pmAspAT was synthesized in rabbit reticulocyte lysate as previously described [2]. pmAspAT–H_6Y (an engineered version of the authentic precursor that contains a C–terminal LGH_6Y tail) was expressed in *E. coli*, purified, and denatured and refolded by published procedures [11]. The purified protein was radiolabeled with $Na^{125}I$ using Pierce's Iodogen. GroEL and GroES were overexpressed in *E. coli* MZ1 using the pRE1–GroESL plasmid [12] and purified as described earlier [11]. The conditions used for limited proteolysis and analysis of the digestion have been previously published [13].

Mitochondria were isolated from rat liver according to published procedures [2] and suspended at 5 mg/ml in MESH (220 mM mannitol, 70 mM sucrose, 20 mM HEPES, 0.1 mM EDTA pH 7.5). ^{125}I pmAspAT–H_6Y, labeled to $\approx 10^6$ cpm/µg, was denatured at room temperature for 30 min at 1 mg/ml in 4 M guanidinium chloride, 100 mM HEPES, 10 mM DTT pH 7.5. The denatured protein was diluted by addition of 40–volumes of ice cold 0.4 mg/ml GroEL in 100 mM HEPES pH 7.5. One part of the resulting GroEL-^{125}I pmAspAT–H_6Y complex was added to 9 parts of reticulocyte lysate or MESH. When indicated, GroES was added at a 2:1 molar ratio relative to GroEL. Import reactions consisted of equal volumes of the diluted GroEL-^{125}I pmAspAT–H_6Y complex and mitochondria. After 30 min at 30°C, the import reaction was sampled and the remaining mitochondria were reisolated by centrifugation. The reisolated mitochondria were washed three times with 150 µl MESH then resuspended at a nominal 2.5 mg/ml. Part of the reisolated mitochondria were digested at 0°C for 30 min with 20 µg/ml trypsin, part were digested in the presence of 0.1% Triton X100, and part were untreated. The mitochondria and the unfractionated reaction sample were then denatured and analyzed by SDS PAGE and PhosphorImager autoradiography.

Results and Discussion

When purified mitochondrial aspartate aminotransferase precursor (pmAspAT) is transferred from denaturing to native conditions, it rapidly collapses to a compact state sometimes described as a "molten globule" and then slowly refolds to the native, dimeric conformation [14]. Adding this partially folded protein to mitochondria results in aggregation rather than import (data not shown). pmAspAT synthesized in reticulocyte lysate is slower to fold and is readily imported into mitochondria [2]. Reticulocyte lysate evidently has factors that slow the folding of pmAspAT and possibly maintain the partially folded translation product in a soluble, import competent conformation. Simply adding the refolding, purified protein to reticulocyte lysate does not confer import competence—it still aggregates (data not shown).

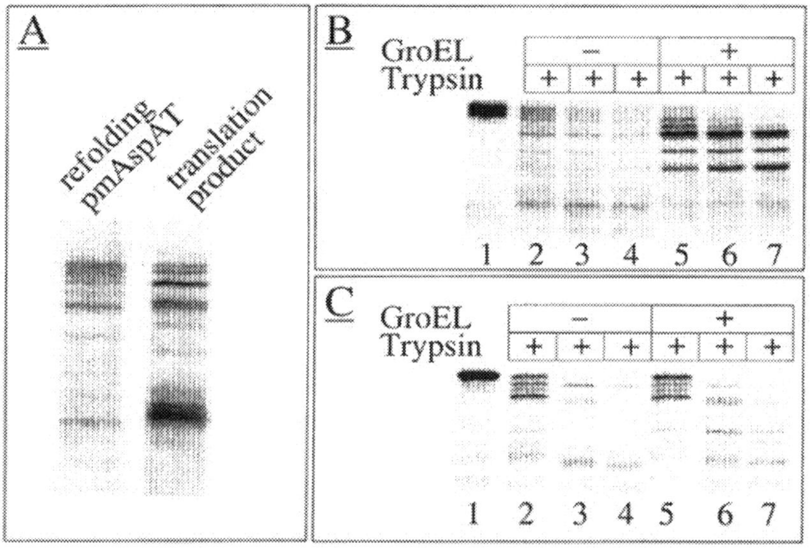

Figure 1. **Limited proteolysis of refolding pmAspAT and the pmAspAT translation product.** *Panel A* shows a comparison of the trypsin digestion products from refolding pmAspAT and the nascent protein synthesized in reticulocyte lysate. Samples were digested with trypsin as described under materials and methods. *Panel B* shows the effect of GroEL on the trypsinolysis of the refolding ^{125}I labeled pmAspAT. Note that the two bands seen in the undigested sample are produced by traces of contaminating endoproteases. Digestion times were 10 min for lanes 2 and 5, 20 min for lanes 3 and 6, and 30 min for lanes 4 and 7. *Panel C* shows the effect of GroEL on the trypsinolysis of nascent translation product. Digestion times were 0.5 min for lanes 2 and 5, 10 min for lanes 3 and 6, and 30 min for lanes 4 and 7.

A flexible conformation is a clear requirement for import, but the extent of this flexibility has yet to be defined [15]. The folding path that a protein follows during and after synthesis in the cytosol may not be the same that a denatured full–length protein experiences as it refolds *in vitro*. A protein synthesized in reticulocyte lysate might have a more flexible and extended conformation than the refolding protein. This could explain why refolding pmAspAT could not be imported into mitochondria either directly or in the presence of reticulocyte lysate. We used limited proteolysis to compare the conformations of the two proteins and test this hypothesis.

Several intermediate species are noted during trypsin digestion of refolding pmAspAT (Figure 1a), though none ever amounts to more than ≈10% of the radiolabel initially present. We assume that these species reflect regions in which potential trypsin cleavage sites are relatively inaccessible, presumably because these regions spend a significant amount of time in a compact conformation. These regions probably fluctuate between compact and unfolded conformations. This would account for the observation that they are eventually digested, given enough time or a higher trypsin concentration. Nonetheless, they may be stable enough to interfere with import. If so, one would not expect to see such protease resistant domains in the translation product. Trypsinolysis of the translation product, however, shows transient species whose sizes correspond to the same intermediates (Figure 1a). This suggests that the translation product and the refolding protein may spend time in roughly similar conformations.

Molecular chaperones can be classified according to the conformation of the unfolded proteins to which they bind. Hsp70, for example, apparently binds to polypeptides that are extensively unfolded [16]. Chaperonins such as GroEL apparently bind to partially folded polypeptides that are in compact "molten globule" conformations [16]. We used binding to GroEL as another probe of the conformation of the pmAspAT translation product. GroEL binds refolding pmAspAT tightly, forming a complex that is stable to native gel electrophoresis, glycerol gradient centrifugation, and gel filtration chromatography [11]. Trypsinolysis of GroEL–bound pmAspAT produces the same species seen in the digestion of unbound, partially–folded pmAspAT (Figure 1b), though the intermediates are more stable. GroEL similarly binds the translation product, forming a stable complex that somewhat protects the translation product from trypsinolysis (Figure 1c) and inhibits import (Figure 2). GroES releases the translation product from GroEL so that it can then be imported into mitochondria (Figure 2). This observation again suggests that the translation product and refolding pmAspAT have generally similar conformations.

We reasoned that GroEL might also present refolding pmAspAT to mitochondria in an import competent conformation. Addition of ^{125}I pmAspAT–H_6Y-GroEL to an import reaction containing 0.5 X reticulocyte lysate and mitochondria does result in some import (Figure 3). GroES is not required for import—in fact, it inhibits import (data not shown). Reticulocyte lysate, however, is required (Figure 3). We interpret these results to mean that the partially folded pmAspAT is transferred from GroEL to one or more reticulocyte lysate factors that promote the import reaction. When partially folded pmAspAT is added to lysate directly, it apparently aggregates faster than it can bind to these factors. Alternatively, partially folded pmAspAT may not be in exactly the correct conformation for binding. The inhibitory effect of GroES suggests that refolding pmAspAT might be sequestered within a GroES–GroEL cage. GroES apparently does not trap the translation product in such a cage, possibly because reticulocyte factors may remain bound to the translation product forming a complex too large to fit into the GroES–GroEL cage.

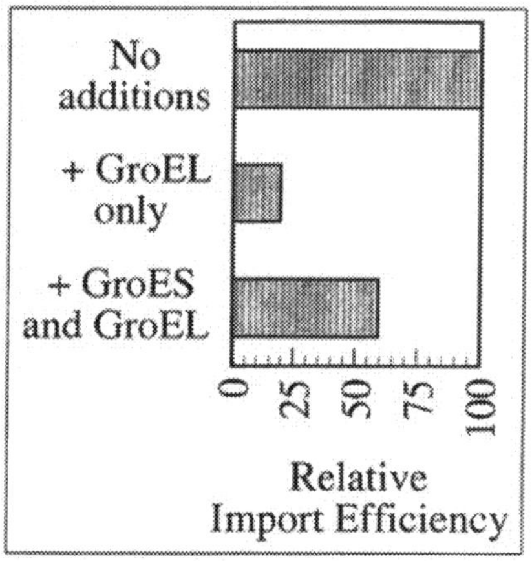

Figure 2. **Inhibitory effect of GroEL on the mitochodrial import of the pmAspAT translation product.** Conditions for import are described under Materials and Methods.

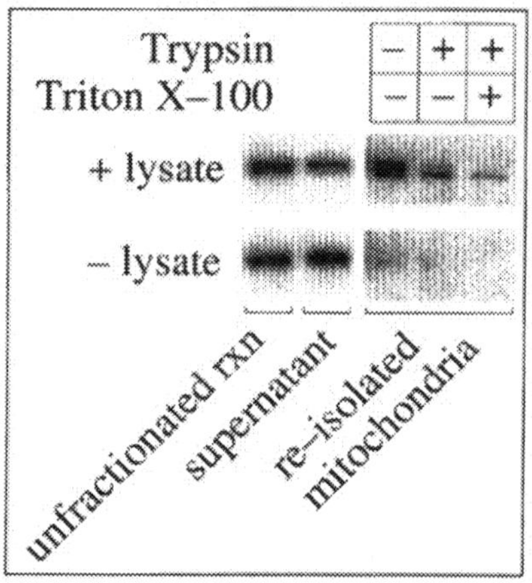

Figure 3. **Mitochondrial import of ^{125}I pmAspAT–H_6Y.** Conditions for import are described under Materials and Methods.

References

1. Neupert, W. (1997) Protein import into mitochondria. *Annual Review of Biochemistry 66*, 863-917.

2. Mattingly, J. R., Jr., Youssef, J., Iriarte, A., and Martinez-Carrion, M. (1993) Protein folding in a cell-free translation system. The fate of the precursor to mitochondrial aspartate aminotransferase. *J Biol Chem 268*, 3925-37.
3. Eilers, M., and Schatz, G. (1986) Binding of a specific ligand inhibits import of a purified precursor protein into mitochondria. *Nature 322*, 228-32.
4. Murakami, K., Amaya, Y., Takiguchi, M., Ebina, Y., and Mori, M. (1988) Reconstitution of mitochondrial protein transport with purified ornithine carbamoyltransferase precursor expressed in Escherichia coli. *J Biol Chem 263*, 18437-42.
5. Deshaies, R. J., Koch, B. D., Werner-Washburne, M., Craig, E. A., and Schekman, R. (1988) A subfamily of stress proteins facilitates translocation of secretory and mitochondrial precursor polypeptides. *Nature 332*, 800-5.
6. Murakami, H., Pain, D., and Blobel, G. (1988) 70-kD heat shock-related protein is one of at least two distinct cytosolic factors stimulating protein import into mitochondria. *J Cell Biol 107*, 2051-7.
7. Murakami, K., and Mori, M. (1990) Purified presequence binding factor (PBF) forms an import-competent complex with a purified mitochondrial precursor protein. *EMBO J 9*, 3201-8.
8. Ono, H., and Tuboi, S. (1990) Presence of the cytosolic factor stimulating the import of precursor of mitochondrial proteins in rabbit reticulocytes and rat liver cells. *Arch Biochem Biophys 277*, 368-73.
9. Sheffield, W. P., Shore, G. C., and Randall, S. K. (1990) Mitochondrial precursor protein. Effects of 70-kilodalton heat shock protein on polypeptide folding, aggregation, and import competence. *J Biol Chem 265*, 11069-76.
10. Hachiya, N., Komiya, T., Alam, R., Iwahashi, J., Sakaguchi, M., Omura, T., and Mihara, K. (1994) MSF, a novel cytoplasmic chaperone which functions in precursor targeting to mitochondria. *EMBO J 13*, 5146-54.
11. Mattingly, J. R., Jr., Iriarte, A., and Martinez-Carrion, M. (1995) Homologous proteins with different affinities for groEL. The refolding of the aspartate aminotransferase isozymes at varying temperatures. *J Biol Chem 270*, 1138-48.
12. Kamireddi, M., Eisenstein, E., and Reddy, P. (1997) Stable Expression and Rapid Purification of Escherichia Coli Groel and Groes Chaperonins. *Protein Expr Purif 11*, 47-52.
13. Mattingly, J. R., Jr., Torella, C., Iriarte, A., and Martinez-Carrion, M. (1998) Conformation of aspartate aminotransferase isozymes folding under different conditions probed by limited proteolysis. *J Biol Chem 273*, 23191-202.
14. Reyes, A. M., Iriarte, A., and Martinez-Carrion, M. (1993) Refolding of the precursor and mature forms of mitochondrial aspartate aminotransferase after guanidine hydrochloride denaturation [published erratum appears in J Biol Chem 1994 Feb 18;269(7):5480]. *J Biol Chem 268*, 22281-91.
15. Schwartz, M. P., Huang, S. H., and Matouschek, A. (1999) The structure of precursor proteins during import into mitochondria. *J Biol Chem 274*, 12759-12764.
16. Bukau, B., and Horwich, A. L. (1998) The Hsp70 and Hsp60 Chaperone Machines. *Cell 92*, 351-366.

Biochemistry and Molecular Biology of Vitamin B_6
and PQQ-dependent Proteins
ed. by A. Iriarte, H.M. Kagan and M. Martinez-Carrion

Engineering of B_6 enzymes

Sandmeier E., Mouratou B., Kasper P., Graber R., de Vries D., Mehta P., Gehring H., Christen P.

Biochemisches Institut der Universität Zürich, CH-8057 Zürich, Switzerland

Summary: To alter the reaction and substrate specificity of B_6 enzymes we substituted, guided by homology modeling, critical residues in the active site of two α-family members. The arginine-shift mutation R100T/V283R transformed tyrosine phenol-lyase into a dicarboxylic amino acid β-lyase that degrades L-aspartate to pyruvate, ammonia and formate. Aspartate aminotransferase was converted into a L-aspartate β-decarboxylase by an arginine-shift mutation (Y225R/R386A) plus a conservative substitution of a substrate-binding residue (R292K); the triple mutation I37Q/Y225Q/R292Y transformed the same enzyme into a cysteine sulfinate desulfinase.

B_6 enzymes that act on amino acid substrates are of multiple evolutionary origin and constitute four families of orthologous proteins of which the α-family is by far the largest (1). Foremost the α family shows that with similar protein scaffolds quite diverse reactions of amino acids can be catalyzed. Thus, alterations of the reaction and substrate specificity of a given enzyme by substitution of a limited number of critical amino acid residues seemed feasible. Tyrosine phenol-lyase (TPL) and aspartate aminotransferase (AspAT), two members of the α family with known 3-D structures, were subjected to oligonucleotide-directed mutagenesis of few residues in or close to the active site.

Conversion of tyrosine phenol-lyase into dicarboxylic amino acid β-lyase. The structures of the active sites of TPL and AspAT are similar; most of the residues that participate in the binding of the coenzyme and the α-carboxylate group of the substrate in AspAT (2) are conserved in the structure of tyrosine phenol-lyase (3).

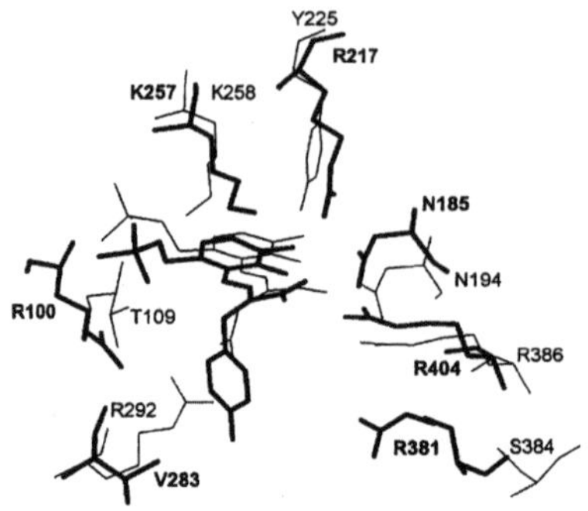

Figure 1. Superposition of the active-site structures of *C. freundii* TPL (thick lines) and *E. coli* AspAT (thin lines). The structures correspond to the quinonoid intermediates with L-tyrosine and L-aspartate, respectively, as obtained by molecular modeling and dynamics simulations (for details, see Ref. 5). The Cα atoms of K257, R217, R404 and V283 of TPL were superimposed with the Cα atoms of the corresponding K258, Y225, R386 and R292, respectively, in AspAT.

To change the substrate specificity of TPL in favor of dicarboxylic amino acids, the active-site structure of this enzyme was compared with that of AspAT (Fig.1). The specificity of AspAT for dicarboxylic amino acids and dicarboxylic 2-oxo acids seems to be based primarily on the salt bridge-hydrogen bond interaction of Arg292 with the distal carboxylate group of these substrates. The sequence identity between AspAT and TPL is too low to allow the use of standard alignment algorithms (23 % between TPL of *Citrobacter freundii* and AspAT of *Escherichia coli)*. However, superposition of their 3-D structures (3, 4) identified Val283 of TPL as the residue corresponding to Arg292 in AspAT. Another significant difference in the active sites of the two enzymes is the replacement of a residue interacting with the phosphate group of the coenzyme. Arg100 in TPL apparently corresponds to Thr109 in AspAT which is conserved among all AspATs. The substitution of Arg100 of TPL with an uncharged residue together with the introduction of an arginine residue into position 283, i.e. the arginine-shift mutation R100T/V283R, was expected to mimic the binding site for dicarboxylic substrates of AspAT. Double-mutant TPL indeed catalyzed β-elimination reactions with L-aspartate and L-glutamate as substrates (5). These new reactions do not occur with the wild-type enzyme and are only one order of magnitude slower than the reaction of L-tyrosine with the wild-type enzyme (Table 1). Apparently, dicarboxylic substrates adopt in the active site of TPL R100T/V283R a similar configuration as L-tyrosine in the wild-type enzyme and interact in a similar way with the critical residues that control reaction specificity. Dicarboxylic amino acid β-lyase is an enzyme that is not found in nature.

Table 1. Kinetic parameters for β-elimination reactions of TPL wild-type and R100T/V283R. Values were determined in 50 mM potassium phosphate pH 8.0 at 25 °C. Pyruvate production was followed with the coupled assay with lactate dehydrogenase and NADH. In control reactions with 50 μM pyridoxal-5'-phosphate but without enzyme, no activity was detected. The concentrations of the amino acid substrates were 10-200 mM for dicarboxylic amino acids and 0.2-2 mM for L-tyrosine.

Substrate	Wild-type enzyme			R100T/V283R		
	k_{cat}	K_m	k_{cat}/K_m	k_{cat}	K_m	k_{cat}/K_m
	s^{-1}	mM	$mM^{-1}s^{-1}$	s^{-1}	mM	$mM^{-1}s^{-1}$
L-Tyrosine	3.7	0.2	1.85×10^4	0.11	0.32	344
L-Aspartate	Activity below detection			0.21	54	3.9
L-Glutamate	Activity below detection			0.10	5.3	19

Conversion of aspartate aminotransferase into aspartate β-decarboxylase. Displacement of a positively charged active-site residue may be expected to alter bond angles and electron distribution, two important determinants of the reaction pathway taken by the coenzyme-substrate adduct. Again, an arginine-shift mutation converted AspAT into an L-aspartate β-decarboxylase. In AspATs, Arg386 binds the α-carboxylate group of the substrate. The triple mutation Y225R/R386A/R292K only slightly changes the topography of the active site (Fig. 2) but drastically alters the reaction specificity of the enzyme (Table 2). The various single- or double-mutant AspATs corresponding to the triple-mutant enzyme showed, with the exception of AspAT Y225R/R386A, no measurable or only very low β-decarboxylase activity. The introduction of the third mutation (R292K) reduces the aminotransferase activity without

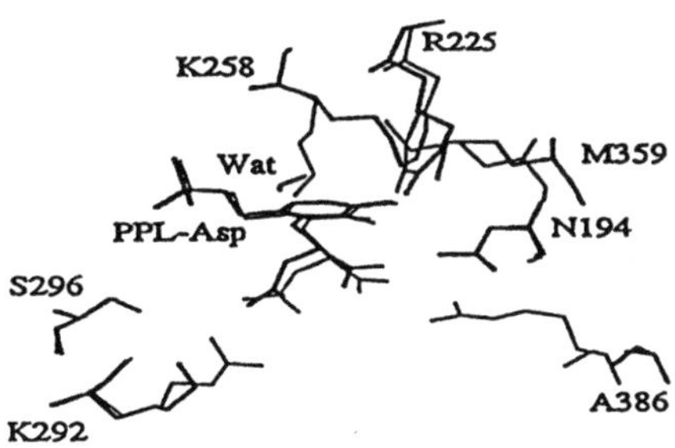

Figure 2. Active site of the 5'-phosphopyridoxyl-L-aspartate (PPL-Asp) complexes of AspAT Y225R/R292K/R386A (*thick lines*) and wild-type AspAT (*thin lines*). Wild-type AspAT assumes the closed conformation, whereas the triple-mutant stays in the open conformation. A water molecule (Wat) is found in the triple-mutant structure close to the position of a water molecule that, in mitochondrial AspAT, is assumed to carry out hydrolysis of the ketimine intermediate (from Ref. 6 and J.N. Jansonius, unpublished data).

Table 2. Aspartate β–decarboxylase activity of mutant and wild-type AspAT. The K_m values of the triple-mutant enzyme for aspartate and 2-oxoglutarate are increased in comparison to the wild-type enzymes by less than one order of magnitude.

AspAT	k_{cat}		Ratio
	β-Decarboxylation	Transamination	β-decarboxylation/ transamination
	s^{-1}	s^{-1}	
Y225R/R386A/R292K	0.08	0.01	8
Y225R/R386A	0.08	0.19	0.42
Y225R	< wild-type	0.45	
R386A	< wild-type	0.33	
R292K	$1.8 \cdot 10^{-3}$	0.5	$3.6 \cdot 10^{-3}$
Wild-type	$6 \cdot 10^{-5}$	180	$3.3 \cdot 10^{-7}$

reducing the β-decarboxylase activity (6). Similar to authentic L-aspartate β-decarboxylase, the triple-mutant enzyme produces L-alanine through direct decarboxylation of aspartate without intermediary formation of pyruvate. In comparison to wild-type AspAT, the ratio of decarboxylase to aminotransferase activity is increased 25 million-fold. AspAT is thus converted into a true aspartate β-decarboxylase which catalyzes transamination as a side reaction. Molecular dynamics simulations on the basis of the high-resolution crystal structures of the double- and triple mutant enzymes suggested that a new hydrogen bond forms between Arg225 and the imine *N* of the coenzyme adduct and reinforces the electron sink capacity of the π system of the imine and the cofactor pyridine ring to such an extent that, even after deprotonation at Cα, it remains effective enough to stabilize the carbanion produced by β-decarboxylation.

Conversion of aspartate aminotransferase into an L-cysteine sulfinate desulfinase. With the triple mutation I37Q/Y225Q/R292Y the active-site structure of AspAT was made to resemble that of dialkylglycine decarboxylase, a pyridoxal-5'-phosphate-dependent enzyme catalyzing α-decarboxylation and/or transamination, depending on the substrate. Both I37 and Y225 in AspAT were replaced by glutamine. Arg292 binds the distal carboxylate group of dicarboxylic substrates. Replacing Arg292 with tyrosine may be expected to increase the motility of the distal part of the substrate. The three mutations I37Q/Y225Q/R292Y profoundly altered the reaction

Table 3. Comparison of k_{cat} values of both β-decarboxylation and transamination with those of desulfination. Transamination of L-cysteine sulfinate was measured under single-turnover conditions.

AspAT	k_{cat} toward L-aspartate		k_{cat} toward L-cysteine sulfinate	
	β-Decarboxylation	Transamination	β-Desulfination	Transamination
	s^{-1}	s^{-1}	s^{-1}	s^{-1}
I37Q/Y225Q/R292Y	0.07	0.13	0.5	0.05
Wild-type	$6x10^{-5}$	530	0.05	512

specificity of AspAT (Table 3). The ratio of β-decarboxylase to transaminase activity was increased by 6 orders of magnitude, while transaminase activity was decreased by 3 orders of magnitude. L-Cysteine sulfinate was desulfinated directly to L-alanine at a high rate that is only 30 times slower than that of authentic cysteine sulfinate desulfinase (7). The rate of desulfination is 10 times faster than that of transamination indicating very efficient reprotonation at Cα instead of C4'.

Concluding remarks. Alterations in bond angles and electron repartition within the pyridoxal-5'-phosphate-substrate adduct as brought about by amino acid substitutions in the active site of B_6 enzymes may leave their catalytic potency at least in part intact while modifying reaction and substrate specificity. The reaction specificity of B_6 enzymes appears to be achieved not only by accelerating the specific reaction but also by preventing potential side reactions (8,9). Similar amino acid substitutions might have occurred in early phases of molecular evolution when the ancestor proto-enzyme of one of the B_6 enzyme lineages diverged into different reaction and substrate specific enzymes. Subsequent amino acid exchanges including non-active-site residues (10) improved its catalytic efficacy.

References

1. Mehta, P.K. and Christen P. (2000) The molecular evolution of pyridoxal-5'-phosphate-dependent enzymes. *Advan. Enzymol.*, in press.
2. Kirsch, J.F., Eichele, G., Ford, G.C., Vincent, M.G., Jansonius, J.N., Gehring, H. and Christen, P. (1984) Mechanism of action of aspartate aminotransferase proposed on the basis of its spatial structure. *J. Mol. Biol.* 174: 497-525.

3. Antson, A.A., Demidkina, T.V., Gollnick, P., Dauter, Z., Von Tersch, R.L., Long, J., Berezhnoy, S.N., Phillips, R.S., Harutyunyan, E.H. and Wilson, K.S. (1993) Three-dimensional structure of tyrosine phenol-lyase. *Biochemistry* 32: 4195-4206.
4. Jäger, J., Moser, M., Sauder, U. and Jansonius, J.N. (1994) Crystal structures of *Escherichia coli* aspartate aminotransferase in two conformations. *J. Mol. Biol.* 239: 285-305.
5. Mouratou, B., Kasper, P., Gehring, H. and Christen, P. (1999) Conversion of tyrosine phenol-lyase to dicarboxylic amino acid β–lyase, an enzyme not found in nature. *J. Biol. Chem.* 274: 1320-1325.
6. Graber, R., Kasper, P., Malashkevich, V.N., Strop, P., Gehring H., Jansonius, J.N. and Christen P. (1999) Conversion of aspartate aminotransferase into an L-aspartate β–decarboxylase by a triple active-site mutation. *J. Biol. Chem.* 274: 31203-31208.
7. Mihara, H., Kurihara, T., Yoshimura, T., Soda, K. and Esaki N. (1997) Cysteine sulfinate desulfinase, a NIFS-like protein of *Escherichia coli* with selenocysteine lyase and cysteine desulfurase activities. *J. Biol. Chem.* 272: 22417-22424.
8. Rétey, J. (1990) Reaction selectivity of enzymes through negative catalysis or how enzymes work with highly-reactive intermediates. *Angew. Chem. Int. Ed. Engl.* 29: 355-361.
9. Vacca, R.A., Giannattasio, S., Graber, R., Sandmeier, E., Marra, E. and Christen P. (1997) Active-site Arg → Lys substitutions alter reaction and substrate specificity of aspartate aminotransferase. *J. Biol. Chem.* 272: 21932-21937.
10. Oue, S., Okamoto, A., Yano, T. and Kagamiyama, H. (1999) Redesigning the substrate specificity of an enzyme by cumulative effects of the mutations of non-active site residues. *J. Biol. Chem.* 274: 2344-2349.

Biochemistry and Molecular Biology of Vitamin B_6
and PQQ-dependent Proteins
ed. by A. Iriarte, H.M. Kagan and M. Martinez-Carrion

Association of Newly Synthesized Mitochondrial Aspartate Aminotransferase with Cytosolic Factors

B. Lain, C. A. Tanase, A. Iriarte, A. E. Johnson* and M. Martinez-Carrion

School of Biological Sciences, University of Missouri-Kansas City, Kansas City, MO 64110
*College of Medicine, Texas A&M University, College Station, TX 77843, USA

Summary

To investigate the possible interaction of mitochondrial aspartate aminotransferase (mAAT) with cytosolic components, a photoreactive probe was incorporated co-translationally into mAAT chains synthesized in a cell-free extract. Irradiation of the translation reaction results in the appearance of four crosslinked species containing radioactive mAAT with apparent molecular weights ranging from 70 to 200 kDa. These results suggest that newly synthesized mAAT is associated with as yet unknown factors from the cell extract. Fractionation of a fresh translation reaction by gel filtration chromatography confirms that indeed a large fraction of mAAT translation product is present as high molecular weight species.

Introduction

Cytosolic (cAAT) and mitochondrial (mAAT) aspartate aminotransferases share a high degree of sequence similarity and almost identical three-dimensional structures. Both of them are encoded by the nuclear DNA and synthesized in the cytoplasm, mAAT as a precursor with an N-terminal extension or presequence peptide (pmAAT) that targets the protein to mitochondria (1). However, following synthesis *in vitro* in a cell-free extract, cAAT folds very rapidly whereas it takes several hours for pmAAT to acquire protease resistance (2). This different behavior may result from a selective interaction of each isozyme with the members of the network of cytosolic proteins that participate in folding and translocation processes. This network of cytosolic proteins is formed by molecular chaperones, folding catalysts and targeting factors. Their interaction with nascent chains may prevent aggregation or other undesirable interactions of the growing polypeptides as well as the premature folding of those proteins destined to other cellular compartments.

In this work, we have used two alternative approaches to investigate the possible association of newly synthesized pmAAT with cytosolic components. One of them involves the incorporation of a photocrosslinker during translation of pmAAT in a cell-free extract. Formation of covalent crosslinked products between nascent pmAAT and cytosolic proteins facilitates the detection of potentially short-lived or low affinity complexes. We also describe the partial purification of high molecular weight species containing newly translated full-length pmAAT using conventional gel filtration chromatography. The fractions obtained were used to dissect the requirements for translocation of these putative complexes into mitochondria.

Materials and Methods

Yeast $tRNA^{Lys}$ was purified and aminoacylated with ^{14}C-Lys as previously described (3). Lys- tRNA was selectively modified at the ε amino group of Lys with the N-hydroxysuccinimide ester of 5-azido-2-nitrobenzoic acid (NOS-ANB) (3, 4) to yield photoreactive εANB-^{14}C-Lys-tRNA. *In vitro* translation of pmAAT mRNA was performed using Promega Flexi-Lysate in the presence of 2 μCi/μl of ^{35}S-methionine and 30 pmoles of εANB-Lys-tRNA. After incubation for 30 min at 30° C in the dark, cycloheximide was added to stop protein synthesis, and the samples were irradiated for 15 min using a 500 W mercury arc lamp (Oriel) through a filter combination that provides a 300-400 nm band pass. After photolysis, samples were analyzed by gradient SDS-PAGE (10-15% acrylamide) either directly or after immunoprecipitation using antibodies raised against the proteins of interest. The radioactivity on the dried gels was detected using a Molecular Dynamics PhosphorImager™.

Samples of ^{35}S-labeled pmAAT freshly translated in rabbit reticulocyte lysate (RRL) were fractionated in a Superose 12 column (1 x 30 cm, Pharmacia) equilibrated and eluted with 50 mM potassium phosphate buffer, pH 7.5 containing 150 mM NaCl. Elution of ^{35}S-pmAAT was monitored by analysis of the collected fractions by SDS-PAGE and autoradiography.

Results and Discussion

We previously reported that pmAAT interacts with the cytosolic chaperone hsp70 very early during translation (5). The co-immunoprecipitation technique that we used relies on the affinity of hsp70 for mAAT and makes difficult the detection of transient and/or very short-lived complexes. To investigate whether there are other cytosolic components that associate with nascent pmAAT, we have used a photo-crosslinking approach. The photoreactive crosslinker ANB has been successfully incorporated during *in vitro* translation of several proteins (3, 4, 6). The modified Lys-tRNA containing ANB, when added to a cell-free extract such as RRL or wheat germ extract, allows for the incorporation of the crosslinker at any of the lysyl residues within the sequence of the protein being translated. Upon irradiation at 330 nm, ANB generates a nitrene that will covalently react with any protein that associates closely with the translated chain and is within reach of the probe.

Full-length pmAAT has 439 amino acids with 12 methionine and 29 lysine residues (1). Translation of pmAAT mRNA in RRL in the presence of ^{35}S-methionine (to facilitate detection of the translation product) and εANB-Lys-tRNA yielded a prominent radioactive band of the correct molecular weight (45 kDa). Although εANB-Lys-tRNA must compete for incorporation with endogenous Lys-tRNA in the translation, we found that approximately 1 of every 4 lysines incorporated into pmAAT was εANB-Lys (results not shown; 4). Upon irradiation of full-length pmAAT, we detected the presence of four crosslinked species with apparent molecular masses ranging from 70 to 200 kDa (Fig.1, lane 3). In addition, some radioactive material was always found at the top of the separating gel suggesting the presence of even larger crosslinked species. By contrast, no crosslinked radioactive adducts were observed when ANB was chemically inactivated by addition of 100 mM dithiothreitol before photolysis or when the samples were not irradiated (Fig. 1, lane 2). These crosslinked species contained pmAAT since they could be immunoprecipitated using anti-mAAT antibodies (Fig. 1, lane 3) but not with preimmune serum (not shown). Antibodies against mammalian hsp70 or yeast hsp40 (YDJ1P) did not immunoprecipitate the crosslinked products (Fig. 1, lanes 4 and 5). These negative results suggest that these chaperones are probably not associated with full-length pmAAT translation product, although we can not discard the possibility that they might be present but not accessible to the reactive crosslinking probe.

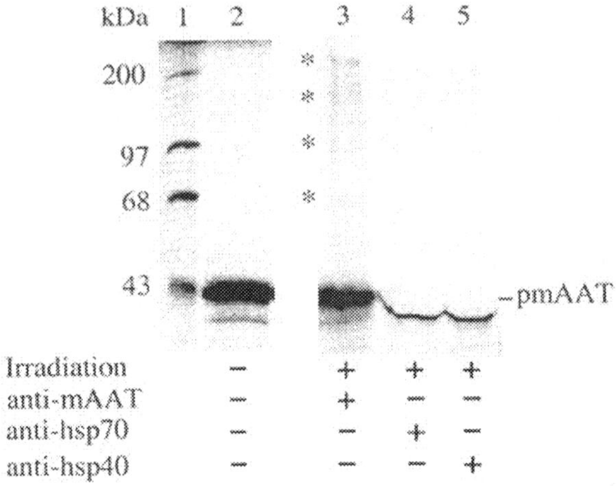

Figure 1. Crosslinking of *in vitro* synthesized pmAAT. pmAAT was synthesized in RRL in the presence of εANB-Lys-tRNA and ^{35}S-methionine and then irradiated as described under Materials and Methods. Samples were analyzed by SDS-PAGE and autoradiography. Asterisks indicate the position of the putative crosslinked species. Lane 1, ^{14}C-labeled molecular weight markers.

Several additional observations support the association of a newly-translated ^{35}S-pmAAT chain with proteins in the lysate implied by the results from the crosslinked experiments. Most of the radiolabeled protein was recovered in the retentate after ultrafiltration of the translation reaction on microcon YM-100 (100,000 molecular weight cut-off). Further fractionation of this retentate by gel filtration chromatography revealed the presence of two pmAAT peaks. The elution volume of the main radioactive protein peak was 10 ml, which corresponds roughly to an apparent molecular weight of 400 kDa. The material recovered in these fractions could be imported into mitochondria, although only after supplementation with RRL. A small fraction of pmAAT chains was found in a later peak eluting at 13 ml, which coincides with the elution volume of native dimeric pmAAT from the same column (97-kDa molecular weight). Only a very small portion of this material was competent for import into mitochondria. Together these findings suggest that this 13-ml fraction probably represents partially folded dimers that have escaped their initial interactions with the lysate chaperones. Similar chromatographic results were reported earlier for the fractionation of chicken pmAAT expressed in RRL or synthesized in chicken embryo fibroblasts (7).

In summary, newly synthesized pmAAT chains appear to associate with other proteins in the lysate forming complexes that can be detected by crosslinking and are stable enough to withstand isolation by size exclusion chromatography. However, import into mitochondria of pmAAT in these putative complexes still needs the presence of additional proteins, most likely import factors that are present in the cytosolic-like cell extract used for translation of the protein. The identification and characterization of these components are in progress.

Acknowledgments

This work was supported by funds from NIH Grants GM-38341 (M.M.C.) and GM-26494 (A.E.J.), and the Robert A. Welch Foundation (A.E.J.).

References

1. Mattingly, J. R., Jr., Rodriguez-Berrocal, F. J., Gordon, J., Iriarte, A. and Martinez-Carrion, M. (1987) Molecular Cloning and *In Vivo* Expression of a Precursor to Rat Mitochondrial Aspartate Aminotransferase. *Biochem. Biophys. Res. Commun. 149*: 859-865.
2. Mattingly, J. R., Jr., Iriarte, A. and Martinez-Carrion, M. (1993) Structural Features Which Control Folding of Homologous Proteins in Cell-free Translation Systems. *J. Biol. Chem. 268*: 26320-26327.
3. Crowley, K. S., Reinhart, G. D. and Johnson, A. E. (1993) The Signal Sequence Moves through a Ribosomal Tunnel into a Noncytoplasmic Aqueous Environment at the ER Membrane Early in Translocation. *Cell 73*: 1101-1115.
4. Krieg, U. C., Johnson, A. E. and Walter, P. (1987) Protein Translocation across the Endoplasmic Reticulum Membrane: Identification by Photocross-Linking of a 39-kD Integral Membrane Glycoprotein as Part of a Putative Translocation Tunnel. *J. Cell Biol. 109*: 2033-2043.
5. Lain, B., Iriarte, A. and Martinez-Carrion, M. (1994) Dependence of the Folding and Import of the Precursor to Mitochondrial Aspartate Aminotransferase on the Nature of the Cell-free Translation System. *J. Biol. Chem. 269*: 15588-15596.
6. Do, H., Falcone, D., Lin, J., Andrews, D. W. and Johnson, A. E. (1996) The Cotranslational Integration of Membrane Proteins into the Phospholipid Bilayer Is a Multistep Process. *Cell 85*: 369-378.
7. Behra, R. and Christen, P. (1986) In Vitro Import into Mitochondria of the Precursor of Mitochondrial Aspartate Aminotransferase. *J. Biol. Chem. 261*:257-263.

Biochemistry and Molecular Biology of Vitamin B_6
and PQQ-dependent Proteins
ed. by A. Iriarte, H.M. Kagan and M. Martinez-Carrion

Expression and Purification of Serine Palmitoyltransferase

Hiroko Ikushiro, Hideyuki Hayashi, and Hiroyuki Kagamiyama.

Department of Biochemistry, Osaka Medical College, Takatsuki, Osaka 569-8686, Japan

Summary

Serine palmitoyltransferase (SPT, EC 2.3.1.50) is the key enzyme in sphingolipid biosynthesis, and catalyzes the PLP-dependent condensation of L-serine and palmitoyl coenzyme A to 3-ketodihydrosphingosine. Recently, two different genes, *LCB1* and *LCB2*, have been isolated, and it has been suggested that the SPT enzyme is composed of two subunits named LCB1 and LCB2. We constructed a co-expression plasmid for LCB1 lacking N-terminal and His6-tagged LCB2 lacking N-terminal. Both proteins were expressed in *Escherichia coli* and LCB1 was co-purified with the His6-tagged LCB2 on metal-immobilized affinity resin. This indicates that LCB1 and LCB2 expressed in *E. coli* form a heterooligomer.

Introduction

Sphingolipids are ubiquitous eukaryotic membrane components with a hydrophobic segment consisting of a long-chain base (LCB) that is N-fatty acylated and linked to various polar head groups. Sphingolipids are known to play roles in general membrane functions, cell-to-cell recognition, regulation of cell growth, differentiation, oncogenesis, and immune functions. Recent results implicate sphingolipid breakdown products, sphingosine, sphingosine-1-phosphate, lysosphingolipids, and ceramide, as second messengers in incompletely characterized signal transduction pathways (1).

Serine palmitoyltransferase (SPT; EC 2.3.1.50) catalyzes the condensation reaction of L-serine with palmitoyl coenzyme A to generate 3-ketodihydrosphingosine. This reaction is the first unique and committed step in sphingolipid biosynthesis, utilizing substrates that are shared by other pathways, and has a lower activity than other enzymes involved in sphingolipid biosynthesis. Therefore, SPT is thought to be a rate-determining enzyme in the sphingolipid synthetic pathway and, accordingly, a key enzyme regulating the cellular sphingolipid content. However, the actual mechanisms for the regulation remain unclear because of the incomplete knowledge of the properties of the biosynthetic enzymes. The eukaryotic SPT has been known as a membrane-bound protein enriched in the endoplasmic reticulum (2). Recently, genetic studies have shown that two different genes, *LCB1* and *LCB2*, are required for SPT activity in the yeast *Saccharomyces cerevisiae* (3, 4, 5), and the isolation of the mammalian homologs of the *LCB* genes from CHO, mouse and human cells have been reported (6, 7, 8). It has been suggested that the SPT enzyme is composed of two subunits named LCB1 and LCB2.

Snell and co-workers recognized that SPT requires pyridoxal phosphate (PLP) for activity (9). LCB2 has been implicated to be responsible for catalysis because its amino acid sequence is similar to aminolevulinate synthase and it carries a lysine residue that is believed to form a Schiff base with PLP. There is a high sequence homology between LCB1 and LCB2, but LCB1 doesn't contain the PLP-binding motif in its protein sequence. LCB1 is thought to be necessary for SPT catalytic activity, regulation of SPT activity, self-interaction, or interaction with other subunit(s). As a first step to elucidate the reaction mechanism of SPT and the functions of LCB1 and LCB2, we tried to construct an expression system of SPT in *E. coli*.

Material and Methods

Cloning of the mouse *LCB1* and *LCB2* cDNAs. Reverse-transcription PCR of mouse mRNA was performed with the oligonucleotides mLCB1s (5'-GACTGTACGAAGCTTCA-TATGGCGACAGTGGCGGAGCAGTGGGTGCTGGTG-3') and mLCB1as (5'-GACTG-TACGGTCGACCTACAACAGCACAGCCTGGGCCGCCTCCCTGAT-3') for the mouse LCB1 gene and mLCB2s (5'-GACTGTACGGAATTCCATATGCGGCCGGAGCCCGGA-GGCTGCTGCTGCCGC-3') and mLCB2as (5'-GACTGTACGGTCGACTCAGTCTTCT-GTCTCTTCATAGGTAGTCTC-3') for the mouse LCB2 gene. These PCR fragments were subcloned into pUC118 vectors and sequenced.

Plasmid constructs for expression in *E. coli*. *LCB1* and *LCB2* cDNA were excised with *Nde*I/*Sal*I from the respective pUC vectors and ligated into pET vectors (Qiagen). To construct the expression plasmid carrying both *LCB1* and *LCB2* genes and the plasmids with single truncated *LCB* gene, PCR-based methods were employed. The vectors for glutathione S-transferase (GST)-fusion LCB2 proteins were also constructed in a similar way using a pGEX vector (Pharmacia Biothech).

Western Blot Analysis. Purified inclusion bodies of each LCB protein were used as antigens to immunize rabbits. The specific antibodies were purified from the antisera by affinity chromatography on the antigen-immobilized columns. For Western blot analysis, protein samples were incubated in SDS-sample buffer at 90 °C for 3 min, separated by SDS-polyacrylamide gel electrophoresis (10% acrylamide), and transferred to a Immobilone^{TM}-P^{SQ} transfer membrane (MILLIPORE). After blocking the blotted membrane, the LCB1 and LCB2 proteins on the membrane were detected by using the rabbit anti-mLCB1 antibody and anti-mLCB2 antibody, respectively, as the primary antibody and an alkaline phosphatase-conjugated goat anti-rabbit IgG (Bio-Rad) as the secondary antibody.

Purification of recombinant protein. Immobilized metal affinity chromatography was performed by using TALON^{TM} metal affinity resin (CLONTECH) according to the manufacturer's protocol. The recovered fractions were analyzed for distribution of the LCB1 and LCB2 proteins by Western blotting with anti-His6 IgG (CLONTECH) or anti-FLAG IgG (Kodak), and assayed for SPT activity.

Assay. Incorporation of L-[3-^{14}C] serine (2.07GBq/mmol; Amersham) into 3-ketodihydrosphingosine was assayed as described below. The assay was performed in a 0.1

ml reaction mixture containing 50 mM HEPES (pH 7.5), 0.1 mM EDTA, 5 mM dithiothreitol, 10 mM PLP, 2 mM serine (10 kBq), 0.4 mM palmitoyl coenzyme A. The reaction was started by the addition of substrates and incubated at 37 °C for 30 min. The reaction was terminated by the addition of 0.1 M NH_4OH. Lipids were extracted twice with 0.3 ml $CHCl_3/CH_3OH$ (4:1 (v/v)) and separated by TLC with a solvent system $CHCl_3/CH_3OH$/0.1 M NH_4OH (40:10:1 (v/v)). Radioactive lipids on the TLC plates were visualized by autoradiography using a BAS2000 Image Analyzer (Fuji).

Results and Discussion

Expression of SPT. Both SPT genes, *LCB1* and *LCB2*, were cloned by reverse-transcription PCR of mouse brain mRNA. Overexpression of the entire coding region of *LCB1* formed inclusion bodies of the protein. Another construct, LCB1 lacking N-terminal membrane anchor region (Δ3-42; His6-tagged) was well expressed in a soluble form and purified by a metal affinity column chromatography. Only one construct of LCB2 lacking N-terminal 125 amino acids that contains membrane anchor region (Δ2-125; His6-tagged), was expressed in partially soluble form. All the GST-fusion constructs of LCB2 made inclusion bodies. Next, we constructed a co-expression plasmid for LCB1 (Δ3-42; FLAG-tagged) and LCB2 (Δ2-125; His6-tagged) (Figure 1-A). Both proteins were expressed together in soluble forms and LCB1 without His6-tag was co-purified with His6-tagged LCB2 on metal affinity resin (Figure 1-B). This indicates that LCB1 and LCB2 expressed in *E. coli* interact with each other. Although the spectrum of the purified protein showed Schiff-base formation, the SPT activity could not be detected under this condition.

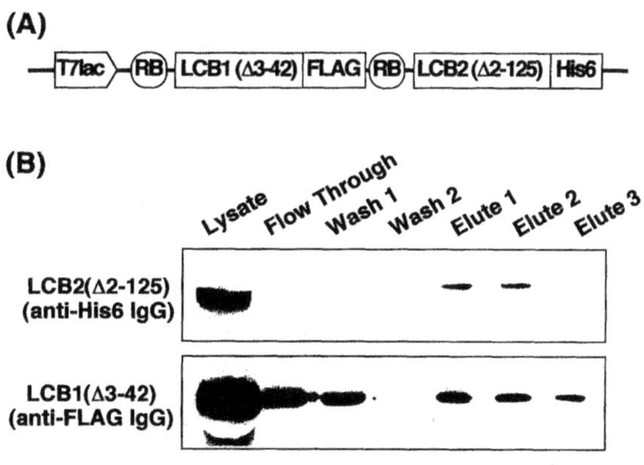

Figure 1: (A) A co-expression plasmid for N-terminal lacking LCB1 protein with a FLAG-tag and N-terminal lacking LCB2 protein with a His6-tag. In this vector, a ribosome binding site and the *LCB2* gene were ligated immediately after the stop codon of the *LCB1* gene. (B) Co-purification of the LCB proteins by a metal affinity column chromatography. An equivalent volume of each fraction was loaded onto each lane of SDS-PAGE, and were analyzed by Western blotting with an anti-His6-IgG (*upper panel*) or an anti-FLAG IgG (*lower panel*).

Preparation of polyclonal antibodies against LCB1 and LCB2. LCB1 (full-length; His6-tagged) and LCB2 (Δ2-125; His6-tagged), were used as antigens to immunize rabbits. After three times boosts at two-week intervals with 300 μg recombinant protein, the antibody titer reached 1:100,000. Using purified specific antibodies, Western blot analysis of the microsomes from mouse tissues were carried out. Both LCB proteins were found in all tissues examined. The tissue distribution of the LCB1 and LCB2 proteins showed an essentially identical pattern in mouse. Relative high levels of the LCB proteins were found in uterus, testis, lung, and kidney as shown in Figure 2-A. The activities of SPT were paralleled by the expression level of the enzyme (Figure 2-B). The exact relationship between the level of the LCB proteins and the specific activity of SPT in tissues requires further study.

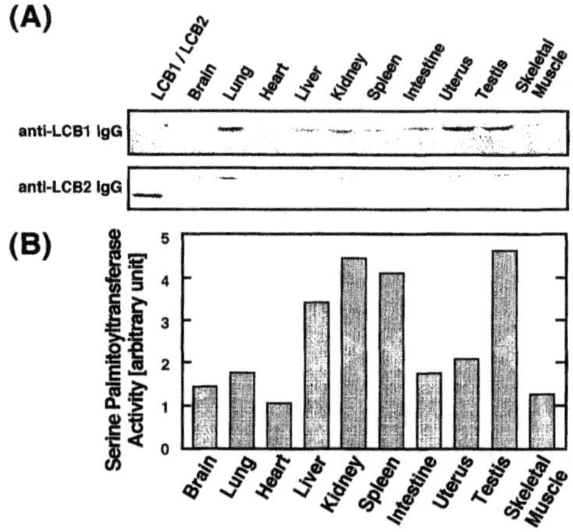

Figure 2: Expression of SPT in mouse tissues. (A) Western blot analysis of LCB proteins in the microsomes from mouse tissues (*upper panel*; anti-LCB1-IgG, *lower panel*; anti-LCB2 IgG). The left lane in each panel showed LCB1 (*upper*) or LCB2 (*lower*) expressed in *E. coli*. (B) SPT activities in the microsomes from mouse tissues.

References

1. Merrill, A. H., Jr., and Jones, D. D. (1990) *Biochim. Biophys. Acta* **1044**, 1-12
2. Mandon, E. C., Ehses, I., Rother, J., van Echten, G., and Sandhoff, K. (1992) *J. Biol. Chem.* **267**, 11144-11148
3. Buede, R., Rinker-Schaffer, C., Pinto, W. J., Lester, R. L., and Dickson, R. C. (1991) *J. Bacteriol.* **173**, 4325-4332
4. Nagiec, M.M., Lester, R. L., and Dickson, R. C. (1996) *Gene (Amst.)* **177**, 237-241
5. Nagiec, M.M., Baltisberger, J.A., Wells, G.B., Lester, R. L., and Dickson, R. C. (1994) *Proc. Natl. Acad. Sci. U. S. A.* **91**, 7899-7902
6. Hanada, K., Hara, T., Nishijima, M., Kuge, O., Dickson, R. C., and Nagiec, M. M. (1997) *J. Biol. Chem.* **272**, 32108-32114
7. Weiss, B., and Stoffel, W., (1997) *Eur. J. . Biochem.* **249**, 239-247
8. Hanada, K., Hara, T., Fukusawa, M., Yamaji, A., Umeda, M., and Nishijima, M. (1998) *J. Biol. Chem.* **273**, 33787-337947.
9. Braun, P., E., and Snell, E. E. (1967) *Proc. Natl. Acad. Sci. U. S. A.* **58**, 298-303

DEHYDRATASES, LYASES AND OTHER PLP-DEPENDENT PROTEINS

Biochemistry and Molecular Biology of Vitamin B_6
and PQQ-dependent Proteins
ed. by A. Iriarte, H.M. Kagan and M. Martinez-Carrion

5-Aminolevulinate Synthase: Pre-Steady State Reaction and Functional Role of Specific Active Site Residues.

Gloria C. Ferreira

Department of Biochemistry and Molecular Biology, College of Medicine, Institute for Biomolecular Science, and H. Lee Moffitt Cancer Center and Research Institute, University of South Florida, Tampa, Florida 33612, USA.

Summary

5-Aminolevulinate synthase (ALAS), the first enzyme of the heme biosynthetic pathway in non-plant eukaryotes and some bacteria, catalyzes the condensation of glycine and succinyl-CoA. Rapid scanning stopped flow experiments of murine erythroid ALAS demonstrate that reaction with glycine plus succinyl-CoA results in a pre-steady state burst of quinonoid intermediate formation. Reaction of either glycine or 5-aminolevulinate (ALA) with ALAS is slow and approximates k_{cat}. The turnover appears to be limited by release of the ALA product or a conformational change associated with ALA release. The function of specific active site residues (*e.g.*, K313, D279 and R439) has been established in relation to the ALAS catalytic mechanism.

Introduction

Heme, an iron-containing tetrapyrrole, is the prosthetic group for a wide range of proteins. In animals, heme is synthesized in all nucleated cells, but predominantly in differentiating erythrocytes, where requirements are greater due to the synthesis of hemoglobin. The heme biosynthetic pathway in animal cells comprises eight enzymes, with the first and the three terminal enzymes being mitochondrial while the intervening four are cytosolic. 5-Aminolevulinate synthase (ALAS; E.C. 2.3.1.37), a pyridoxal 5'-phosphate (PLP)-dependent enzyme, catalyzes the first and regulatory step of the heme biosynthetic pathway, the condensation of glycine with succinyl-coenzyme A (CoA) to yield 5-aminolevulinate (ALA), carbon dioxide and CoA [1-3]. There are two ALAS isozymes encoded by separate genes. ALAS1, the housekeeping isoform, is responsible for providing heme for respiratory cytochromes and other hemoproteins, whereas the erythroid-specific ALAS2 isoform ensures that the large heme requirements in hemoglobin synthesis are met during erythroid cell differentiation [4]. In humans, the ALAS1 gene is located on chromosome 3, while the ALAS2 gene is on X chromosome. Molecular defects in the ALAS2 gene are associated with X-linked sideroblastic anemia, a hypochromic and microcytic anemia characterized by iron deposition in erythroblast mitochondria [5]. In this article the emphasis is on the ALAS reaction mechanism in relation to the architecture of the active site.

ALAS Catalytic Mechanism: Pre-Steady State Reaction

The mechanism of ALAS catalysis appears to be similar to those of other PLP-dependent enzymes [6,7]. The PLP cofactor is bound covalently to an active site lysine, forming an internal aldimine (Fig. 1 - **I**) [8], which is subsequently displaced by the incoming amino acid substrate to form an external aldimine (Fig. 1 - **II**). The first step following formation of the external aldimine is the specific, heterolytic cleavage of one of the bonds to the α-carbon of the substrate, leading to the resonance stabilized quinonoid intermediate (Fig. 1 - **III**).

Initial steady state kinetic analysis of the reaction catalyzed by *Rhodobacter spheroides* ALAS revealed an ordered bi-bi mechanism, in which glycine binds before succinyl-CoA, and ALA is dissociated after carbon dioxide and CoA [9]. However, the mechanistic and stereochemical details of the ALAS-catalyzed reaction only began to be unraveled with the pioneering work of Jordan and colleagues using radiolabeled glycine and ALA with ALAS isolated from *R. spheroides* [10-12]. Their studies established that ALAS catalyzes stereospecific removal of the pro-*R* proton of glycine [10], and that the pro-*S* proton of glycine occupies the S-configuration at C-5 of ALA [11]. The evidence that the pro-*S* proton of glycine remains in the pro-S position of C-5 of ALA supported the proposal that the decarboxylation of the 2-amino-3-keto-adipate intermediate (Fig. 1 - **IV**) occurs on the enzyme surface and not free in solution. Further, the proposal that the 2- amino-3-keto-adipate is an intermediate, which is catalytically decarboxylated, gained support from the fact that ALAS catalyzes the removal of the pro-*R* proton at C-5 of ALA [12]. In essence, the radiolabeling experiments provided evidence for the chemical mechanism proposed of ALAS by Kikuchi *et al.* [13], although the enzyme had not been purified to homogeneity.

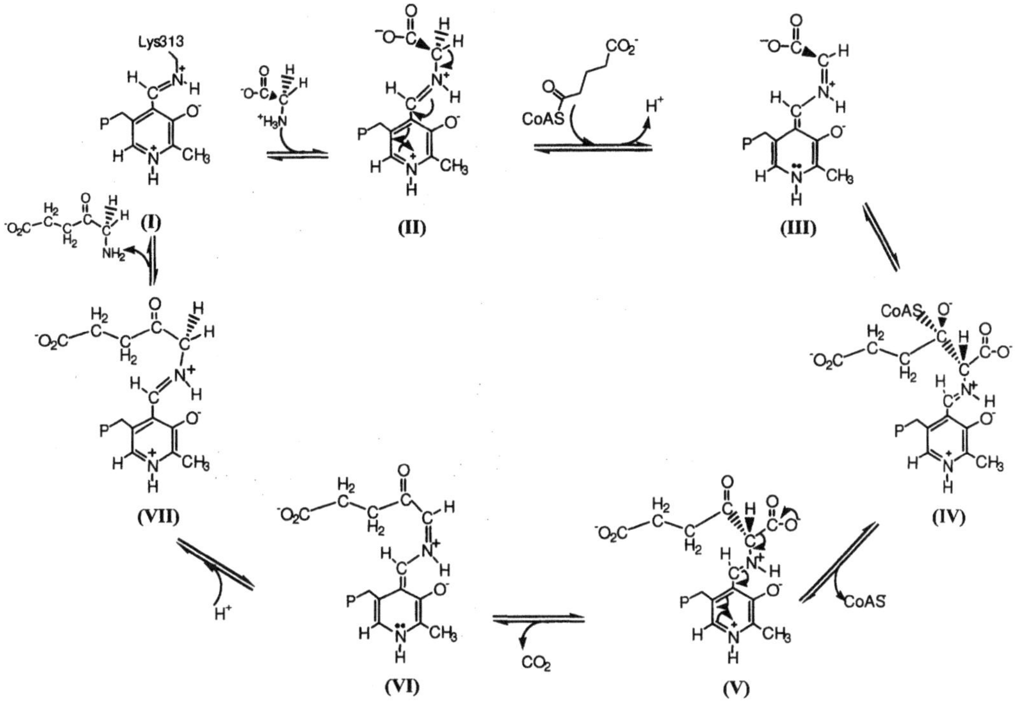

Figure 1: Proposed reaction mechanism for ALAS. Roman numerals represent intermediate transition states. Structures I and III correspond to the quinonoid intermediates.

The development of an overexpression system for murine ALAS2 in *Escherichia coli* [14] made it possible to obtain purified recombinant enzyme with concentrations amenable to initiate detailed studies on the ALAS reaction mechanism. In addition, the ability to identify, spectroscopically, several reaction intermediates permitted Hunter and Ferreira [15] to monitor the rate of formation of these intermediates. The binding reactions of ALAS with glycine and ALA and the initial chemical event in the forward reaction, *i.e.*, removal of the pro-*R* proton of glycine leading to a quinonoid intermediate, were analyzed using pre-steady-state kinetics [15]. Rapid scanning stopped flow experiments of ALAS demonstrated that reaction with glycine plus succinyl-CoA results in a pre-steady-state burst of quinonoid intermediate (Fig. 1 - **III**) formation, indicating that a step following binding of substrates and initial quinonoid intermediate formation is rate-determining. The observation that the steady-state spectrum of ALAS is similar to that formed in the presence of ALA suggests that the rate-determining step is associated with the release of the ALA product.

In the absence of succinyl-CoA the binding reaction of glycine with ALAS occurs in two distinct kinetic steps, *possibly* representing an initial binding step followed by a conformational change of the enzyme-glycine complex [15]. The second step occurs with a rate constant of 0.15 s^{-1}. However, in the presence of succinyl-CoA, the rate of the second step of the reaction of glycine with ALAS is increased by at least 90-fold, suggesting that the binding of succinyl-CoA accelerates the possible conformational change of the enzyme-glycine complex to a catalytically active configuration [15]. Binding of ALA to ALAS is also slow (k_f = 0.15 s^{-1}) and approximates k_{cat}; the second and slow step of binding reaction of ALA to the enzyme is independent of ALA concentration, also suggestive of a slow isomerization of the enzyme [15]. Collectively, the results suggest that turnover is limited by release of ALA or a conformational change associated with ALA release [15].

Architecture of the ALAS Active Site

Although the three-dimensional structure of ALAS remains to be determined, ALAS has been demonstrated to be a homodimer with the active site residing at the subunit interface [16]. Lysine 313 (K313) of murine ALAS2 was identified as the residue involved in the Schiff base linkage with the PLP cofactor [8] and a conserved glycine loop was proposed to form part of the PLP-binding site [17,18]. The common evolutionary origin between ALAS and other members of the α -family of PLP-dependent enzymes made it possible, using sequence homology analyses and homology modeling against PLP-dependent enzymes of known three-dimensional structure (*i.e.*, aspartate aminotransferase [19]), to identify putative ALAS active site residues with critical roles in substrate binding and catalysis. Spectroscopic and biochemical characterization (*e.g.*, absorption and CD spectroscopies, steady-state kinetic parameters, kinetic isotope effect) of recombinant murine ALAS2 variants validated the proposed roles [20-23]. Briefly, K313, the residue involved in the Schiff base linkage to the PLP cofactor, acts as a general base catalyst in the proton abstraction from the external aldimine complex (with the glycine substrate or the ALA product) to form the quinonoid reaction intermediates [20]. During formation of ALA from glycine and succinyl-CoA, K313 is proposed to act as a general base to abstract the pro-*R* proton of glycine, and as a general acid to donate a proton to the quinonoid intermediate formed from decarboxylation of the 2-amino-3-keto-adipate intermediate (Fig. 1) [20]. Aspartate 279 (D279) has been identified as a critical residue in ALAS catalysis by enhancing the electron withdrawing capacity of the PLP cofactor through the stabilization of the protonated form of the cofactor pyridinium ring nitrogen [21]. Arginine 439 (R439) has been reported to be crucial in recognition and binding of substrate by forming a bifurcated coplanar ionic bond with the carboxyl group of the glycine substrate [22]. Tyrosine 121 (Y121) has been demonstrated to play a critical role in cofactor binding by donating a hydrogen bond to the phosphate group of the PLP cofactor [23]. Figure 2 represents a model for the ALAS active site based on the studies summarized above.

In conclusion, research in the past recent years research has provided considerable progress on structural aspects and mechanism of ALAS. The three-dimensional structure of ALAS, however, remains unknown. The recent determination of the three-dimensional structure of 8-amino-7-oxononanoate synthase [7], an enzyme which along with ALAS is a member of the α-oxoamine synthase family and shares a high degree of homology with ALAS, has corroborated the above proposed structural roles for some of the ALAS active site residues [20-23].

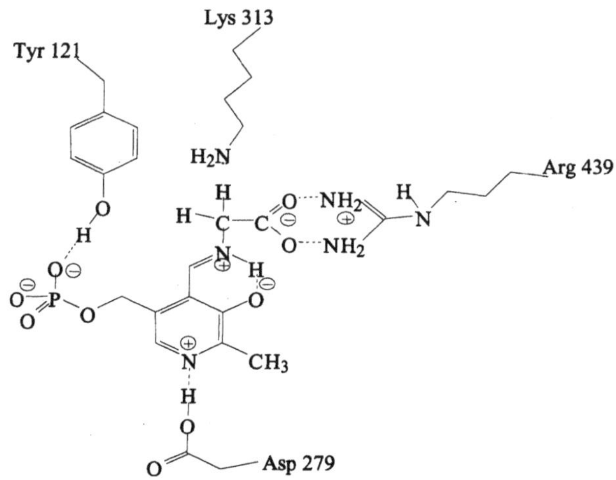

Figure 2: **Model for the ALAS active site.** Residue numbering is according to murine ALAS2.

Acknowledgements

This work was support by a grant from the National Institutes of Health (DK52053).

References

1. Ferreira, G. C. (1999) 5-Aminolevulinate Synthase and Mammalian Heme Biosynthesis. In "Inorganic Biochemistry and Regulatory Mechanisms of Iron Metabolism", (eds. Ferreira, G.C., Moura, J.J.G. and Franco, R.), Wiley-VCH, Weinheim, Germany, 15-34.
2. Ponka, P. (1997) Tissue-specific Regulation of Iron Metabolism and Heme Synthesis: Distinct Control Mechanisms in Erythroid Cells. *Blood 89*:1-25.
3. Sadlon, T. J., Dell'Oso, Surinya, K. H. and May, B. K. (1999) Regulation of Erythroid 5-Aminolevulinate Synthase Expression During Erythropoiesis. *Int. J. Biochem. & Cell Biol. 31*: 1153-1167.
4. Nakajima, O., Takahashi, S., Harigae, H., Furuyama, K., Hayashi, N., Sassa, S. and Yamamoto, M. (1999) Heme Deficiency in Erythroid Lineage Causes Differentiation Arrest and Cytoplasmic Iron Overload. *EMBO J. 18*: 6282-6289.
5. Bottomley, S. S. (1999) Sideroblastic Anemias. In: "Wintrobe's Clinical Hematology" (eds. Lee GR et al.), Williams &Wilkins, Baltimore, 1022-1045.
6. John, R. A. (1995) Pyridoxal Phosphate-dependent Enzymes. *Biochim. Biophys. Acta 1248*: 81-96.
7. Alexeev, D., Alexeeva, M., Baxter, R.L., Campopiano, D.J., Webster, S.P. and Sawyer, L. J. (1998) The Crystal Structure of 8-Amino-7-Oxononanoate Synthase: a Bacterial PLP-dependent, Acyl-CoA-Condensing Enzyme. *J. Mol. Biol. 284*: 401-419.
8. Ferreira, G. C., Neame, P. J., and Dailey, H. A. (1993) Heme Biosynthesis in Mammalian Systems. Evidence of a Schiff Base Linkage Between the Pyridoxal Phosphate Cofactor and a Lysine Residue in Aminolevulinate Synthase. *Protein Science 2*: 1959-1965.
9. Fanica-Gaignier, M. and Clement-Metral, J. (1973) 5-Aminolevulinic-Acid Synthetase of *Rhodopseudomonas spheroides Y.* Kinetic Mechanism and Inhibition by ATP. *Eur. J. Biochem. 40*: 19-24.
10. Zaman, Z., Jordan, P. M., Akhtar, M. (1973) Mechanism and Stereochemistry of the 5-Aminolaevulinate Synthetase Reaction. *Biochem. J. 135*: 257-263.
11. Abboud, M. M., Jordan, P. M., and Akhtar, M. (1974) Biosynthesis of 5-Aminolevulinic Acid: Involvement of a Retention-Inversion Mechanism. *J.S.C. Chem. Comm. 276*: 643-644.
12. Laghai, A. and Jordan, P. M. (1977) An Exchange Reaction Catalyzed by δ-Aminolaevulinate Synthetase from *Rhodopseudomonas spheroides. Biochem. Soc. Trans. 5*: 299-300.
13. Kikuchi, G., Kumar, A., Talmage, P., and Shemin, D. (1958) The Enzymatic Synthesis of δ-Aminolevulinic Acid. *J. Biol. Chem. 233:*1214-1219.
14. Ferreira, G.C. and Dailey, H.A. (1993) Expression of Mammalian 5-Aminolevulinate Synthase in *Escherichia coli*: Overproduction, Purification and Characterization. *J. Biol. Chem*. 268: 584-590.

15. Hunter, G. A. and Ferreira, G. C (1999) Pre-Steady State Reaction of 5-Aminolevulinate Synthase: Evidence for a Rate-Determining Product Release. *J. Biol. Chem. 274*: 12222-12228.

16. Tan, D. and Ferreira, G. C. (1996) Active Site of 5-Aminolevulinate Synthase Resides at the Subunit Interface. Evidence from *In Vivo* Heterodimer Formation. *Biochemistry 35*: 8934-8941.

17. Gong, J. and Ferreira, G.C. (1995) Aminolevulinate Synthase: Functionally Important Residues at a Glycine Loop, a Putative Pyridoxal Phosphate Cofactor-Binding Site. *Biochemistry 34*: 1678-1685.

18. Gong, J., Kay, C. J., Barber, M. J., and Ferreira, G. C (1996) Mutations at a Glycine Loop in Aminolevulinate Synthase Affect Pyridoxal Phosphate Cofactor Binding and Catalysis. *Biochemistry 35*: 14109-14117.

19. Kirsch, J. F., Eichele, G., Ford, G. C., Vincent, M. G., Jansonius, J. N., Gehring, H. and Christen, P. (1984) Mechanism of Action of Aspartate Aminotransferase Proposed on the Basis of its Spatial Structure. *J. Mol. Biol. 174*: 497-525.

20. Hunter, G. A. and Ferreira, G. C (1999) Lysine-313 of 5-Aminolevulinate Synthase Acts as a General Base During Formation of the Quinonoid Reaction Intermediates. *Biochemistry 38*: 3711-3718.

21. Gong, J., Hunter, G. A. and Ferreira, G. C (1998) Aspartate-279 in Aminolevulinate Synthase Affects Enzyme Catalysis Through Enhancing the Function of the Pyridoxal Phosphate Cofactor. *Biochemistry 37*: 3509-3517.

22. Tan, D., Harrison, T., Hunter, G. A. and Ferreira, G. C. (1998) The Role of Arginine 439 in Substrate Binding of 5-Aminolevulinate Synthase. *Biochemistry 37*: 1478-1484.

23. Tan, D., Barber, M. J. and Ferreira, G. C. (1998) The Role of Tyrosine 121 in Cofactor Binding of 5-Aminolevulinate Synthase. *Protein Science 7*: 1208-1213.

Biochemistry and Molecular Biology of Vitamin B_6
and PQQ-dependent Proteins
ed. by A. Iriarte, H.M. Kagan and M. Martinez-Carrion

Function of Pyridoxal-5′-Phosphate in Bacterial Phosphorylases

Reinhard Schinzel[1], Dieter Palm[1], Birgit Böck[1] and Bernd Nidetzky[2]

[1]Theodor-Boveri-Institut, Physiologische Chemie I, Biozentrum der Universität Würzburg, Am Hubland D-97074 Würzburg, Germany

[2]Institut für Lebensmitteltechnologie, Universität für Bodenkultur, Muthgasse 18, A-1190 Wien Austria

Summary

Bacterial phosphorylases differ from mammalian phosphorylases in their physiological functions, reflected by differences in substrate specificity and regulatory mechanisms. By using *E. coli* maltodextrin phosphorylase (MalP) as a model enzyme, the properties of the active-site cofactor pyridoxal-5'-phosphate (PLP) in various bacterial enzymes are compared. The spectroscopic and chemical properties of PLP in most eubacterial phosphorylases appear to be very similar to those of MalP and eucariotic phosphorylases; but strikingly different to thermophilic phosphorylases, which could indicate the adaptation of the active sites of thermozymes to specific environmental requirements. Furthermore, there is convincing evidence that PLP-dependent α-glucan phosphorylases are also present in archaeons.

Introduction

α-Glucan phosphorylases (EC 2.4.1.1) catalyse the reversible cleavage of polysaccharides into α-D-glucose-1-phosphate (Glc-1-P) and hence play a central role in the mobilisation of storage polysaccharides. They are ubiquitous enzymes that are found in many organisms including bacteria, yeast, slime mold, plants and vertebrates. All known α-glucan phosphorylases require pyridoxal-5'-phosphate (PLP) for activity but their catalytic mechanism differs from that of other pyridoxal-containing enzymes. The phosphorylase reaction mechanism does not depend on the formation of a Schiff base between the aldehyde group of the cofactor and the substrate like in aminotransferases, but on a close contact of the cofactor phosphate group to either the inorganic phosphate (in the direction of glycogen degradation) or the phosphate of Glc-1-P (in the direction of glycogen synthesis) [1-3]. Although glycogen is present in many bacteria, only few α-glucan phosphorylases from bacteria have been identified and characterised on the protein or gene level. Like the glucan phosphorylases from higher organisms, all bacterial phosphorylases investigated so far require a short oligosaccharide as a primer molecule for glycogen synthesis, and seem to contain PLP. However, the bacterial enzymes differ from their eucariotic counterparts in terms of substrate specificity, regulation and physiological role [4]. To see whether these differences may reflect different modes of cofactor binding and function, the properties of PLP were examined in three bacterial phosphorylases: the maltodextrin phosphorylase (MalP) from *Escherichia coli* as a representative of the gram-negative bacteria; the starch phosphorylase from the gram-positive soil bacterium *Corynebacterium callunae;* and a thermophilic phosphorylase isolated from the eubacterium *Thermus thermophilus* [5-8].

Cofactor analysis

As anticipated by sequence comparisons [9,10], all three phosphorylases examined contain 1 PLP molecule per subunit [1, 6, 7]. The UV/Vis spectra of the phosphorylases show a typical shoulder at 330 nm (Fig. 1A). A similar spectrum was observed for the maltodextrin phosphorylase from the archaeon *Thermococcus litoralis* indicating the presence of PLP also in archaebacteria [11]. The activity of all phosphorylases is dependent on the presence of the cofactor as shown for the phosphorylases from *E. coli* and *C. callunae* by reversible cofactor dissociation experiments [Palm and Zeier, unpublished results, 6]. In phosphorous NMR, the phosphate group of PLP of phosphorylase from *E. coli* and *C. callunae* shows a similar downfield shift of 3.60 and 3.85 ppm, respectively [6, 12]. The presence of PLP in the active site of a bacterial phosphorylase has been confirmed by the resolution of the MalP structure [13, 14]. As expected, the overall structure of MalP shows a fold which is similar to that of the best characterised phosphorylase, the glycogen phosphorylase from rabbit muscle. While the control sites present in regulated glycogen phosphorylase are missing in MalP, the cofactor-binding site is very similar in both enzymes. The active site structure of the ternary complex indicates identical catalytic mechanisms. Structural data for MalP provide strong support for the suggestion that MalP and glycogen phosphorylase utilise identical catalytic mechanisms[14]. Overall, *E. coli* MalP appears to be a basic archetype of a phosphorylase, which promotes degradation of maltodextrins but lacks regulatory sites [13].

Fluorescence Spectroscopy

The fluorescence emission spectrum of PLP of glycogen phosphorylase shows a large Stokes shift of approx. 200 nm, with a wavelength of maximum emission of 540 nm. Similar fluorescence properties are found for PLP in the enzymes from *E. coli* and *C. callunae* [1, 6]. In marked contrast, an emission maximum was found at 420 nm for the thermophilic enzyme when excited at 330 nm (Fig. 1B) [7].

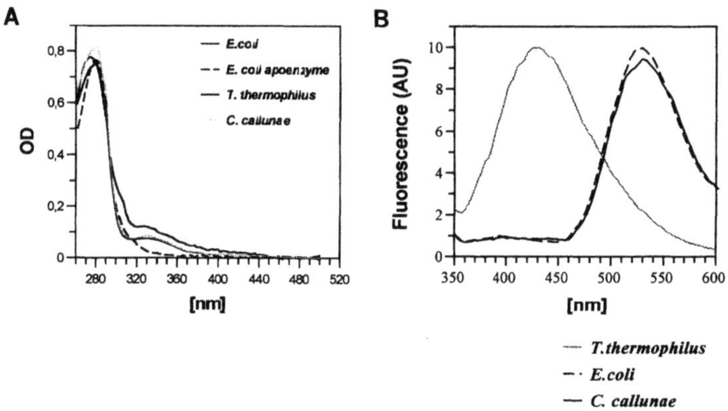

Figure 1. UV-Vis (A) and fluorescence-spectra (B) of bacterial α-glucan phosphorylases.

The fluorescence spectrum of the enzyme from *T. thermophilus* is similar to the spectra measured for model Schiff base compounds such as PLP aminocaproic acid, or other PLP-containing enzymes such as amino transferases as well as for denatured phosphorylases [Palm, unpublished results, 14]. The fluorescence spectrum from the archaebacterial thermophilic phosphorylase has an intermediate maximum at 460 nm (Table 1) [11]. Fluorescence energy transfer from tryptophans (excited at 280 nm) to the cofactor was observed for the mammalian phosphorylases and MalP but not for the enzymes from *C. callunae* and *T. thermophilus* (Table

1). The wavelength of maximum fluorescence emission of PLP is thought to be characteristic of the polarity of the active site of phosphorylase. By that criterion, the differences in the fluorescence properties indicate that the cofactor is slightly more exposed to the aqueous solvent in the phosphorylase from *C. callunae* and even more exposed in the thermophilic phosphorylase. They indicate subtle differences in the active site structure of the phosphorylases, which are also reflected in some of the catalytic properties such as the capability to use arsenate as a slow substrate. Compared with malP, the rate of degradation of oligosaccharides in the presence of arsenate is reduced in the *C. callunae* phosphorylase and absent in the *T. thermophilus* phosphorylase (Table I) [6,7].

Table I. Biochemical properties of different bacterial α-glucan phosphorylases

	Arsenolysis[a)]	Emission Max.[b)]	Energy Transfer[c)]
GP	12%	540	+
E. coli	12%	532	+
C. callunae[d]	8%	512	-
T. therm.[e]	<0.1%	430	-
T. litoralis[f]	n.d.	460	n.d.

a) Percent of residual activity using arsenate as a substrate, compared to the activity with phosphate b) Excitation at 330 nm c) excitation at 330 nm, emission at 540 nm d) from [6] e) from [7] f) from [11]; GP, glycogen phosphorylase from rabbit muscle; n.d., not determined

Archaebacterial α-glucan phosphorylases

Glycogen was found in some archeabacteria - for example in *Halobacterium salinarium*, *Methanobacteria* and *Sulfolobus* [16]. Phosphorylase activity has been shown in cellular extracts of *Methanococcus maripaludis* and *Sulfolobus acidocaldarius* [17, Neubert, unpublished results], recently a maltodextrin phosphorylase was cloned and purified from the hyperthermophilic archaebacterium *Thermococcus litoralis* [11]. Furthermore, the first genomes of archeabacteria are fully or partially sequenced [18,19]. By using the Prosite - program a phosphorylase-PLP-attachment site

E-A-[SC]-G-X-[GS]-X-M-K-X-[LM]-N

was described, which is only found in glycogen phosphorylases [20]. This site was also found in the archaebacterial phosphorylase from *T. litoralis* [11] and in ORFs from the genomes of the archaebacteria *Methanococcus jannaschii* and *Pyrococcus horikoshii*. These genes were aligned to other phosphorylases by the GAP program of the GCG package 9.1. While glucan phosphorylases from prokaryotes and eukaryotes share a more than 40% sequence identity, the archaebacterial genes showed only an about 20% identity to eubacterial and eucariotic phosphorylases (Table II). The enzymes from the closely related strains *T. litoralis* and *P. horikoshii* show a high degree of identity. Hence, PLP-dependent phosphorylases are present in all phylogenetic domains and share many properties despite relative low amino acid sequence similarity.

Table II. Conservation of bacterial glucan phosphorylases in % identity

	S. cerevisiae	*B. subtilis*	*P. horikoshii*	*M. jannaschii*	*Thermococcus litoralis*[a]
Escherichia Coli	47	44	18	19	13
Saccharomyces cerevisiae		42	19	17	15
Bacillus subtilis			21	19	20
Pyrococcus horikoshii				28	66
Methanococcus jannaschii					32

a) from [11]

In conclusion, all bacterial phosphorylases characterised so far seem to follow the same catalytic mechanism, and their activities are dependent on the cofactor pyridoxal 5'-phosphate. However, different physiological functions and the adaptation to extreme environments appear to be reflected in different biophysical and biochemical properties of cofactor binding and function.

References:

[1] Palm, D., Klein, H. W., Schinzel, R., Buehner, M., and Helmreich, E. J. M. (1990) The role of pyridoxal-5'-phosphate in glycogen phosphorylase catalysis. *Biochemistry 29:* 1099-1107.

[2] Johnson. L. N. (1992) Glycogen phosphorylase control by phosphorylation and allosteric effectors. *FASEB J.*: 2274-2282.

[3] Browner, M. F., and Fletterick, R. J. (1992) Phosphorylase: a biological transducer. *Trends Biochem. Sci. 17*: 66-71.

[4] Schinzel, R., and Nidetzky, B. (1999) Bacterial α-Glucan phosphorylases. *FEMS Microbiol. Lett. 171:* 73-79

[5] Weinhaeusel, A., Nidetzky, B., Rohrbach, M., Blauensteiner, B., and Kulbe K. D. (1994) A new maltodextrin-phosphorylase from *Corynebacterium callunae* for the production of glucose-1-phosphate. *Appl. Microbiol. Biotechnol. 41*: 510-516.

[6] Weinhaeusel, A., Griessler, R., Krebs, A., Haltrich, D., Kulbe, K.D., Zipper, P. and Nidetzky, B. (1997) □-1,4-D-Glucan-phosphorylase from gram-positive Corynebacterium callunae: *Biochem. J. 326* 773-778.

[7] Böck, B., and Schinzel, R. (1996) Purification and characterisation of a thermophilic □-glucan phosphorylase from *Thermus thermophilus. Eur. J. Biochem. 238:* 150-155.

[8] Böck, B. and Schinzel, R. (1998) Growth-dependence of alpha-glucan phosphorylase activity in *Thermus thermophilus. Res. Microbiol. 149:* 171-176

[9] Palm, D., Goerl, R., and Burger, K. L. (1985) Evolution of catalytic and regulatory sites in phosphorylases. *Nature 313:* 500-502

[10] Newgard, C. B., Hwang, P., and R. J. Fletterick (1989) The family of glycogen phosphorylases: structure and function. *Crit. Rev. Biochem. Mol. Biol. 24*: 69-99.

[11] Xavier, K. B., Peist, R., Kossmann, M., Boos, W., and Santos, H. (1999) Maltose metabolism in the hyperthermophilic archaeon *Thermococcus litoralis*: Purifcation and characterisation of key enzymes *J. Bacteriol. 181*: 1158-1167

[12] Palm, D., Schächtele, K.H., Feldmann, K, and Helmreich E.J.(1979) Is the active form of PLP in glycogen phosphorylases a 5'-phosphate dianion? *FEBS Lett. 101:* 403-406

[13] Watson, K. A., Schinzel, R., Palm, D., and Johnson, L. N. (1997) The crystal structure of *E. coli* maltodextrin phosphorylase provides an explanation for the activity without control in this basic archetype of a phosphorylase. *EMBO J. 16*: 1-14.

[14] Watson, K. A., McCleverty, C, Geremia, S., Cottaz, S., Driguez, H., and Johnson L. N. (1999) Phosphorylase recognition and phosphorolysis of its oligosaccharide substrate. *EMBO J. 18:* 4619-4632

[15] Shaltiel, S., Cortijo, M., and Zaidenzaig, Y. (1972) In *Metabolic Interconversion of Enzymes* (Wieland O., Holzer H. and Helmreich, E.; eds), pp. 73-101, Springer Verlag, Berlin

[16] König, H., Skorko, R., Zillig, W and Reiter, W-D. (1982) Glycogen in thermoacidophilic archaebacteria . *Arch. Microbiol. 132:* 297-303

[17] Yu, J.P., Ladapo J., and Whitman, W. B. (1994) Pathway of glycogen metabolism in *Methanococcus maripaludis. J. Bacteriol. 176:* 325-332.

[18] Bult C.J., et. al. (1996), Complete genome sequence of the methanogenic archaeon, *Methanococcus jannaschii Science 273*:1058-1073

[19] Kawarabayasi et al., (1998) Complete sequence and gene organization of a hyper-thermophilic archaebacterium, *Pyrococcus horikoshii* OT3. *DNA Research 5*:55-76

[20] Hofmann K., Bucher P., Falquet L., and Bairoch A. (1999). The PROSITE database. *Nucl. Acids Res. 27*: 215-219.

Biochemistry and Molecular Biology of Vitamin B_6
and PQQ-dependent Proteins
ed. by A. Iriarte, H.M. Kagan and M. Martinez-Carrion

Inhibition and Structural Changes of *O*-Acetylserine Sulfhydrylase-A from *Salmonella typhimurium* Upon Binding Sulfate and Chloride Anions

Chia-Hui Tai[1], Peter Burkhard[2], Johan N. Jansonius[2] and Paul F. Cook[1]

[1]Department of Chemistry and Biochemistry, University of Oklahoma, 620 Parrington Oval, Norman, OK 73019, U.S.A.
[2]Department of Structural Biology, Biozentrum, University of Basel, Klingelbergstrasse70, CH-4056 Basel, Switzerland

Summary

Two anion binding sites, for sulfate and chloride ions, on *O*-acetylserine sulfhydrylase-A were determined using X-ray data. The sulfate ion binding site is similar to the carboxylate moiety of the substrate analogue methionine in the closed conformation, while the chloride ion binding site is located at the interface of the dimer. Both anions inhibit enzyme activity, but with different inhibition patterns, reflecting differences in the binding sites for the two anions.

Introduction

O-Acetylserine sulfhydrylase-A (OASS-A) is a dimeric pyridoxal 5-phosphate (PLP) dependent enzyme, with one PLP per subunit, which is responsible for the synthesis of L-cysteine in bacteria [1] and plants [2]. The biosynthesis of L-cysteine in enteric bacteria, such as *Salmonella typhimurium* and *Escherichia coli*, and plants proceeds via a two-step pathway (Scheme 1).

OASS-A belongs to the β family of PLP-dependent enzymes that catalyze β-replacement or β-elimination reactions. The amino acid precursor of L-cysteine is L-serine, which is first acetylated at its β-hydroxyl by acetyl-CoA to give *O*-acetyl-L-serine (OAS), catalyzed by the enzyme serine acetyltransferase (SAT) (E. C. 2.3.1.30) [3]. The final step in cysteine synthesis, replacement of the acetate side chain by a thiol, is catalyzed by *O*-acetylserine sulfhydrylase (OASS) (E. C. 4.2.99.8) with inorganic sulfide as the thiol [1].

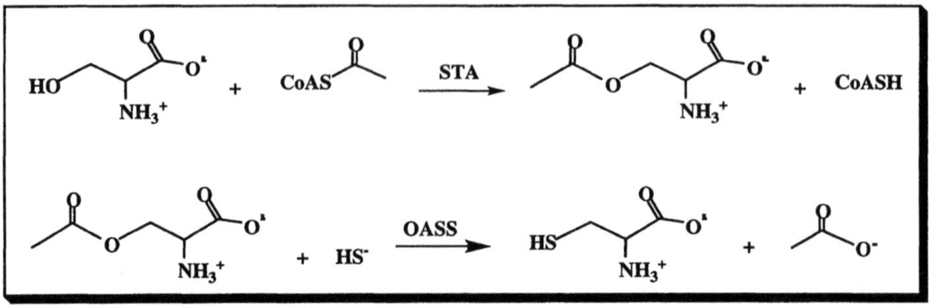

Scheme 1

Tryptophan synthase, one of the β family of PLP-dependent enzymes, and others have been shown to have monovalent cation sites that regulate their enzyme activity [4]. In this paper, we report that OASS-A has two different anion binding sites that regulate the enzyme activity and change enzyme conformation.

Materials and Methods

Enzyme. Recombinant OASS-A was purified from a plasmid-containing overproducing strain by conventional methods of sonication, streptomycin sulfate precipitation, ammonium sulfate fractionation [5], and high performance liquid chromatography [6].

Crystallization. Crystals of OASS-A with either sulfate or chloride ions were grown by vapor diffusion using the hanging-drop method [7]. The crystal form of OASS-A with chloride ion was grown in the presence of 0.18 M Li_2SO_4 and at pH 6.7. All crystal forms are orthorhombic.

Enzyme Assay. Initial velocities were measured by varying one reactant, OAS, over a range of concentrations and several fixed concentrations of the second reactant, 5-mercapto-2-nitrobenzoic acid (TNB). The disappearance of TNB at 412 nm resulting of formation of S-(3-carboxy-4-nitrophenyl)-L-cysteine was monitored continuously as the reaction proceeds. Inhibition patterns were obtained by measuring the initial velocity at varying concentrations of one reactant with the second maintained at K_m at several different fixed concentrations of inhibitor including zero [6].

Results and Discussion

Dead-end Inhibition. For a ping pong mechanism competitive inhibition pattern is expected between a substrate and its analog since they bind to the same enzyme form, while an uncompetitive inhibition pattern is expected for the analog versus the other substrate since they bind to different enzyme forms. Inhibition patterns and constants for sulfate and chloride ions are listed in Table 1. The inhibition pattern with sulfate is uncompetitive vs OAS and competitive vs TNB, suggesting that sulfate ion binds to α-aminoacrylate intermediate much more tightly than to the free enzyme form. Inhibition patterns with chloride are noncompetitive vs both OAS and TNB, suggesting that chloride ion binds to both α-aminoacrylate intermediate and free enzyme form with similar affinity.

Table1. Inhibition Constants for *O*-Acetylserine Sulfhydrylase-A

Inhibitor	Variable Substrate	Type of Inhibition	K_{is} (mM)	K_{ii} (mM)
[c]SO_4^{2-}	OAS[a]	UC		20 ± 8
	TNB[b]	C	21 ± 2	
[c]Cl^-	OAS[a]	NC	21 ± 8	39 ± 28
	TNB[b]	NC	37 ± 5	48 ± 13

[a]Data were obtained at 0.1 mM TNB
[b]Data were obtained at 2 mM OAS.
[c]Data were obtained with 300 μg OASS at pH 6.5 and 25 °C, 100 mM Mes.

Location of the Sulfate Ion Binding Site. The stereo view of the active site region with the bound sulfate ion is shown in Figure 1, the asparagine loop adopts a new inhibited conformation compared to the open and closed conformations and the residues N69 has three different positions (N69o, N69i and N69c where o, i and c represent open, inhibited, and closed) [8,9]. The sulfate ion is involved in many ionic interactions similar to the carboxylate moiety of the substrate analogue methionine in the "closed" conformation [9]. A total of ten interatomic distances shorter than 3.3 Å can be observed between the sulfate ion and the protein and two water molecules. However, there is no positively charged residue in hydrogen bonding distance to the sulfate ion, but rather the helix dipole of helix 2 [8] is compensating for the negative charge of the sulfate ion.

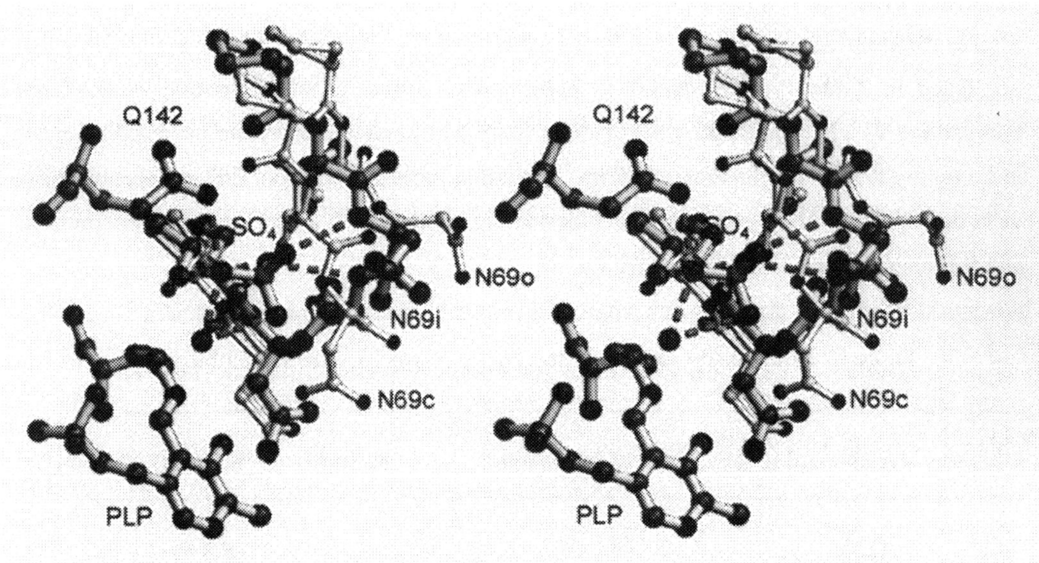

Figure 1. Stereo representation of sulfate ion bound to the active site region. The picture was drawn with the programs MOLSCRIPT [10] and Raster3D [11].

Location of the Chloride Ion Binding Site. This chloride ion is involved in five ionic interactions as shown in Figure 2, three with protein residues and two with water molecules with interatomic distances of about 3.2 Å. The chloride ion is located at the interface of the dimer. The chloride atom interacts with the amide nitrogen of Asn 7 of one monomer and the peptide nitrogen of Leu 268' of the other monomer while the positive charge of the guanidinium group of Arg 34' of the other monomer compensates the negative charge of the chloride ion. In addition, two water molecules are involved in complex formation with the chloride ion. These two water molecules are part of a hydrogen bonding network around the chloride binding site. One water molecule forms hydrogen bonds to the carbonyl moiety of Leu 268' and the side chain oxygen of Ser 33', while the other water molecule forms hydrogen bonds to the peptide nitrogen of Ser 37 and to another water molecule which then is further hydrogen bonded to the side chain oxygen of Ser 33'.

The location of the chloride atom brings it in close neighborhood to the carbonyl oxygen of Pro 36. The carbonyl moiety of this residue in the "open" conformation is only a few Å apart from the chloride ion (dotted line) and therefore undergoes a conformational change that induces a cascade of side chain reorientations, resulting in a movement of the whole "movable domain"

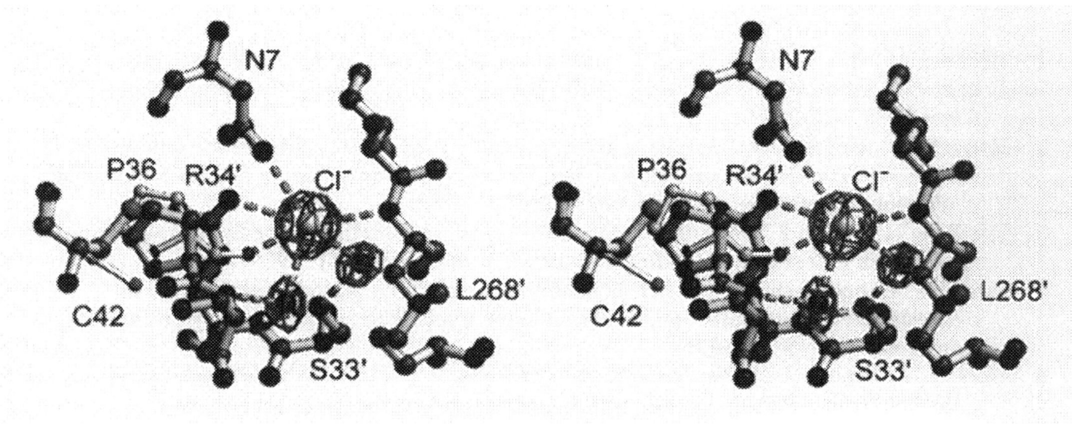

Figure 2. Stereo representation of the allosteric binding site with a bound chloride ion. The picture was drawn with the programs MOLSCRIPT [10] and Raster3D [11].

to a new location. This new conformation of the "movable domain" represents the structure of an inhibited conformation of the protein

Conclusions

Sulfate ion binds to the α-carboxyl subsite of the active site and causes a partial closing of the active site, and as a result the position of the moveable domain is between that found in the opened and closed conformations. From the inhibition patterns obtained with sulfate, sulfate ion also binds to the allosteric inhibitory site in the α-aminoacrylate form of the enzyme. The chloride binding site is buried in the dimer interface, generates a new stable conformation of the enzyme. The chloride site is most likely the sulfide inhibitory binding site, i.e. the site for competitive substrate inhibition by sulfide. In contrast to tryptophan synthase, which has an allosteric cation activation site, OASS has an allosteric anion inhibition site. This represents a new subclass for type II PLP enzymes.

References

1. Becker, M. A., Kredich, N. M., and Tomkins, G. M. (1969) The purification and characterization of *O*-acetylserine sulfhydrylase A from *Salmonella typhimurium*. *J Biol Chem* 244: 2418-2427.
2. Rolland, N., Ruffet, M.-L., Job, D., Douce, R., and Droux, M. (1996) Spinsch chloroplast O-acetylserine (thio)-lyase exhibits two catalytically non-equivalent pyridoxal 5'phosphate-containing active sites. *Eur. J. Biochem.* 236: 272-282.
3. Kredich, N. M. and Tomkins, G. M. (1966) The enzymatic synthesis of L-cysteine in *Escherichia coli* and *Salmonella typhimurium*. *J Biol Chem* 241: 4955-4965.
4. Woehl, E. U. and Dunn, M. F. (1995) Monovalent metal ions play an essential role in catalysis and intersubunit communication in the tryptophan synthase bienzyme complex. *Biochemistry* 34: 9466-9476.
5. Hara, S., Payne, M. A., Schnackerz, K. D., and Cook, P. F. (1990) A rapid purification procedure and computer-assisted sulfide ion electrode assay for *O*-acetylserine sulfhydrylase from *Salmonella typhimurium*. *Protein Expression Purif* 1:70-76.
6. Tai, C.-H., Nalabolu, S. R., Jacobson, T. M., Minter, D. E., Cook, P. F. (1993) Kinetic mechanisms of *O*-acetylserine sulfhydrylases A and B from *Salmonella typhimurium* with natural and alternate substrates. *Biochemistry* 32: 6433-6442.
7. Rao, J. G. S., Goldsmith, E. J., Mottonen, J., Cook, P.F. (1993) Crystallization and preliminary x-ray data for the A-isozyme of *O*-acetylserine sulfhydrylase from *Salmonella typhimurium*. *J Mol Biol* 231: 1130-1132.
8. Burkhard, P., Rao, G. S. J., Hohenester, E., Cook, P. F., Jansonius, J. N. (1998) Three dimensional structure of *O*-acetylserine sulfhydrylase from *Salmonella typhimurium* at 2.2 Å. *J Mol Biol* 283: 111-120.
9. Burkhard, P., Cook, P. F., Jansonius, J. N. (1999) Structure of an external aldimine intermediate in the *O*-acetylserine sulfhydrylase reaction: the K41A mutant enzyme. *J Mol Biol* 291: 941-953.
10. Kraulis, P. J. (1991) MOLSCRIPT: a program to produce both detailed and schematic plots of protein structure. *Journal of Applied Crystallography* 24:946-950.
11. Merritt, E. A. and Bacon, D. J. (1997) Raster3D: Photorealistic molecular graphic. *Methods in Enzymology* 277 part B: 505-524.

Biochemistry and Molecular Biology of Vitamin B_6
and PQQ-dependent Proteins
ed. by A. Iriarte, H.M. Kagan and M. Martinez-Carrion

Structure-Function Relationships of Porcine Pyridoxal Kinase †

Y. C. Leung, H. Y. Wong, J. E. Churchich, S. C. L. Lo and F. Kwok

Department of Applied Biology and Chemical Technology, The Open Laboratory of Asymmetric Synthesis, The Hong Kong Polytechnic University, Hung Hom, Kowloon, Hong Kong, China

Summary

Pyridoxal kinase (PK) catalyzes the formation of pyridoxal-5-phosphate (PLP) from pyridoxal (PL), ATP and a divalent cation (Zn^{2+}). So far, there is no three-dimensional structure of PK available. Site-directed mutagenesis was carried out to study the importance of three conserved residues: Tyr137, Gly242 and Gly244. The mutants (Y137F, G242A and G244A) were constructed, expressed, purified and analyzed. The results demonstrated that the mutants had much less activity but with no dramatic variation in protein stability. Tyr137 residue was shown to be involved in PL binding but not ATP binding. For the G244A mutant, the absence of enzyme activity was probably due to the deficiency in PL binding rather than the lack of ATP binding. In the case of the G242A mutant, it did not bind to ATP or PL.

Introduction

Pyridoxal 5-phosphate (PLP) is the metabolically active form of vitamin B_6 and acts as an essential coenzyme in the synthesis, catabolism, and interconversion of amino acids in cellular metabolism. Pyridoxal kinase (PK) catalyzes the formation of PLP from pyridoxal (PL), ATP and a divalent cation (Zn^{2+}) [1]. However, the amino acid residues in PK that are responsible for this important biochemical reaction have not been determined. In addition, PK from porcine has been shown to be a dimer of identical subunits [2]. Each subunit, with a molecular mass of about 40 kDa, may contain 2 domains responsible for ATP and PL binding.

Recently, the porcine PK gene has been cloned in our laboratory [3]. With the gene in hand, it is now possible to use the powerful site-directed mutagenesis method to change the potentially important amino acids in this enzyme so that their structural and functional relationships can be studied in a more logical manner. Residues that are highly conserved in PK

†This paper is dedicated to the memory of Professor Jorge Churchich.

were searched and potentially important mutants (Y137F, G242A and G244A) were constructed for detailed studies.

Materials and Methods

For both site-directed mutagenesis and gene expression, plasmid pAED4 containing the porcine PK gene was used. Site-directed mutagenesis was performed using QuikChange™ site-directed mutagenesis kit (Stratagene). Mutant and wild-type PK were expressed in *E.coli* BL21 (DE_3) expression system, in which the PK gene was put under the control of a T7 phage promoter. The expressed proteins were purified using 0-40% ammonium sulfate fractionation followed by Q-sepharose chromatography. Enzyme purity was tested by SDS-polyacylamide gel electrophoresis. Protein concentration was determined by the method of Bradford [4]. Enzyme kinetic parameters were measured in 70 mM potassium phosphate buffer at pH 6.5 at 25 °C by following the change in absorbance at 388 nm, at which PLP has an absorption maximum with an absorption coefficient of 4900 cm^{-1} M^{-1}. Emission fluorescence spectra were recorded in a Perkin-Elmer LS-50B spectrofluorometer. The absorbance of the samples at the excitation wavelengths were around 0.1 (1-cm cuvette).

Results and Discussion

Activity assays confirmed that the mutants G242A and G244A did not show any detectable activities. Unfolding studies indicated that the replacement of the Tyr137, Gly242, or Gly244 residue did not cause a dramatic conformational change of the enzyme (data not shown). The kinetic parameters of the purified Y137F mutant were measured and are shown in Table 1.

Table 1. The steady-state kinetic parameters were determined from the initial rate. Reactions were performed in 70 mM potassium phosphate buffer (pH 6.5) at 25°C. k_{cat} and k_{cat}/K_m values were calculated by using MW = 35.4 kDa for PK.

Enzyme	Substrate	K_m (μM)	k_{cat} (s^{-1})	k_{cat}/K_m ($mM^{-1}s^{-1}$)
Wild-type	PL	236 ± 46	0.185 ± 0.014	0.78 ± 0.21
	ATP	47 ± 10	0.054 ± 0.004	1.15 ± 0.33
Y137F	PL	2504 ± 1419	0.056 ± 0.025	0.022 ± 0.023
	ATP	34 ± 5	0.003 ± 0.0001	0.088 ± 0.017

The K_m value of Y137F for ATP was similar to that of the wild-type, whereas for PL, K_m was about 10 times higher than that of the wild-type. The k_{cat} values for PL and ATP of Y137F were both smaller than those of the wild-type. As a result, the k_{cat}/K_m values for PL and ATP of Y137F were 35 and 13 times smaller than those of the wild-type enzyme respectively.

The binding of 2 fluorescent substrate analogs, i.e., TNP-ATP and pyridoxaloxime, to the catalytic sites of wild-type and mutants were also examined. The fluorescence enhancement on binding to the enzymes was studied by varying the concentration of TNP-ATP (from 0 to 4 μM). All of them showed an increase in fluorescence enhancement with increasing TNP-ATP concentration, except mutant G242A (Fig. 1).

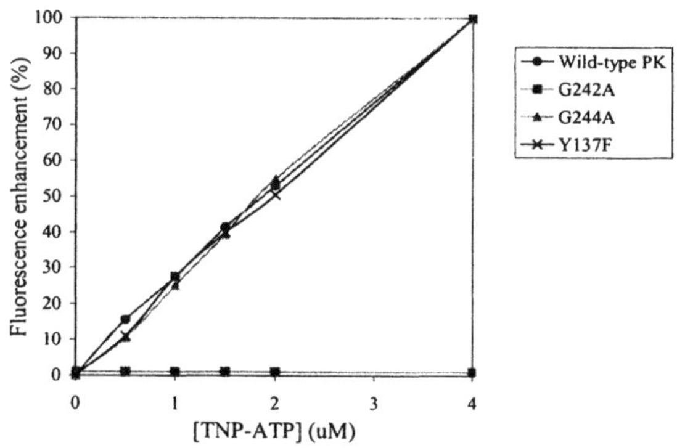

Fig. 1. Fluorescence enhancement by wild-type and mutants of PK (2 μM) at emission wavelength 540 nm in various concentrations of TNP-ATP. Excitation wavelength was set at 460 nm.

Increasing the concentration of pyridoxaloxime (from 0 to 2 μM) in the presence of the wild-type or mutant Y137F caused an increase in the quenching of fluorescence intensity (Fig. 2). At 1 μM pyridoxaloxime concentration, a 40% and 20% quenching in fluorescence was observed for the wild-type and the Y137F mutant respectively. However, no fluorescence quenching was demonstrated by the mutants G242A and G244A, suggesting that these mutants did not bind pyridoxaloxime.

From the ATP- and PL-analogs binding experiments, similar fluorescence enhancement was obtained when TNP-ATP was added to the wild-type and the Y137F mutant (Fig. 1). This indicated that the replacement of the Tyr137 residue by phenylalanine did not affect the binding between ATP and PK. On the other hand, the quenching of pyridoxaloxime fluorescence intensity on binding to the Y137F mutant was lower than that on binding to the wild-type (Fig.

2), indicating that the binding ability towards the substrate PL might be affected by the mutation. Taken together, our data suggested that Tyr137 was not required for ATP binding but was involved in PL binding. The decrease in enzyme activity after the mutation was probably due to the poor binding of PL to the PK protein.

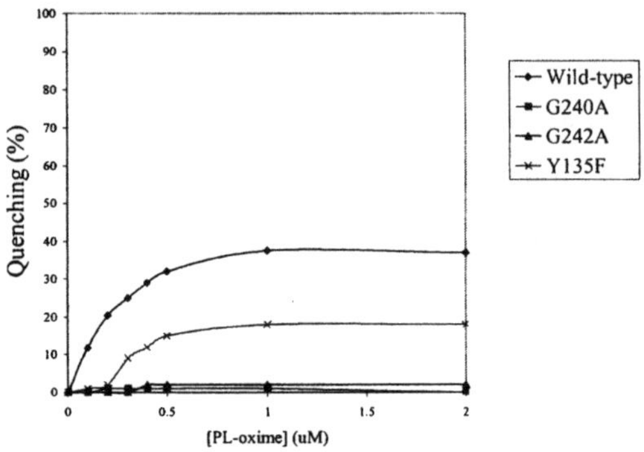

Fig. 2. Quenching effect on fluorescence upon addition of various concentrations of pyridoxaloxime to a fixed concentration of enzyme (4 μM) at the emission wavelength 450 nm. Excitation wavelength was set at 360 nm.

Although both Gly242 and Gly244 were located in the Gly-rich loop that has been suggested as an unique feature of the ATP-binding site [5, 6], our results did not show a direct critical role of Gly244 in ATP binding. Rather than direct involvement in catalysis, Gly242 and Gly244 were proposed to be taking part in maintaining the flexibility of the Gly-rich loop and thus facilitating the substrate binding in PK.

Acknowledgment
The authors would like to thank Dr. M. S. Wong for useful discussion. This work was supported by the Area of Strategic Development Grant (A016).

References

[1] McCormick, D.B. and Snell, E. E. (1959). *Proc. Natl. Acad. Sci. USA*, 45, 1371-1379.
[2] Kerry, J. A., Rohde, M. and Kwok, F. (1986). *Eur. J. Biochem.* 158, 581-585.
[3] Gao, Z.G., Lau, C. K., Lo, S. C. L., Choi, S. Y., Churchich, J. E. and Kwok, F. (1998). *Int. J. Biochem. & Cell. Bio.* 30, 1379-1388.
[4] Bradford, M. M. (1976). *Anal. Biochem.* 72, 248-254.
[5] Walker, J. E., Saraste, M., Runswick, M. J. and Gay, N. (1982). *EMBO J.* 1, 945-951.
[6] Traut, T. W. (1994). *Eur. J. Biochem.* 222, 9-19.

Biochemistry and Molecular Biology of Vitamin B_6
and PQQ-dependent Proteins
ed. by A. Iriarte, H.M. Kagan and M. Martinez-Carrion

The Contribution of a Conformationally-Mobile, Active-Site Loop to the Reaction Catalysed by Glutamate Semialdehyde Aminomutase.

R. Contestabile, S. Angelaccio, R. Maytum[§], F. Bossa and R.A. John[§].

Dipartimento di Scienze Biochimiche "A. Rossi Fanelli" and Centro di Biologia Molecolare del CNR, Università di Roma "La Sapienza", Piazzale Aldo Moro 5, 00185 Roma, Italy and [§]Cardiff School of Biosciences, University of Wales, Cardiff CF1 3US, UK

Summary

The behaviour of glutamate semialdehyde aminomutase is markedly affected by the extent to which the central intermediate in the reaction, 4, 5-diaminovalerate, is allowed to dissociate from the enzyme. The kinetic properties of the wild-type enzyme are compared with those of a mutant form in which a flexible loop, that reversibly plugs the entrance to the active site, has been deleted by site-directed mutagenesis. The deletion has three distinct effects: i) the dissociation constant for diaminovalerate is increased 100-fold; ii) the catalytic efficiency of the enzyme is lowered 30-fold; iii) during the aminomutase reaction, the mutant enzyme undergoes absorbance changes not seen with the wild-type. These are proposed to be due to an abortive reaction of the substrate.

Introduction

Glutamate semialdehyde aminomutase (GSAM) catalyses the isomerisation of L-GSA into aminolevulinate (ALA). The enzyme is clearly related to the aminotransferases in both tertiary structure and reaction mechanism. However, GSA, the only substrate in the aminomutase reaction, is the source of both carbonyl and amino groups. Therefore, in principle, the initial reaction in the mechanism could be external imine formation with either the amino- (E_M) or the imine-form (E_L) of the enzyme. Depending on this first step, the mechanism entails the formation of either a dioxo or a diamino central intermediate, which is, at the same time, the product of the first half-transamination and the substrate of the second half. Although it is demonstrated (1) that the enzyme starts the catalytic cycle in the E_M-form and that 4, 5-diaminovalerate (DAVA) is the central intermediate, it is not clear why and how the other alternative is precluded. Dissociation of the product of the first half-transamination, essential to the continuation of the reaction in the aminotransferase ping-pong mechanism, appears instead detrimental to the catalytic efficiency of GSAM. In fact, DAVA is released from the enzyme, at least to some extent (1), thereby leaving the cofactor partially as an internal aldimine of PLP. The crystal structure of the homodimeric enzyme from *Synechococcus* shows significant asymmetry in the conformation of a polypeptide chain loop which apparently controls the access to the substrate binding pocket (2). To assess the contribution of the loop to the enzyme's mechanism, we have produced a mutant form (GSAMdel) in which a portion of the loop that forms an helical active site plug (residues 159-172) has been deleted and replaced with a glycine residue. The study compares the steady-state kinetic properties of wild-type and mutant enzymes.

Materials and methods

All reactions were carried out in 0.1 M Na-tricine buffer at pH 7.9 and at 25 °C. [ALA] was measured as previously described (3). Curve-fitting and statistical analyses were performed using the data manipulation software, Scientist (Micromath, Salt Lake City, UT). The dependence of the steady state velocity on total enzyme concentration ($[E_0]$) was fitted by combining Eq. 1 with the quadratic relationship (Eq. 2) that predicts how the steady-state concentration of the enzyme-substrate complex [ES] increases non-linearly as a function of total enzyme concentration.

$$v = k_{cat}[ES] \qquad \text{(Eq. 1)}$$

$$\left(1+\frac{Ks}{[S]}\right)^2[ES]^2 - \left\{2\left(1+\frac{Ks}{[S]}\right)[E_0]+K_d\right\}[ES]+[E_0]^2 = 0 \qquad \text{(Eq. 2)}$$

Results and discussion

Changes in cofactor absorbance at 420 nm and the ALA produced were measured after mixing the wild-type enzyme (0.3 μM) in the E_M-form with L-GSA (115 μM), in absence or in presence of DAVA (350 μM), until all the substrate was virtually consumed (Fig 1a). Best global fitting of the data to the Michaelis-Menten relationship gave estimates of K_m and k_{cat} (Table 1). The deletion mutant did not have detectable catalytic activity in the same conditions. However, at higher enzyme concentration (14.5 μM), ALA was formed at a rate similar to that seen for the wild-type enzyme. In the presence of DAVA, the kinetic behaviour of GSAMdel conformed closely to the Michaelis relationship with k_{cat} being decreased 13-fold and K_m increased 2.5 fold relative to the wild-type enzyme. However, when DAVA was not added, a slow increase in A_{420} continued until the substrate was almost exhausted, whereupon A_{420} decreased to a final value that was significantly higher than that at the start (Fig 1b). The slow increase in A_{420} was not present at all in the wild-type nor in the mutant enzyme reaction when DAVA was included. This suggests that it is a reaction of GSAMdel in the E_L-form, arising from DAVA dissociation. The formation of this chromophore stops abruptly when GSA is depleted, consistently with the hypothesis that the

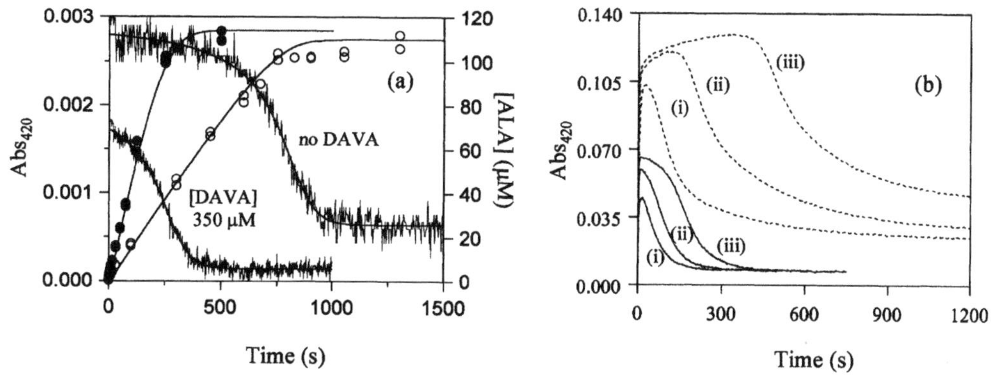

Fig. 1 *Panel a) Correlation of changes in cofactor absorbance with product formation (wild-type enzyme).* Changes in cofactor absorbance were observed at 420 nm (continuous noisy lines) after mixing the enzyme with the substrate. The observed initial Abs_{420} is actually formed during the transient phase of the reaction. The ALA produced in the reaction was measured in presence (●) or in absence (O) of DAVA. The smooth lines through both sets of experimental data are those predicted by a model based on numerical integration of the Michaelis relationship. *Panel b) Mutant enzyme.* Changes in Abs_{420} observed upon mixing GSAMdel with GSA at three concentrations (i, 60 μM, ii, 120 μM and iii 240 μM) in presence of 10 mM DAVA (——) or in absence of added DAVA (-----).

reaction is with GSA itself. Therefore, we propose that GSAMdel in the E_L-form combines with GSA to form an abortive complex (intermediate III in Scheme 1). This hypothesis would also explain why the reaction mechanism begins by reaction of the substrate carbonyl group with E_M rather than reaction of the amine with E_L. In the wild-type enzyme, the loop apparently prevents the formation of this abortive complex. In the absence of added DAVA, the steady-state rate of the reaction catalysed by low concentrations of GSAM increases non-linearly with increasing enzyme concentration. This unusual behaviour is attributed (1) to partial dissociation of DAVA during the transient phase that precedes the steady state. Addition of sufficiently high concentrations of DAVA results in a linear dependence (Fig 2, inset). The reaction catalysed by GSAMdel also showed the same behaviour, but non-linearity continued to much higher enzyme concentrations (Fig 2). Comparison of the dissociation constants estimated using Eqs 1 and 2

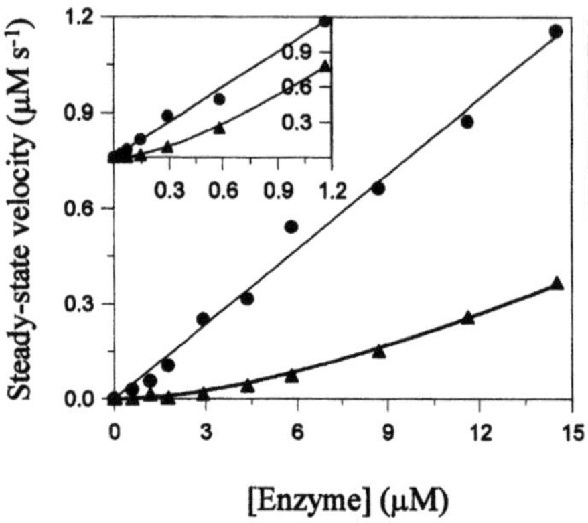

Fig. 2 The rate of ALA formation was measured at constant [GSA] (125 μM), in the absence (▲) or in presence (●) of 10 mM DAVA. The best fit value for DAVA K_d with the wild-type enzyme (inset) was estimated to be 0.29 ± 0.007 μM, while K_d with GSAMdel was 33 ± 3 μM.

Table 1

Enzyme form	K_m (μM)	k_{cat} (s^{-1})	k_{cat}/K_m (mM^{-1} s^{-1})
Wild-type	9	0.55	61
Wild-type + DAVA	29	1.99	69
GSAMdel + DAVA	71	0.15	2.1

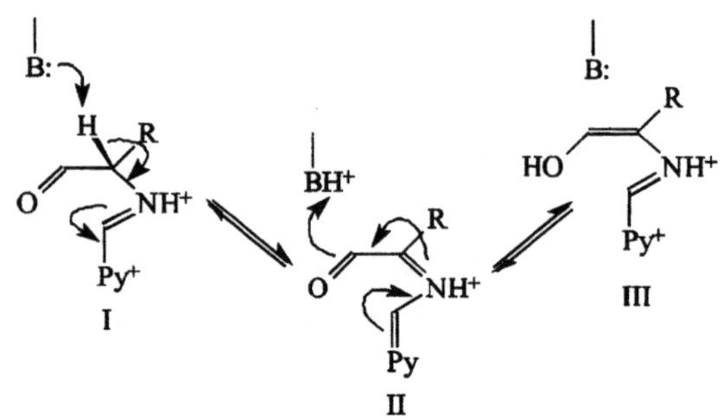

Scheme 1 Proposed abortive reaction between GSA and the E_L-form of the enzyme.

indicates that, during the steady state, DAVA binds approximately 100 times less tightly to GSAMdel (Fig 2). Therefore, a further contribution of the loop to catalysis appears to be that of preventing as far as possible the dissociation of DAVA, which however still occurs to a limited extent and may be regarded as an ancestral reminiscence of the common evolutionary origins with aminotransferases.

Acnowledgements
This work was supported by the UK B.B.S.R.C. and by the Italian M.U.R.S.T.

References

1. Tyacke, R. J. Harwood, J. L. and John, R. A. (1993) *Biochem. J* 293:, 697-701
2. Hennig, M., Grimm, B., Contestabile, R., John, R. A. & Jansonius, J. N. (1997) *Proc Natl. Acad Sci.* USA 94: 4866-4871
3. Pugh, C.E., Nair, S.P., Harwood, J.L. and John, R.A. (1991) *Anal. Biochem.* 198: 43 - 46

Biochemistry and Molecular Biology of Vitamin B_6
and PQQ-dependent Proteins
ed. by A. Iriarte, H.M. Kagan and M. Martinez-Carrion

The Reactions of Glutamate 1-Semialdehyde Aminomutase with *(R)* and *(S)* Enantiomers of a Novel, Mechanism-Based Inhibitor, 2, 3-Diaminopropyl Sulfate.

R. A. John, K. Khayer[#], T. Jenn[#], M. Akhtar[#], D. Gani[#] and R. Contestabile[§]

School of Biosciences, University of Wales, Cardiff, PO Box 911, Cardiff, UK; [#]School of Chemistry, University of Birmingham, Birmingham UK and [§]Istituto di Chimica Biologica, University of Rome "La Sapienza", Piazza Aldo Moro 5, 00185 Roma, Italy

Summary

Glutamate semialdehyde aminomutase is a recognised target for selective herbicides and antibacterial agents because it provides the aminolevulinate from which tetrapyrroles are synthesised in plants and bacteria but not in animals. The reactions of the enzyme with *(R)* and *(S)* enantiomers of a novel compound, 2, 3 diaminopropyl sulfate, designed as a mechanism-based inhibitor of the enzyme, are described. The *(S)*-enantiomer undergoes transamination without significantly inactivating the enzyme. The *(R)*-enantiomer inactivates the enzyme rapidly. Inactivation is accompanied by the formation of a 520 nm absorbing chromophore and by elimination of sulfate.

Introduction

In plants, the 5-aminolevulinate (ALA) from which tetrapyrroles are assembled is made from glutamate 1-semialdehyde in a reaction catalysed by the PLP-dependent enzyme glutamate semialdehyde aminomutase (GSAM) (1). This enzyme is absent from mammals, which instead synthesize ALA from glycine and succinyl-CoA in a reaction catalysed by ALA synthase. Because no counterpart of GSAM exists in animals, the enzyme is a potential target for safe, selective herbicides and anti-bacterial agents. Although GSAM catalyses the exchange of amino and carbonyl functions within the same molecule, it is clear that the enzyme is structurally homologous to the aminotransferases (2) and that the reaction mechanism is analogous to classical transamination. Successful mechanism-based inhibitors of aminotransferases have been designed by including a good leaving group at the β-carbon of an amino acid, thereby producing a structure which resembles the true substrate sufficiently to bind at the active site and to undergo all of the catalytic steps of the natural reaction. Elimination of the β-substituent during the reaction produces an intermediate which reacts covalently with the enzyme thereby inactivating it (3). GSAM-catalysed reaction begins and ends with the enzyme in the pyridoxamine form (E_M) (4). However, the catalytic mechanism requires the formation of a pyridoxaldimine form of the enzyme. The intermediate 4,5-diaminovalerate (DAVA) dissociates from this form of the enzyme to some extent (4), to give significant concentrations of a form of the enzyme (E_L) in which the cofactor is present as an internal aldimine of PLP. Because free DAVA reacts rapidly with E_L, the structurally analogous 2, 3 diaminopropyl sulfate (DAPS) might be expected to react but to undergo β-elimination (Scheme 1) with possible inactivation of the enzyme.

Materials and methods

Details of the synthesis of the enantiomers of DAPS will be reported elsewhere (5). All reactions were carried out in 0.1 M tricine buffer pH 7.9 at 25 °C. GSAM activity was measured in the presence of DAVA as described earlier (4). DAPS was quantified by its reaction with *o*-phthaldialdehyde. Solutions containing DAPS were mixed with an equal volume of 60 mM *o*-phthaldialdehyde, containing 230 mM 2-mercaptoethanol, in 0.1 M $NaHCO_3$ at pH 10. After 5 min, absorbance at 452 nm was measured and concentrations were calculated using a value of $\varepsilon_{452} = 6\ \mu M^{-1}\ cm^{-1}$ established in separate experiments. Sulfate was measured by turbidimetry (6).

Results and discussion

Both *(R)*- and *(S)*-enantiomers of DAPS reacted with the enzyme to produce rapid changes in the absorption spectrum of the enzyme-bound cofactor. The spectral changes were distinctly different for each enantiomer. The major change observed with *(S)*-DAPS was the conversion of the chromophore absorbing maximally at 420 nm, due to the E_L-form, to one absorbing maximally at 340 nm. Addition of succinic semialdehyde (SSA) converted the enzyme back to its initial 420 nm absorbing E_L-form. Assay of enzyme activity in samples taken from the reaction mixtures, including those to which SSA had been added, showed no detectable loss of enzyme activity at any stage in the reaction. The reaction of E_L with *(R)*-DAPS produced very different changes in the enzyme absorption spectrum. The major change was a large, initially rapid, increase in absorbance with a maximum at 520 nm. Inclusion of SSA in the reaction mixture almost doubled the amplitude of the increase in A_{520} and the reaction became monophasic. Measurement of enzyme activity in samples withdrawn at time intervals from identical solutions, showed that inactivation paralleled the increase in 520 nm absorbance (Fig 1a). The reaction of *(R)*-DAPS with

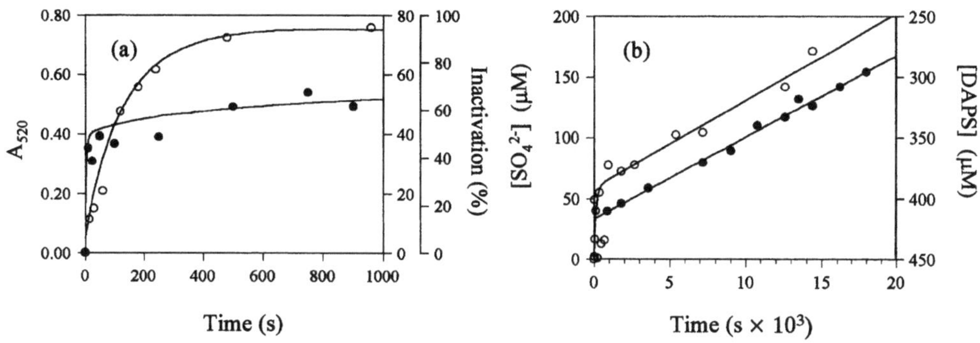

Fig. 1 *Panel (a).* Correlation of changes in absorbance at 520 nm with loss of enzyme activity during reaction with *(R)*-DAPS. The enzyme (15 μM) was mixed with *(R)*-DAPS (10 mM). The continuous lines are A_{520}. The symbols show loss of enzyme activity in the absence of succinic semialdehyde (●) and in the presence of 5 mM SSA (O). Percentage inactivations on full scale are for the experiment with SSA. *Panel (b).* Production of sulfate and depletion of substrate. Same conditions as in panel (a) with 450 μM *(R)*-DAPS. Inorganic sulfate (●) and DAPS (O, shown on an inverted scale) were measured in samples removed at time intervals.

the enzyme resulted in a burst, followed by a slow, almost constant rate of sulfate production. The rate of depletion of *(R)*-DAPS was also monitored and found to coincide closely with sulfate production except that a burst of greater amplitude was observed (Fig 1b). The observed bursts are likely to be due to slowing of the β-elimination reaction as a result of the inactivation, while part of the substrate is also consumed in a half-transamination reaction (Scheme 1, react. 3a). The inactivated enzyme was found to recover 98.3 % of the original activity when, after 1 hr reaction with 1 mM *(R)*-DAPS and 5 mM SSA, it was separated from DAPS by gel filtration on Sephadex G-25. The activity returned in an exponential process governed by a rate constant of 0.54 hr^{-1}. The difference between the absorption spectra produced when GSAM reacts with *(S)*- or *(R)*-DAPS is clearly due to the different stereochemistry at C-2. The absorbance change observed with (*S*)-DAPS is the same as that occurring with the natural intermediate, DAVA. It seems certain that these changes are due to reversible conversion of the enzyme to the E_M-form and we

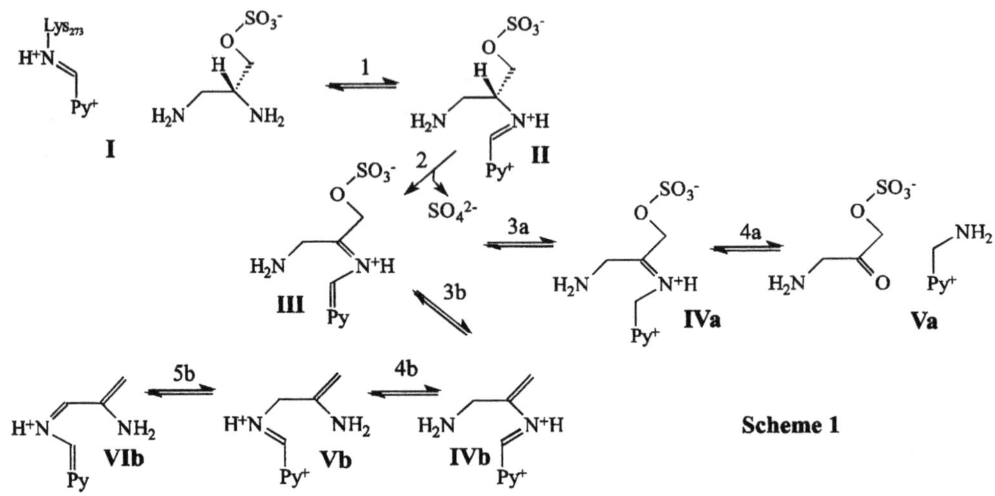

Scheme 1

can conclude that *(S)*-DAVA acts predominantly as a transamination substrate, with the reaction probably occurring at the non-chiral C-3. The production of the inactive 520nm-absorbing chromophore by *(R)*-DAPS shows that a reaction, other than transamination, predominates with this enantiomer. Elimination of sulfate by the mechanism of Scheme 1 results in the intermediate IVb, from which a quinonoid structure could be formed by attack of an enzyme nucleophile at Cβ. However, there seems to be no reason for this quinonoid structure to be stable. A further quinonoid structure (VIb) would be possible after transaldimination between intermediates IVb and Vb (react. 4b), followed by labilisation of a C3 proton. The extensive conjugation in the resulting intermediate would account for the stability and high λ_{max} of the inactive chromophore.

Acknowledgements
This work was supported by the UK B.B.S.R.C. and by the Italian M.U.R.S.T. Roberto Contestabile was the recipient of a fellowship from Istituto Pasteur-Fondazione Cenci Bolognetti.

References

1. Kannangara, G., Gough, S. P., Bryant, P., Hoober, J. K., Kahn, A. and von Wettstein, D. (1988) *Trends Biochem. Sci.* 13: 139-143
2. Hennig, M., Grimm, B., Contestabile, R., John, R. A. And Jansonius, J. N. (1997) *Proc. Natl. Acad. Sci.* U.S.A. 94: 4866-4871
3. John, R. A. and Fasella, P. (1969) *Biochemistry* 8: 4477-4482
4. Tyacke, R., Harwood, J. and John, R. A. (1993) *Biochem. J.* 293: 697-701
5. Gani D. et al. (2000), unpublished results
6. Sorbo, (1987) *Methods in Enzymol.* 143: 3-6

Biochemistry and Molecular Biology of Vitamin B_6
and PQQ-dependent Proteins
ed. by A. Iriarte, H.M. Kagan and M. Martinez-Carrion

STRUCTURE AND FUNCTION OF *E. COLI* PYRIDOXINE PHOSPHATE OXIDASE

Martino di Salvo, Emily Yang, Martin Safo, Faik Musayev, and Verne Schirch

Department of Biochemistry and Molecular Biophysics and Institute for Structural Biology and Drug Discovery, Virginia Commonwealth University, Richmond, Virginia 23298

SUMMARY:

Escherichia coli pyridoxine 5'-phosphate oxidase was purified from an overexpressing plasmid containing the *pdxH* gene. The enzyme is a 53 kDa homodimer with 1 FMN bound non-covalently to each subunit. A method was developed for removing the bound FMN to generate an apoenzyme with no absorbance maxima above 300 nm. Complete activity could be restored by addition of FMN. The apoenzyme was observed to bind pyridoxal 5'-phosphate tightly at a non-catalytic site. The tightly bound pyridoxal 5'-phosphate did not dissociate during size exclusion chromatography, and did not affect the maximum catalytic activity of the enzyme with pyridoxine 5'-phosphate as substrate. A crystal structure of the enzyme has been determined to 1.8 Å.

Introduction

Pyridoxal 5'-phosphate (PLP) is a required cofactor for most enzymes in amino acid metabolism. In *E. coli* the last step in the *de novo* biosynthesis of PLP is shown in Eq. 1 and is catalyzed by pyridoxine 5'-phosphate oxidase (PNPOx) (1). In addition to using PNP as a substrate it can also form PLP from pyridoxamine 5'-phosphate (PMP). In mammalian cells, as well as in most other cells, PNPOx serves as a part of the salvage pathway for reusing PLP after

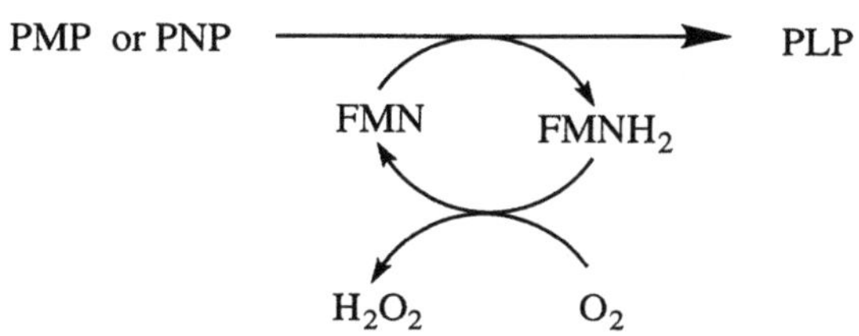

(Eq. 1)

protein degradation (2). This enzyme has been purified from rabbit liver, sheep and pig brain, and *E. coli* and in each case it is a homodimer of subunits that are in the range of 25 to 27 kDa (2 and Ref. therein).

Currently there are 11 amino acid sequences for PNPOx in the protein database. The core of all sequences shows a high degree of homology with most differences in size and sequence occurring at the N-terminus.

Materials and Methods

PNPOx was purified from an overexpressing pET22b plasmid in *E. coli* strain HMS174(DE3). Purification involves ammonium sulfate and protamine sulfate fractionation and chromatography on CM-Sepharose and hydroxylapatite. Thirty mg of pure enzyme are obtained from 2 liters of cells (2).

The enzyme is yellow when it is purified due to the bound FMN. The cofactor can be removed by chromatographing on a phenyl Sepharose column in 100 mM potassium monobasic phosphate, pH 5, with 20% saturation of ammonium sulfate. The column is washed with this buffer until no more FMN is being eluted. The apoenzyme is then eluted from the column with 20 mM potassium phosphate, pH 7.3, containing 5 mM dithiotreitol.

Catalytic activity is followed by observing the rate of formation of PLP as determined by its absorbance at 388 nm. Increased sensitivity can be obtained by using 50 mM Tris, pH 8.0, as the buffer, which forms a 414nm absorbing aldimine with the PLP.

Protein concentration was determined from the number of Trp, Tyr, and Cys residues, and the spectra of apoenzyme in both 6 M guanidine hydrochloride and 20 mM potassium phosphate, pH 7.3 (4).

Results and Discussion

Stoichiometry of FMN binding-Previous investigations had shown that there is only one FMN molecule bound per dimer of PNPOx. When the amino acid composition of the enzyme became available it was clear that the large number of Arg and Trp residues would make determining the true concentration of the enzyme very difficult by either absorbance or dye binding assays. We used three methods to reexamine the stoichiometry of FMN binding to PNPOx (2). The first method determined the concentration of FMN from its absorbance at 450 nm and the concentration of protein from the absorbance of the apoenzyme at 278 nm. The molar extinction coefficient was determined from the number of Trp, Tyr, and Cys residues in the protein (4). In the second method, we compared the FMN concentration, as determined from absorbance at 450 nm, to the concentration of Cys residues, determined by titration with dithionitorbenzoic acid. There is only a single Cys residue in a PNPOx subunit. The third method took advantage of the greater than 90%

quenching of fluorescence of FMN when bound to PNPOx. We titrated apoPNPOx with known concentrations of FMN and plotted the fluorescence against the ratio of apoPNPOx/FMN. All three methods showed that there was one FMN bound per subunit of PNPOx.

Kinetic properties of PNPOx-The K_m for PNP is 2 μM in Tris, phosphate, and Hepes buffers, at 37 °C and pH 7.6. The value of k_{cat} in Tris buffer is 0.3 s^{-1} and in phosphate and Hepes buffers it is 0.2 s^{-1}. These values are in agreement with previously determined values for *E. coli* PNPOx (3). However, they differ significantly from *E. coli* PNPOx purified as a maltose binding fusion protein (5).

Evidence for tight binding of PLP- Previous investigators had shown that the product PLP had high affinity for the enzyme and served as a product inhibitor (3). However, the K_i values were in the low μM range and this concentration of free PLP is unlikely to exist in the cell because PLP is very reactive and would form complexes with many other nucleophiles and proteins. In an attempt to understand the binding of PLP to PNPOx, we looked at PLP binding to the apoenzyme where the absorbance of FMN would not interfere with the spectral properties of PLP. A shift in the 388 nm to 398 nm was observed when a stoichiometric amount of PLP was added to apoPNPOx. Passing the complex down a Bio-Gel P6-DG desalting column resulted in complete retention of the PLP with the protein peak (6).

The stoichiometry of the PNPOx•PLP complex was determined from the known absorbance extinction coefficient of the protein at 278 nm and the extinction coefficient of PLP after it was released from the protein by 0.1 M NaOH. There was one PLP tightly bound per subunit of apoPNPOx. The experiment was repeated with holoPNPOx, and again we found one PLP tightly bound per subunit of the enzyme.

The tightly bound PLP was not at the active site since assaying the enzyme with PNP resulted in normal catalytic activity. This suggested that PLP was bound tightly at a non-catalytic site, which raises the question of how this PLP gets transferred to other B_6 requiring apoenzymes. This was tested by adding the PNPOx•PLP complex to a solution of aposerine hydroxymethyltransferase (SHMT) and determining the rate of return of catalytic activity of this enzyme. The results showed that the PLP is transferred completely to apoSHMT and the rate is the same as if we added an equivalent amount of free PLP. This experiment was repeated in a buffer containing 300 mg per ml of an *E. coli* extract that lacked any SHMT activity. This was to simulate in vivo conditions where PLP would be expected to react with other nucleophiles, such as thiols, and proteins in the extract and be trapped from forming a productive complex with apoSHMT. Under these conditions apoSHMT was activated several fold faster with PNPOx•PLP than with free PLP, suggesting the tightly bound PLP on PNPOx is targeted to apoSHMT.

Crystal structure of E. coli PNPOx-Crystals of PNPOx were grown, data collected, and the diffraction pattern solved to 1.8 Å resolution (7). The structure of the yeast PNPOx is available from the protein data bank, but there has been no report about the properties of this enzyme. The fold of the *E. coli* enzyme is very similar to that of the yeast PNPOx, as both are characterized by an eight-stranded β-sheet surrounded by five α-helical structures. At the dimer interface are two grooves that extend from the top to the bottom of the dimer. In each of these two grooves is a molecule of FMN bound with the ribityl and phosphate moieties buried deeply at the narrow end of the groove and the isoalloxazine ring bound at the more open end. There is evidence that the open end, which is near N5 of FMN, is the site of PNP binding.

Each subunit also contains a bound molecule of inorganic phosphate that is too far from the active site to be the location of the phosphate moiety of PNP. This phosphate may occupy the location of the phosphate moiety of PLP at the non-catalytic tight binding site discussed above. PLP was modeled into this site and fits well with several groups on the enzyme being within hydrogen bonding distance to the functional groups on the PLP ring.

Acknowledgements

This work was supported by NIH grants GM28143 and HL32793. We thank Dr. Irimpan Mathews and Dr. Daniel J. Thiel for there help in collecting data and interpretation of the results.

References

1. Dempsey, W. B. (180) Synthesis of Pyridoxal Phosphate, in *Escherichia coli* and *Salmonella typhimurium. Cellular and Molecular Biology* (Neidhardt, F. C., Ingraham, J. L., Los, K. B., Magasanik, B., Schaechter, M., and Umbarger, H. E., Eds.) pp. 539-543. *Am. Soc. for Microbiol.*, Washington, DC.

2. Di Salvo, M., Yang, E., Zhao, G., Winkler, M. E., and Schirch, V. (1998) Expression, Purification, and Characterization of Recombinant *Escherichia coli* Pyridoxine 5'-Phosphate Oxidase. *Prot. Express. Purif.* 13: 349-356.

3. Zhao, G., and Winkler, M. E. (1995) Kinetic Limitation and Cellular Amount of Pyridoxine (Pyridoxamine) 5'-Phosphate Oxidase of *Escherichia coli* K-12. *J. Bacteriol.* 177: 883-891.

4. Gill, S. C., and von Hippel, P. H. (1989) Calculation of Protein Extinction Coefficients from Amino Acid Sequence Data. *Anal. Biochem.* 182: 319-326.

5. Notheis, C., Drewke, C., and Leistner, E. (1995) Purification and Characterization of the Pyridoxal 5'-Phosphate:Oxygen Oxidoreductase (Deaminating) from *Escherichia coli. Biochim. Biophys. Acta.* 1247: 265-271.

6. Yang, E. S., and Schirch, V.: Tight Binding of Pyridoxal 5'-Phosphate to Recombinant *Escherichia coli* Pyridoxine 5'-Phosphate Oxidase. submitted.

7. Safo, M., Mathews, I., Musayev, F. N., di Salvo, M. L., Thiel, D. J., Abraham, D. J., and Schirch, V.: X-ray Structure of Escherichia coli Pyridoxine 5'-Phosphate Oxidase Complexed with FMN at 1.8 Å Resolution. submitted

MOLECULAR PATHOLOGY AND MEDICINE

Biochemistry and Molecular Biology of Vitamin B_6
and PQQ-dependent Proteins
ed. by A. Iriarte, H.M. Kagan and M. Martinez-Carrion

Role of Branched Chain Aminotransferase Isoenzymes in the Central Nervous System

Susan Hutson, Erich Lieth[*], Kathryn F. LaNoue[#]

Department of Biochemistry, Wake Forest University School of Medicine, Winston-Salem, NC 27157, USA, [*]Departments of Neuroscience and Anatomy, [#]Cellular and Molecular Physiology, Pennsylvania State University College of Medicine, Hershey, PA 17033, USA

Summary

A novel hypothesis for the function of branched chain amino acid (BCAA) transamination in regulating levels of the major excitatory neurotransmitter glutamate in the central nervous system is described. The BCAA shuttle hypothesis predicts that BCAA are required for *de novo* glutamate synthesis in brain. Specifically, the branched chain aminotransferase isoenzymes (mitochondrial BCATm and cytosolic BCATc), which are localized in different cells types, operate in series to provide nitrogen for optimal rates of glutamate synthesis.

Introduction

The first step in catabolism of the nutritionally essential branched chain amino acids (BCAA), leucine, isoleucine, valine, is transamination catalyzed by the branched chain aminotransferase (BCAT) isoenzymes (mitochondrial BCATm and cytosolic BCATc) to form the branched chain α-keto acids (BCKA). BCATm is found in most tissues (1) whereas BCATc is expressed almost exclusively in the central nervous system (2). Additionally, BCATc, but not BCATm, is a target of the anticonvulsant and neuroprotective drug, and leucine analog, gabapentin (3). The second catabolic step, irreversible oxidative decarboxylation of the transamination products catalyzed by the mitochondrial BCKA Dehydrogenase, commits the BCAA to oxidative degradation and results net transfer of nitrogen from the BCAA to nonessential amino acids such as glutamate.

The BCAA, particularly leucine, are known to cross the blood-brain barrier rapidly. In brain slices, BCAA are metabolized more rapidly than they are incorporated into protein (5). On the other hand, the ratio of BCAT/BCKA dehydrogenase activity in brain tissue is high (3) suggesting that within the brain BCAA must be needed for something other than fuel. Recent studies suggest BCAA transamination provides nitrogen for the formation of glutamate in the glia (6), needed because nitrogen sources such as aspartate are limited in glia (7,3). Recently, we have proposed that BCAA participate in a nitrogen shuttle that is obligatory for optimal rates of *de novo* glutamate synthesis in brain (3).

Methods

Male Sprague Dawley rats (200-400g) rats were anesthetized with-nembutal, eyes enucleated, and retinas dissected in ice-cold buffer. Half-retinas were preincubated at 37°C for 3 minutes in Krebs buffer (pH 7.4) also containing 20 mM HEPES, 5 mM glucose, 0.05 mM NH_4Cl, 25 mM $NaH^{14}CO_3$, 0.2 mM pyruvate. To measure CO_2 fixation, incubations were initiated by addition of 10 μCi/ml $H^{14}CO_3$. Vessels were then closed to the atmosphere. At 20 min the half retinas were removed and placed in 2% perchloric acid. The buffer was then acidified with perchloric acid and unreacted $^{14}CO_2$ was allowed to diffuse out of all the samples. [1-^{14}C]Leucine (100 μM) or [1-^{14}C]KIC (100 μM) metabolism, with and without 1.0 mM gabapentin, was measured at 30 min as described in Hutson et al (3). Carbon 14 metabolites were separated and quantified by Dowex-1 acetate chromatography as described (7). The mass amount of the glutamate was measured fluorometrically (7) and glutamine was measured after chromatographic separation from glutamate followed by a luminometric determination (7).

Coronal sections (30 μm) of paraformaldehyde fixed rat brain were used. Free floating sections were stained after pretreatment with 10% methanol, 0.3% H_2O_2 in 0.1M PBS. Sections were rinsed in PBS, blocked in 3% normal goat serum (Vector Laboratories, Burlingame, CA), and incubated overnight at 4°C with immunoaffinity purified rabbit anti-rat BCATc peptide antibody (1.2 μg/ml). The secondary antibody was goat anti-rabbit IgG 1:250 (Jackson Immuno), and samples were developed using a standard diaminobenzidine reaction. With the immunoaffinity purified rabbit anti-human BCATm antibody, fresh frozen sections were used. For the control, antibody was preabsorbed overnight with a 50-fold excess of purified BCATc peptide, and tissue slices were incubated with the peptide for 30 min prior to incubation with preabsorbed antibody.

Results and Discussion

Several drugs commonly used to treat brain diseases, including gabapentin (8), influence glutamate turnover by unknown mechanisms. Since glutamate is the most widely utilized excitatory neurotransmitter in the nervous system, it is critical to understand how neurons replenish the glutamate they release during neurotransmission. The brain glutamate/glutamine cycle (Fig. 1) is a metabolic pathway which involves the synaptic release of glutamate from neurons, rapid and efficient glutamate uptake from the synaptic space by glia, conversion of glutamate to glutamine by astrocytic glutamine synthetase followed by release of glutamine to the interstitium and uptake by the neurons for conversion back to glutamate (9). This process efficiently prevents excessive accumulation of glutamate in the interstitium that would induce excitotoxic neurodegeneration.

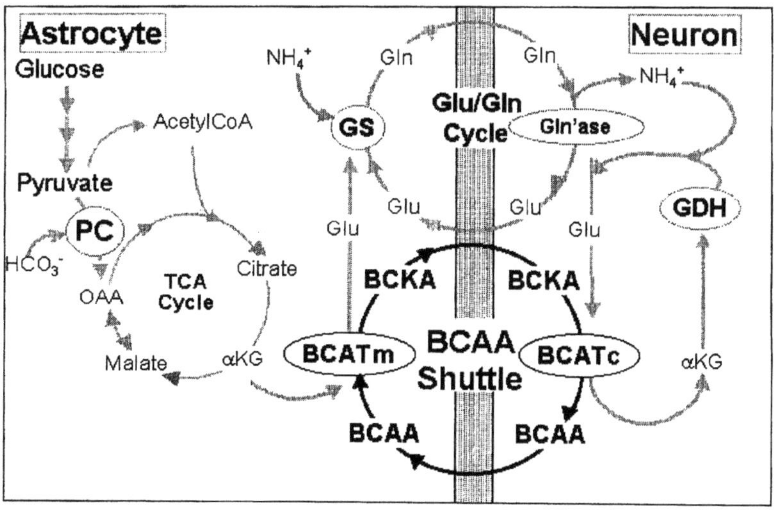

Fig. 1. BCAA Shuttle and brain glutamate metabolism. Gln'ase = glutaminase; PC = pyruvate carboxylase, αKG = α-ketoglutarate, GDH = glutamate dehydrogenase.

Studies (7, 10) of glutamate metabolism in cultured neonatal astrocytes indicate that up to 30% of the glutamate taken up by the astrocytes is not converted directly to glutamine but is actually metabolized to lactate via conversion of glutamate to α-ketoglutarate and subsequent citric acid cycle decarboxylation (Pyruvate/Malate Cycle, Fig. 1). If glial citric acid cycle decarboxylation occurs, equivalent rates of anaplerosis must take place to recover the lost carbon of glutamate, i.e., *de novo* glutamate synthesis. The anaplerotic enzyme, pyruvate carboxylase, is expressed abundantly in the brain, and localized to glia (11). Our previous studies of the rate of anaplerosis in neonatal cultured astrocytes and the resultant accumulation of citric acid cycle intermediates showed that pyruvate carboxylase was very active and dependent on levels of added pyruvate (3,7). The study also indicated that the conversion of the excess tricarboxylic acid cycle intermediates, formed by pyruvate carboxylase, to glutamate and glutamine could be severely limited by lack of a source of nitrogen. Since NH_3 could not supply the needed nitrogen (7), we proposed that BCAA might be the physiological source for the α amino nitrogen of glutamate (3).

Based on the lack of effect of the BCATc-specific drug gabapentin on glutamate metabolism in astrocyte cultures, immunocytochemical results, and results of western blot analysis of BCATm and BCATc localization in primary rat brain cultures, we proposed the BCAA shuttle which involves two BCAT isoenzymes proceeding in opposite directions, one in the glia and the other in neurons (Fig. 1). In the shuttle hypothesis BCAA enter the astrocyte where transamination is catalyzed by BCATm producing glutamate and BCKA. Since BCKA are poorly oxidized in astrocytes (3, 12), they exit the astrocyte and are taken up by the neuron. Neuronal BCATc catalyzes transamination of the BCKA with glutamate. The product, BCAA, exit from the neuron and return to the astrocyte. The α-ketoglutarate product in the neurons may undergo reductive amination to glutamate via neuronal glutamate dehydrogenase. Although reversible, the cycle will be driven in the direction shown in Fig. 1 by anaplerotic synthesis of α-ketoglutarate in astrocytes and by the higher activity of neuronal as opposed to astrocytic glutamate dehydrogenase under physiological ADP concentrations (7).

The model predicts that there is cell-specific localization of BCATm and BCATc in the central nervous system. Using immunoaffinity purified BCATc peptide antibodies, we have observed a neuronal pattern of staining for BCATc in several brain areas. The example in Fig. 2, BCATc immunostaining of neurons is observed in all layers of the somatosensory cortex with staining found primarily in the glutamatergic pyramidal neurons. BCATm staining in the cortex appeared to be in the astroglia. Both patterns are consistent with the shuttle hypothesis.

To determine whether BCAT isoenzymes are important in maintaining levels of glutamate in glutamatergic tissue, the *ex vivo* retina was employed as a model, and the effect of the competitive BCATc inhibitor, gabapentin (3), on metabolism of added leucine as well as rates of *de novo* glutamate and glutamine synthesis (no added leucine) was examined. Addition of gabapentin, resulted in about a 50-60% inhibition of added leucine transamination and oxidation (Fig. 3A). In a separate experiment, addition of gabapentin to the retina resulted in ~30% inhibition in the de novo synthesis of glutamine and glutamate from $^{14}HCO_3^-$ (Fig. 3B). These results are the first demonstration that inhibition of BCATc can affect de novo glutamate synthesis in an intact neural preparation. Therefore, results suggest that the BCAT isoenzymes operate in series in a shuttle that is required for optimal rates of *de novo* glutamate synthesis. Experiments to test this hypothesis in an *in vivo* model are in progress.

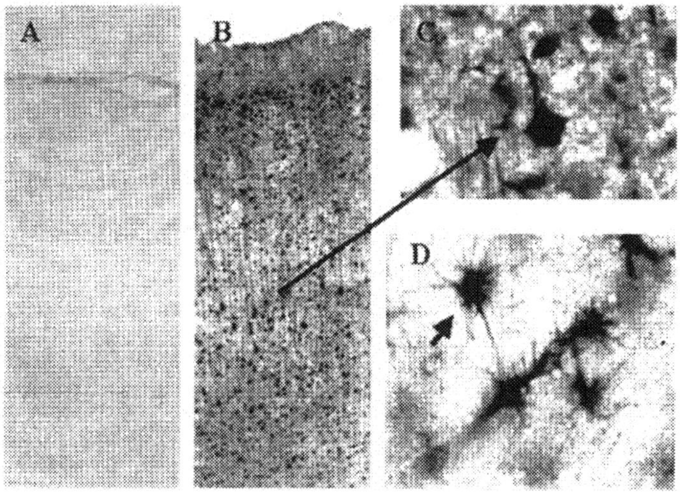

Fig. 2. BCATc (A-C) and BCATm (D) staining in rat brain somatosensory cortex. (A) Preabsorbed control. (B) BCAT+ cells and (D) pyramidal neuron. (D) BCATm+ astroglia.

Fig. 3 Influence of Gabapentin (GP) on leucine metabolism (Left Panel) and de *novo* glutamine and glutamate synthesis (Right Panel) in isolated rat retinas. See Methods. Abbreviations are: TA = transamination, Oxid = oxidation. Values are means ± SD in Left Panel (n = 4) and means ± SEM (n = 6-7, * p<0.05) in Right Panel.

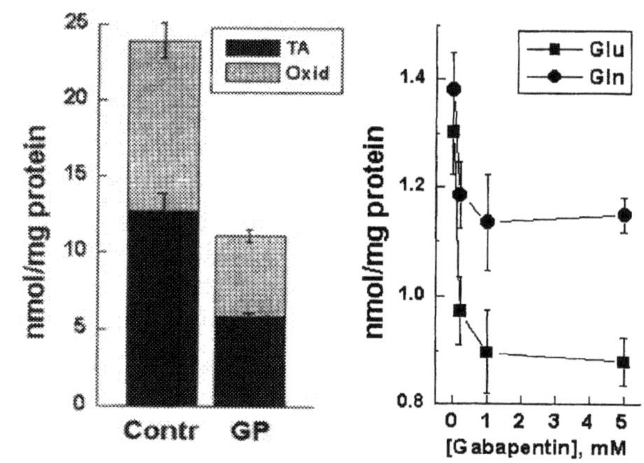

Acknowledgements

Supported by NIH DK34738 (SMH) and in part by grants#197038 and 1-199-678 from the Juvenile Diabetes Foundation International (EL).

References

1. Hutson, S. M., Wallin, R., and Hall, T. R. (1992) Identification of mitochondrial branched chain aminotransferase and its isoforms in rat tissues. *J. Biol. Chem. 267*:15681-15686.
2. Hall, T. R., Wallin, R., and Hutson, S. M (1993) Branched chain aminotransferase isoenzymes: purification and characterization of the rat brain isoenzyme. *J. Biol. Chem. 268*:3092-3098.
3. Hutson, SM, Berkich, DA, Drown, P, Xu, B, and LaNoue, KF: Role of branched-chain aminotransferase isoenzymes and gabapentin in neurotransmitter metabolism. J. Neurochem. **71**:863-874,1998.
4. Oldendorf, W. H. (1971) Brain uptake of radiolabeled amino acids, amines, and hexoses after arterial injection. *Am. J. Physiol. 221*:1629-1639.
5. Chaplin, E. R., Goldberg, A. L., and Diamond, I. (1976) Leucine oxidation in brain slices and nerve endings. *J. Neurochem. 26*:710-717.
6. Yudkoff, M., Daikhin, Y., Grunstein, L., Nissim, I., Stern, J., Pleasure, D., and Nissim, I. (1996) Astrocyte leucine metabolism: significance of branched-chain amino acid transamination. *J. Neurochem. 66*:378-385.
7. Gamberino, W. C., Berkich, D. A., Lynch, C. J., Xu, B., and LaNoue, K. F. (1997) Role of pyruvate carboxylase in facilitation of synthesis of glutamate and glutamine in cultured astrocytes. *J. Neurochem. 69*:2312-2325.
8. Taylor, C. P., Gee, N. S., Su, T. Z., Kocsis, J. D., Welty, D. F., Brown, J. P., Dooley, D. J., Boden, P., Singh, L. (1998) A summary of mechanistic hypotheses of gabapentin pharmacology. *Epilepsy Res. 29*:233-249.
9. Shank, R. P. and Aprison, M. H. (1981) Present status and significance of the glutamate cycle in neural tissue. *Life Sci. 28*:837-842.
10. Sonnewald, U., Westergaard, N., Peterson, S. B., Unsgärd, G., and Schousboe, A. (1993) Metabolism of [U-^{13}C]glutamate in astrocytes studied by ^{13}C NMR spectroscopy: incorporation of more label into lactate than into glutamine demonstrates the importance of the tricarboxylic acid cycle. *J. Neurochem. 61*:1179-1182.
11. Shank, R. P., Bennet, G. S., Freytag, S. O., and Campbell, G. L. (1985) Pyruvate carboxylase: an astrocyte-specific enzyme implicated in the replenishment of amino acid neurotransmitter pools. *Brain Res. 329*:364-367.
12. Bixel, M. G. and Hamprecht, B. (1995) Generation of ketone bodies from leucine by cultured astroglial cells. *J. Neurochem. 65*:2450-2461.

Biochemistry and Molecular Biology of Vitamin B_6
and PQQ-dependent Proteins
ed. by A. Iriarte, H.M. Kagan and M. Martinez-Carrion

Modulation of Gene Expression by Vitamin B_6

Yasuo Natori, Tatsuzo Oka and Masashi Kuwahata

Department of Nutrition, School of Medicine, The University of Tokushima, Kuramoto, Tokushima 770-8503, Japan

Summary

Gene expression of liver-specific proteins, such as serum albumin and aspartate aminotransferase, was found to be modulated by vitamin B_6 through a novel mechanism that involves inactivation of tissue-specific transcription factors by direct interaction with pyridoxal phosphate. We also found that the growth of human hepatoma HepG2 cells in culture and transplanted MH-134 hepatoma cells in C3H/He mice were inhibited by a large dose of pyridoxine. Possible mechanism and application of the supraphysiological dose of vitamin B_6 as an antineoplastic therapy was discussed.

Introduction

Vitamin B_6 has long been recognized for its importance as a cofactor for many enzymes involved particularly in amino acid metabolism. Apart from its role as coenzyme, recent studies are unveiling a new role of vitamin B_6 as a modulator of gene expression. Vitamin B_6 modulates gene expression by interaction with not only steroid hormone receptors but also with many tissue-specific transcription factors such as HNF-1 and C/EBP. Based on the new concept of vitamin B_6 action, we are now working on a possible application of supraphysiological dose of vitamin B_6 in the control of neoplastic cell growth.

Vitamin B_6 and Nucleic Acid-Metabolizing Enzymes

Pyridoxal 5'-phosphate (PLP) has been known, for some time, to be an effective inhibitor of many enzymes that have binding sites for phosphate-containing substrates or effectors, including RNA polymerase (1), reverse transcriptase (2) and DNA polymerase (3). In all these cases, PLP is far more effective than its analogues. It typically interacts with proteins by forming a Schiff base between its aldehyde group and primary amino groups, most commonly the ε -amino groups of lysine residues (4). These effects have been considered unphysiological because the inhibitory action required the presence of several mM PLP in the *in vitro* assay systems.

However, recent studies in our laboratory (5) showed that the activities of RNA polymerase I and II were increased in the liver of vitamin B_6-deficient rats. This finding indicates that the earlier demonstration *in vitro* of PLP as an inhibitor of RNA polymerase (1) may have some physiological significance.

Vitamin B_6 and Steroid Hormone Action

Another interesting feature of vitamin B_6 is its possible involvement in the enzyme induction by steroid hormones. Allgood et al. (6) have demonstrated that vitamin B_6 modulates transcriptional activation by the human glucocorticoid receptor in HeLa cells. Allgood and Cidlowski (7) have subsequently shown that the modulatory effect of vitamin B_6 on transcription activation is not limited to glucocorticoid receptor but extends to multiple members of steroid hormone superfamily. These results indicate that variations in the intracellular concentration of PLP can have pronounced modulatory effects on steroid-induced gene expression. Specifically, elevation of intracellular PLP levels leads to decreased transcriptional responses to glucocorticoid, progesterone, androgen, or estrogen hormones. Conversely, cells in a vitamin B_6-deficient state exhibit enhanced responsiveness to steroid hormones. An excellent review on this subject by Cidlowski and colleagues (8) has appeared recently.

Vitamin B_6 and Gene Expression of Liver Enzymes

The evidence for an interaction between PLP and steroid hormone receptors, described above, has been obtained with the use of cultured cells. However, indications of possible interaction between vitamin B_6 and steroid hormones in the whole animal have existed for some time. For instance, aspartate aminotransferase in rat-liver cytosol (cAspAT) is a PLP-dependent enzyme and the activity of the enzyme is induced by the administration of glucocorticoid hormone. Kondo and Okada (9) found that the induction of the enzyme in the liver of adrenalectomized vitamin B_6-deficient rats by hydrocortisone was suppressed by the administration of pyridoxine.

We recently found that the level of cAspAT mRNA in the liver of vitamin B_6-deficient rats was several fold higher than that of the control rats (10). The administration of hydrocortisone induced expression of the hepatic cAspAT mRNA and the induction was suppressed by the simultaneous administration of pyridoxine. Since the 5' regulatory region of the rat cAspAT gene contains several sequences showing homology to glucocorticoid-responsive elements (GRE), we synthesized an oligonucleotide probe of GRE sequence and assayed the binding activity of nuclear extract to the oligonucleotide by gel mobility shift analysis. We found that the binding activity of nuclear extract prepared from the liver of vitamin B_6-deficient rats was far greater than that of the control rats, indicating that the DNA-binding activity of glucocorticoid receptor was enhanced by vitamin B_6 deficiency. We further found that preincubation of the nuclear extract with PLP brought about a rapid and extensive decrease in the binding of the extract to GRE. Analogues of PLP, such as pyridoxamine 5'-phosphate (PMP), pyridoxal (PL), pyridoxamine (PM) or pyridoxine (PN), did not show an inhibitory effect. These observations suggest that PLP modulates cAspAT gene expression by inactivating the binding activity of glucocorticoid receoptor to GRE.

Glycogen phosphorylase is another PLP-dependent enzyme which catalyzes the first step in the intracellular degradation of glycogen. We observed that the level of phosphorylase mRNA was also increased several fold in the liver of vitamin B_6-deficient rats, compared with vitamin B_6-supplemented control (11). Probably, vitamin B_6 works in a similar way as in the case of cAspAT.

In the course of the study on the effect of vitamin B_6 deficiency on the expression of liver enzymes, we came across unexpected observations. cAspAT and phosphorylase, two enzymes so far described, are PLP-dependent enzymes and one may somehow rationalize the involvement of vitamin B_6 in the control of expression of their genes. However, examination of additional liver enzymes revealed that gene expression of many enzymes, not related to PLP or steroid hormones, were also influenced by vitamin-B_6 nutritional conditions. Thus, mRNA levels of apolipoprotein A-1, phenylalanine hydroxylase, glyceraldehyde-3-phosphate dehydrogenase, and β-actin were elevated in the liver of vitamin B_6-deficient rats (5). The enhanced expression of several vitamin B_6-independent enzymes in the liver of vitamin B_6-deficient rats might be explained, at least in part, by the elevation of RNA polymerase described earlier.

Vitamin B_6 and Expression of Albumin Gene

Serum albumin represents the most abundant protein synthesized in and excreted by the liver. Among the vitamin B_6-independent enzymes and proteins whose gene expressions are enhanced by vitamin B_6 deficiency, we noted that the level of albumin mRNA was increased 7-fold over the control level (12). The magnitude of the increase in albumin mRNA is far greater than can be explained in terms of RNA polymerase activity.

Studies of transcription of the rat albumin gene have shown that the 170-nucleotide region immediately upstream of the transcription initiation is sufficient for tissue-specific expression of the gene. Several *cis*-acting elements have been identified in this region that interact with various transcription factors including HNF-1, C/EBP, DBP, CTF/NF1 and NFY (13). However, there is a hierarchy of importance of the various factors for albumin gene expression; the HNF-1- and C/EBP-binding sites activate transcription more strongly than the other sites (14). We synthesized two oligonucleotides, which interact with HNF-1 and C/EBP respectively, and determined the binding activities of liver nuclear extracts to each of these oligonucleotides by mobility-shift analysis. We found that the activities of the extract prepared from the liver of vitamin B_6-deficient rats were greater than those of controls. As the concentrations of C/EBP in nuclear extracts from control and vitamin-deficient rats, estimated by Western-blot analysis, were essentially the same, the lower binding activity of the extract from control liver is probably due to inactivation of tissue-specific transcription factors by PLP and/or its analogues. We therefore examined the effect of PLP and its analogues on the binding activity of nuclear extract *in vitro* and found that only PLP effectively inhibited the binding. These observations are analogous to the inactivation of glucocorticoid receptor by PLP, described above, and indicate that vitamin B_6 modulates albumin gene expression through a novel mechanism that involves inactivation of tissue-specific transcription factors by direct interaction with PLP. (Figure 1).

In order to elucidate the molecular mechanism whereby DNA-binding activity of tissue-specific transcription factors is inhibited by PLP, we examined the site of attachment of PLP. Determination of the amino acid sequence of PLP-containing peptide revealed that PLP was bound to lysine 197 of HNF-1 molecule (Figure 2). Inasmuch as lysine 197 lies in the homeodomain of HNF-1, PLP binding to this lysine residue would render HNF-1 molecule less accesible to the HNF-1-binding site to albumin gene.

Vitamin B_6 and Cancer

We recently found that the growth of a human hepatoma HepG2 cells was completely inhibited in the medium supplemented with several mM PN. The growth inhibition of HepG2 cells was accompanied by a marked inhibition of protein synthesis and secretion of serum proteins, particularly albumin. PL was as effective as PN, but PM showed no inhibitory action (15). We also found that the growth of MH-134 hepatoma cells, transplanted into C3H/He mice, was significantly retarded by the administration of large doses of pyridoxine. The expression of oncogenes, such as c-*fos* and c-*myc*, was considerably reduced in the tumors probably by a similar mechanism to the inhibition of albumin gene expression by vitamin B_6 (unpublished results).

The antitumor effect of vitamin B_6 is not limited to liver cancer; recent studies have shown that dietary supplementation of vitamin B_6 markedly suppresses azoxymethane-induced colon carcinogenesis in mice (personal communication from N. Kato, Hiroshima University). Application of the supraphysiological doses of vitamin B_6 as an antineoplastic therapy will be a very promising area of research in the future.

References

1. Venegas, A., Martial, J. and Valenzuela, P. (1973) Active site-directed inhibition of E.coli DNA-dependent RNA polymerase by pyridoxal 5'-phosphate. *Biochem. Biophys. Res. Commun.* *55*:1053-1059.

2. Basu, A., Tirumalai, R.S. and Modak, M.J. (1989) Substrate binding in human immunodeficiency virus reverse transcriptase. *J. Biol. Chem.* *264*:8746-8752.

3. Modak, M.J. (1976) Observations on the pyridoxal 5'-phosphate inhibition of DNA polymerases. *Biochemistry* *15*:3620-3626.

4. Fischer, E.H., Kent, A.B., Snyder, E.R. and Krebs, E.G. (1958) The reaction of sodium borohydride with muscle phosphorylase. *J. Am. Chem. Soc.* *80*:2906-2907..

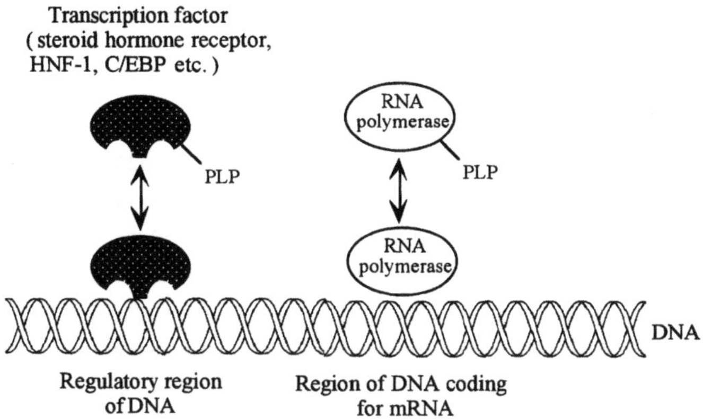

Figure 1. Diagram of binding of transcription factors and RNA polymerase to a regulatory region and a coding region of DNA, respectively. DNA-binding activities are inhibited by interaction of PLP to transcription factors and RNA polymerase.

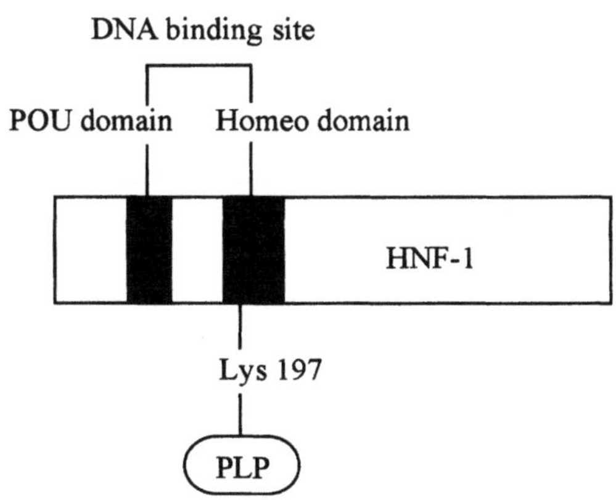

Figure 2. Binding of PLP to transcription factor, HNF-1. PLP binds to lysine-197 in the homeodomain of HNF-1.

5. Oka, T., Komori, N., Kuwahata, M., Sassa, T., Suzuki, I., Okada, M. and Natori, Y. (1993) Vitamin B6 deficiency causes activation of RNA polymerase and general enhancement of gene expression in rat liver. *FEBS Lett.* *331*:162-164.

6. Allgood, V.E., Powell-Oliver, F.E. and Cidlowski, J.A. (1990) Vitamin B6 influences glucocorticoid receptor-dependent gene expression. *J. Biol. Chem.* *265*:12424-12433.

7. Allgood , V.E. and Cidlowski, J.A. (1992) Vitamin B6 modulates transcriptional activation by multiple members of the steroid receptor superfamily. *J. Biol. Chem.* *267*:3819-3824.

8. Tully, D.B., Allgood, V.E. and Cidlowski, J.A. (1994) Modulation of steroid receptor-mediated gene expression by vitamin B6. *FASEB J.* *8*: 343-349.

9. Kondo, T. and Okada, M. (1985) Effect of pyridoxine administration on the induction of cytosolic aspartate aminotransferase in the liver of rats treated with hydrocortisone. *J. Nutr. Sci. Vitaminol.* *31*:509-517.

10. Oka, T., Komori, N., Kuwahata, M., Hiroi, Y., Shimoda, T., Okada, M. and Natori, Y. (1995) Pyridoxal 5'-phosphate modulates expression of cytosolic aspartate aminotransferase gene by inactivation of glucocorticoid receptor. *J. Nutr. Sci. Vitaminol.* *41*:363-375.

11.Oka, T., Komori, N., Kuwahata, N., Suzuki, I., Okada, M. and Natori, Y. (1994) Effect of vitamin B6 deficiency on the expression of glycogen phosphorylase mRNA in rat liver and skeletal muscle. *Experientia* *50*: 127-129.

12. Oka, T., Komori, N., Kuwahata, N., Okada, M. and Natori, Y. (1995) Vitamin B6 modulates expression of albumin gene by inactivating tissue-spesific DNA-binding protein in rat liver. *Biochem. J.* *309*: 242-248.

13. Cereghini, S., Raymondjean, M., Carranca, A.G., Herbomel, P. and Yaniv, M. (1989) Factors involved in control of tissue-specific expression of albumin gene. *Cell* *50*:627-638.

14. Marie, P., Wuarin, J. and Schriber, U. (1989) The role of cis-acting promotor elements in tissue-specific albumin gene expression in rat liver. *Science* *244*:343-346.

15. Molina, A., Oka, T., Munoz, S.M., Chikamori-Aoyama, M., Kuwahata, M. and Natori, Y. (1997) Vitamin B6 suppresses growth and expression of albumin gene in a human hepatoma cell line HepG2. *Nutr. Cancer* *28*: 206-211.

Biochemistry and Molecular Biology of Vitamin B_6
and PQQ-dependent Proteins
ed. by A. Iriarte, H.M. Kagan and M. Martinez-Carrion

PYRIDOXAL 5- PHOSPHATE AND CALCIUM CHANNELS

Krishnamurti Dakshinamurti, Kovvuri J. Lal, Naranjan S. Dhalla *, Sorin Musat* and Xi Wang*
Departments of Biochemistry and Molecular Biology and *Physiology, Institute of Cardiovascular Sciences, St. Boniface General Hospital Research Centre, Faculty of Medicine, University of Manitoba, Winnipeg, Manitoba, Canada R3E 0W3

SUMMARY

The BAY K 8644 – induced influx of ^{45} Calcium into intracellular compartment of artery segments of normal rats was blocked by pyridoxal phosphate (PLP) as well as by the dihydropyridine-sensitive calcium channel antagonists (CCA). PLP also decreased binding *in vitro* of CCA to membrane preparations from normal vascular tissue. PLP antagonized the ATP – induced positive inotropic effect in isolated perfused normal rat hearts without affecting similar action of isoproterenol. PLP prevented the ATP – induced increase of $[Ca^{2+}]_i$ in cardiomyocytes in a concentration – dependent manner and specifically inhibited the specific binding of ATPγS at both high and low affinity sites in cardiac sarcolemmal membranes.

Introduction

Feeding a vitamin B_6- deficient diet to adult rats results in significant hypertension. Administration of a single dose of vitamin B_6 (pyridoxine, 10mg/kg body weight) to vitamin B_6-deficient hypertensive rats (B_6DHT) reversed the hypertension within 24 hrs [1]. Elevated peripheral resistance is the hallmark of hypertension. The vascular smooth muscle (VASM) of caudal artery from B_6DHT had a higher resting tone and spontaneous contractions compared to control normotensive rats and these were abolished by nifedipine or EGTA. Furthermore, when artery segments from B_6DHT rats were incubated *in vitro* with $^{45}Ca^{2+}$, the uptake of calcium was significantly increased compared to that of tissues from control rats. The increased calcium uptake was blocked by nifedipine. These data suggest that the increased intracellular concentrations of calcium $[Ca^{2+}]_i$ in B_6DHT rats might be responsible for the higher tension in the VASM. The higher cytosolic free Ca^{2+} could be caused by an abnormal increase in the permeability of dihydropyridine (DHP)-sensitive Ca^{2+} channel of the VASM plasma membrane.

Since Ca^{2+} is known to play an important role in cellular function in health and disease, we examined the Ca^{2+} handling ability of cardiomyocytes obtained from vitamin B_6 deficient rats. $[Ca^{2+}]_i$ was monitored in both resting and KCl-depolarized cardiomyocytes. In this study we have examined the influence of PLP on the ATP-induced increase in contractile activity of the isolated rat heart and the ATP-mediated increase in $[Ca^{2+}]_i$ in freshly isolated adult cardiomyocytes as well as on the specific binding of ATP to cardiac sarcolemmal membrane in order to examine if PLP is an effective antagonist of ATP receptors in the myocardium.

Materials and Methods

Animals: Adult male Sprague-Dawley rats were used in these experiments. They were randomly divided into two groups, one of which was fed a pyridoxine-deficient diet *ad libitum* and the other pair-fed a pyridoxine-supplemented diet. Systolic blood pressure (SBP) was recorded weekly using tail-cuff plethysmography [1]. Pyridoxine-deficient and corresponding control rats were used in the following experiments.

Calcium uptake: The cold lanthanum-resistant ^{45}Calcium uptake of the caudal artery segments obtained from vitamin B_6-deficient or control rats was determined as described [2]. We studied the effects on ^{45}Calcium uptake of additions to the incubation medium such as pyridoxal phosphate, the calcium channel agonist BAY K 8644 and the calcium channel antagonist nifedipine, singly and in combination at doses indicated in the experiments.

[^{3}H] Nitrendipine binding assay: Modification by PLP of nitrendipine binding sites was measured by binding of [^{3}H] nitrendipine to control and treated membrane preparations isolated from normal rat caudal artery, as described by Costa et al [3].

Isolated heart preparations: Rat hearts were isolated according to Xuet al [4] in the Langendorf perfusion mode. Experiments were started 30 minutes after stabilization of heart function. The left ventricular developed pressure (LVDP) as well as the rate of change in contraction (+dP/dt) and the rate of change in relaxation (-dP/dt) were measured via a transducer which was connected with a latex balloon inserted into the left ventricle.

Isolation of adult cardiomyocytes and measurement of intracellular calcium: Cardiomyocytes were isolated by the method described by Xu et al [4] and incubated with ATP in presence or absence of PLP [5] and $[Ca^{2+}]_i$ was determined according to Grynkiewicz et al [6].

Preparation of sarcolemmal membranes and [^{35}S]ATPγs binding assay The hypotonic shock – LiBr method of Dhalla et al [7] was used to isolate the rat heart purified heavy sarcolemmal preparation. The [^{35}s] ATPγs binding assay was done as described by Zhao and Dhalla [8].

Results

Calcium uptake by caudal artery segments

We investigated the possibility that, pyridoxal phosphate could directly modulate the cellular calcium uptake process. BAY K 8644 stimulated $^{45}Ca^{2+}$ entry into artery segments from control (normal) rats (Fig. 1). Pyridoxal Phosphate dose-dependently (0.1- 10μM)) reduced the BAY K 8644 – stimulated calcium uptake by control artery segments. As seen earlier [5] the basal uptake of $^{45}Ca^{2+}$ by caudal artery segments from vitamin B_6-deficient hypertensive rats was at least twice the uptake by artery segments from control (normal) rats. Pyridoxal phosphate or nifedipine added to the incubation medium reduced significantly the $^{45}Ca^{2+}$ uptake by the artery segments from the deficient hypertensive rats (Fig. 2). However, in presence of BAY K 8644 in the incubation medium both pyridoxal phosphate and nifedipine were much less effective in attenuating the $^{45}Ca^{2+}$ uptake by artery segments from the deficient hypertensive rats.

Fig. 1

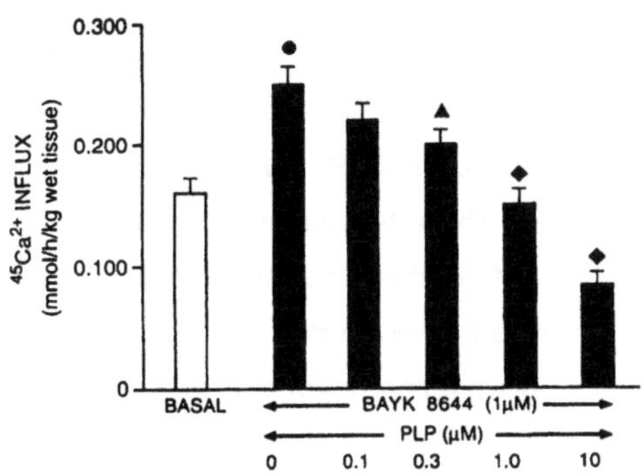

In vitro effect of pyridoxal phosphate addition to the incubation medium on the BAY K 8644 induced $^{45}Ca^{2+}$ influx into caudal artery segments from control rats. Basal, no addition to the incubation medium. Bars represent standard error of mean values for 9 rats in each group. ● $p < 0.05$ with respect to basal; ▲ $p < 0.05$, ◆ $p < 0.01$ with respect to BAY K 8644 alone. (From Lal *et al.* [48]; reprinted with permission of publisher).

Fig. 2

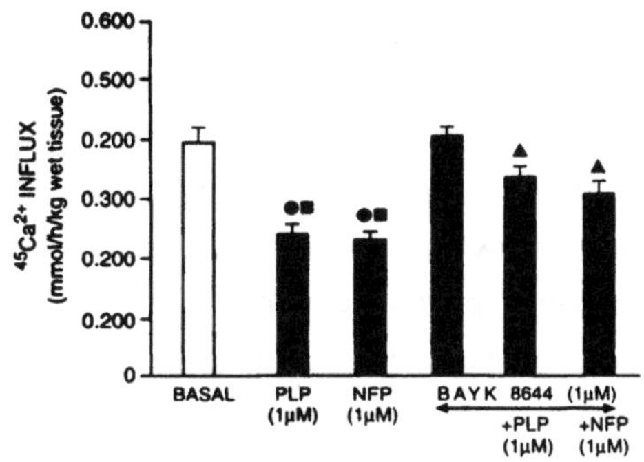

Effect of *in vitro* addition of dihydropyridine and/or pyridoxal phosphate on $^{45}Ca^{2+}$ influx into caudal artery segments from vitamin B_6-deficient hypertensive rats. Basal – no addition to incubation medium; PLP – pyridoxal phosphate; NFP – Nifedipine. Bars represent standard error of mean values for 9 rats in each group. ● $p < 0.05$ with respect to basal; ▲ $p < 0.5$ with respect to BAY K 8644 alone; ■ $p < 0.05$ with respect to presence of BAY K 8644, in addition. (From Lal *et al.* [48]; reprinted with permission of publisher.)

Nitrendipine binding to caudal artery membranes

Figure 3 shows the Scatchard plots of the specific binding of [5-methyl^{3}H]-nitrendipine to crude membrane preparations from caudal artery in the presence or absence of pyridoxal phosphate. The equilibrium dissociation constant (Kd) value for [^{3}H] nitrendipine binding was 0.57 $\pm$0.03 nM and the total number of binding sites (B max) was 150$\pm$7 *f*mol/ mg protein in control membranes. [^{3}H] nitrendipine showed a Kd value of 0.69$\pm$0.04 nM and a Bmax value of 98$\pm$6 *f*mol/mg protein in PLP-treated membranes. The results indicate that PLP treatment of membranes caused a decrease in the number of [^{3}H] nitrendipine binding sites.

Fig. 3

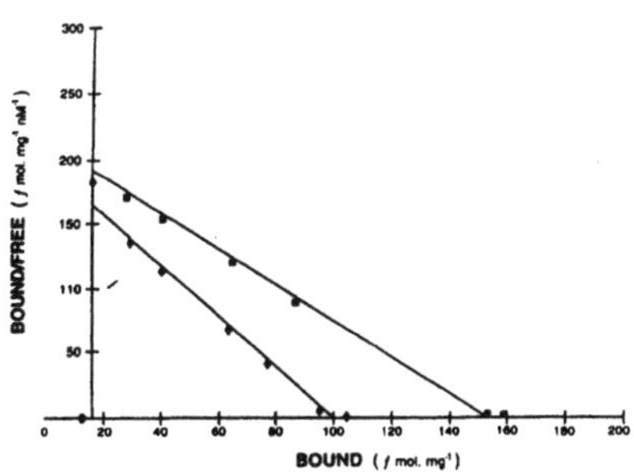

Scatchard plot of [5-methyl ^{3}H] nitrendipine binding to crude membrane preparations from caudal artery of normal rats: ■, in the absence and ◆, in the presence of pyridoxal phosphate (15 μM).

$[Ca^{2+}]_i$ in cardiomyocytes from vitamin B_6-deficient hypertensive rat cardiomyocytes

The results in Fig. 4 indicate that the basal level of $[Ca^{2+}]_i$ in vitamin B_6 deficient cardiomyocytes was not different from the control preparations. However, exposure of the vitamin B_6-deficient cardiomyocytes to 30 nM KCl resulted in a greater increase in $[Ca^{2+}]_i$ in comparison with the control preparations. Administration of vitamin B_6 to vitamin B_6 deficient animals abolished the KCl-induced augmentation in $[Ca^{2+}]_i$.

Effect of PLP on ATP-induced positive inotropic action in the isolated heart

The contractile activity of the isolated perfused rat heart was monitored upon infusing 50 μM ATP in the presence and absence of 50μM PLP. The infusion of ATP caused an immediate increase (within a few seconds) in LVDP, +dP/dt and –dP/dt. This effect was completely blocked in hearts pretreated with PLP for 10 min. The effect of treating the heart with different concentrations (1-50 μM) of PLP on the ATP (50 μM)- induced increase in contractile activity is shown in Fig. 5A. The antagonistic effect of PLP was concentration dependent: ED_{50} for PLP was in the range of 10-15 μM. In another set of experiments, we studied the effect of 15μM PLP on changes in LVDP due to different concentrations (10- 100 μM) of ATP. The results in Fig. 5B reveal that the depressant action of PLP was evident at all concentrations of ATP used in this study. It should be pointed out that the maximal inotropic response on the isolated heart was elicited by 50 μM ATP. The marked increase (more than two-fold) in contractile activity upon infusing 1 μM isoproterenol was not affected by perfusing the heart with 50 μM PLP which prevented the positive inotropic action of ATP. Likewise, 10 μM propranolol, which prevented the positive inotropic action of 1 μM isoproterenol, showed no effect on the positive inotropic action of 50 μM ATP.

Fig. 4

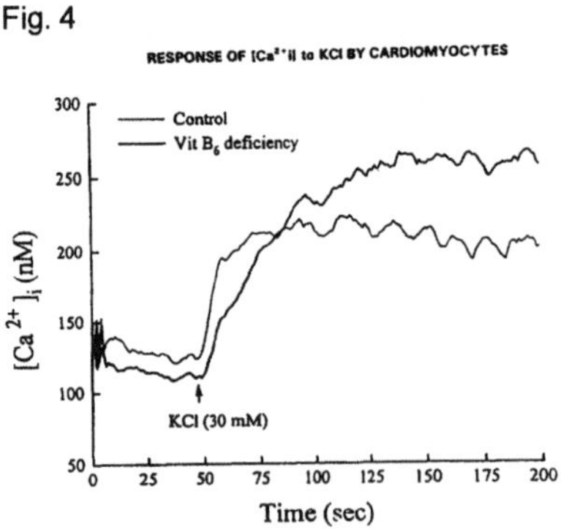

FURA-2 LOADED CARDIOMYOCYTES (10⁶/ml) FROM VITAMIN B6 DEFICIENT AND CONTROL RATS WERE INCUBATED IN K-H BUFFER FOR MEASURING INTRACELLULAR CALCIUM CONCENTRATION. KCl (30 mM) WAS ADDED AT 50 SECONDS OF THE TRACING.

Fig. 5

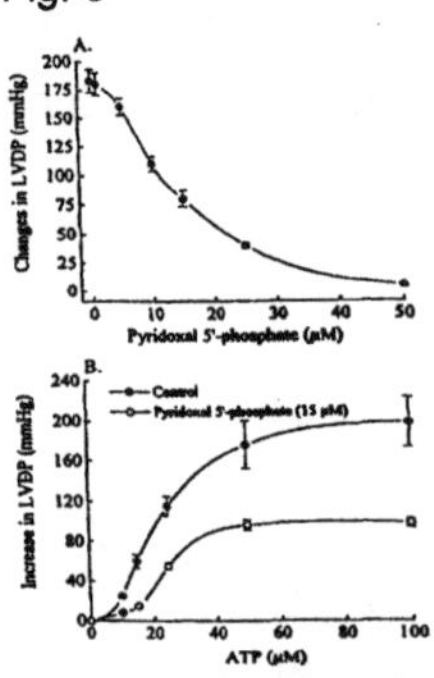

Concentration-dependent effects of pyridoxal 5'-phosphate on ATP-induced increase in left ventricular developed pressure (LVDP) by the isolated perfused rat heart. Panel A: ATP (50 μM) induced increase in LVDP in the absence or presence of different concentrations (1–50 μM) of PLP. Panel B: Increase in LVDP due to different concentrations of ATP (10–100 μM) in the absence (control) and presence of PLP (15 μM). Values are expressed as mean ± SEM of four hearts for each point.

Effect of PLP on ATP-induced increase in $[Ca^{2+}]_i$ in cardiomyocytes

The incubation of cardiomyocytes with 15 or 50 μM of PLP for 10 min did not alter the basal $[Ca^{2+}]_i$. However, the ATP-induced increase in $[Ca^{2+}]_i$ was significantly decreased in PLP-treated cardiomyocytes. The effect of PLP (1-50 μM) on the ATP (50 μM)-induced increase in $[Ca^{2+}]_i$ was (Fig.6A). The effective concentration of PLP was as low as 1μM, whereas 50 μM PLP blocked the action of ATP almost completely (Fig. 6A): ED_{50} for PLP was in the range of 10 –20 μM. In addition, we studied the effect of 15 μM PLP on changes in $[Ca^{2+}]_i$ at different concentrations of ATP (10-100 μM). As shown in Fig. 6B, the inhibitory effect of PLP was evident at all concentrations of ATP employed in this study.

Fig. 6

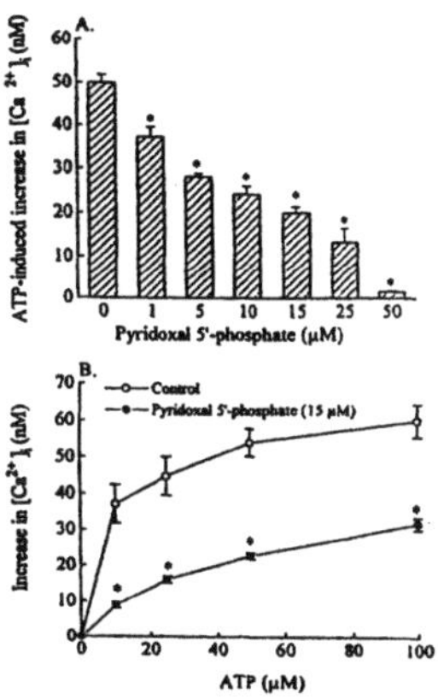

Effect of pyridoxal 5′-phosphate on ATP-induced increase in $[Ca^{2+}]_i$ in cardiomyocytes. Panel A: Fura-2 loaded cardiomyocytes were incubated with different concentrations of pyridoxal 5′-phosphate (1–50 μM) for 10 min and then exposed to 50 μM ATP at 50 s after starting the recording for $[Ca^{2+}]_i$. The increase in $[Ca^{2+}]_i$ at various concentrations of PLP is presented in the histogram. Panel B: Fura-2 loaded cardiomyocytes were incubated with 15 μM pyridoxal 5′-phosphate or same amount of Krebs-Henseleit buffer (control) for 10 min. Different concentrations of ATP (10–100 μM) were added. The increase in $[Ca^{2+}]_i$ is plotted as a function of the concentration of ATP. Values are expressed as mean ± SEM of 6–8 preparations. * $P<0.05$.

Fig. 7

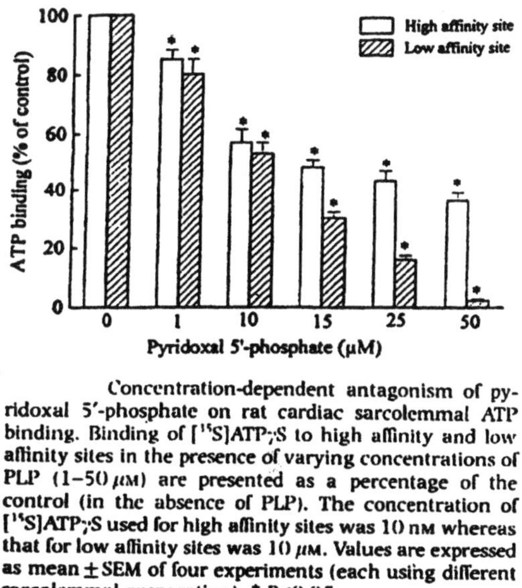

Concentration-dependent antagonism of pyridoxal 5′-phosphate on rat cardiac sarcolemmal ATP binding. Binding of [^{35}S]ATPγS to high affinity and low affinity sites in the presence of varying concentrations of PLP (1–50 μM) are presented as a percentage of the control (in the absence of PLP). The concentration of [^{35}S]ATPγS used for high affinity sites was 10 nM whereas that for low affinity sites was 10 μM. Values are expressed as mean ± SEM of four experiments (each using different sarcolemmal preparation). * $P<0.05$.

Effect of PLP on ATP-binding in cardiac sarcolemma

Since cardiac sarcolemmal membrane has been reported to contain both high and low affinity ATP-binding sites, we studied the effects of PLP on the high affinity and low affinity binding sites by employing 1-10 nM and 1-10 μM [^{35}S]ATPγS, respectively. The maximal ATP-binding at both high and low affinity sites were inhibited by both 50μM PLP and 100 μM suramin without any change in their respective K_6 values. PLP almost completely blocked the low affinity binding. Binding at the high affinity sites was depressed by about 60% of the control value (Fig. 7). Suramin (100 μM) had effects comparable to 50 μM PLP on both high and low affinity sites. Cibacron blue and DIDS at 50 μM concentrations inhibited ATP-binding at high affinity and low affinity sites by 40-45% and 65-70% respectively. The ED_{50} for the inhibitory effects of PLP on high and low affinity ATP-binding sites were about 10 μM and 15 μM, respectively. Agents such as propranolol (β -adrenoceptor blocker), prazosin (α-adrenoceptor blocker), verapamil (L-type Ca^{2+}-channel blocker) and ryanodine (sarcoplasmic reticulum Ca^{2+}- release channel blocker) did not show any significant effect on the high or low affinity ATP-binding sites.

Discussion

Pyridoxal phosphate *in vitro* attenuates the influx of extracellular calcium. This effect is achieved through modulation of ligand binding. This is analogous to the effect of pyridoxal phosphate on steroid hormone activity [9]. Voltage-sensitive calcium channels undergo long-term modulation by neuro-transmitters and a variety of second messengers. Activation of the channels is enhanced by cAMP and AMP-dependent protein kinase [10]. In common with the pharmacological receptors, calcium channels are regulated by homologous and heterologous factors. Chronic channel activation, chronic drug exposure, hormonal influence and specific diseases are all associated with altered expression of calcium channel function and numbers [11]. The action of drugs at the calcium channels would indicate that endogenous factors or ligands might serve as physiological regulators, a function which is mimicked by calcium channel agonists or antagonists.

The KCl-induced increase in $[Ca^{2+}]_i$ was augmented in cardiomyocytes from the vitamin B_6-deficient rats. The administration of vitamin $B_6$to the vitamin B_6-deficient animals was found to abolish this augmentation. It is possible that the observed augmentations of KCl-induced increase in $[Ca^{2+}]_i$ is related to an increase in Ca^{2+}influx through the sarcolemmal Ca^{2+} channels. An increase in Ca^{2+} influx in smooth muscle cells cause an increase in smooth muscle tone and hypertension in vitamin B_6 deficiency [1,12]. The increase in $[Ca^{2+}]_i$ in cardiomyocytes from vitamin B_6 deficient animals may contribute towards heart dysfunction and increased susceptibility to myocardial infarction [13]. These observations can explain the high incidence of hypertension and coronary artery disease in vitamin B_6-deficient patients [14], as well as the beneficial effects of vitamin B_6 in patients with hypertension [15] and myocardial infarction [13]. The antagonistic effect of PLP may not be of a generalized depressant nature. Since ATP is considered to influence cellular functions by acting on purinoceptors [16], it is possible that PLP may affect the cardiac action of ATP by blocking these receptors in the myocardium. This view is consistent with pharmacological studies showing the antagonistic effect of PLP on ATP-induced changes in rat vagus and vas deferens [17]. Furthermore, PLP decreased the specific binding of ATP to both high and low affinity binding sites similar to purinoceptor antagonists such as suramin, Cibacron blue and DIDS. PLP may be a selective antagonist of purinoceptors because other receptor or channel blocking agents, such as propranolol, prazosin, verapamil and ryanodine, did not show any effect on the specific binding of ATP to sarcolemma. Further studies are needed to establish the exact type of purinoceptors for the action of PLP on the myocardium.

It is unlikely that PLP is capable of functioning as an inhibitor of ATP-induced changes in the heart under physiological conditions. However, in view of the suspected availability of extracellular ATP in large amounts in ischemic situations, PLP may prove to be a valuable therapeutic intervention under pathological conditions. Further research is needed to provide experimental support for this viewpoint.

Acknowledgements

This work was supported by grants from the Medical Research Council of Canada and the Heart and Stroke Foundation of Canada to KD and NSD.

References:

1. Paulose CS, Dakshinamurti K, Packer S, Stephens NL. Sympathetic stimulation and hypertension in the pyridoxine-deficient adult rat. Hypertension. 1988; 11: 387-391.
2. Viswanathan M, Bose R, Dakshiamurti K. Increased calcium influx in caudal artery of rats made hypertensive with pyridoxine deficiency. Am. J. Hypertens. 1990; 4: 252-255.
3. Costa B, Giusti L, Martini C, Lucacchini A. Chemical modification of the dihydropyridine binding sites by lysine reagent, pyridoxal 5-phosphate. Neurochem. Int. 1998; 32: 361-364.
4. Xu Y-J, Panagia V. Shao Q, Wang X, Dhalla NS. Phosphatidic acid increases intracellular free Ca^{2+} and cardiac contractile force. Am. J. Physiol. 1996; 271:H651-H650.
5. Dakshinamurti K, Wang X, Musat S, Dandekar M, Dhalla NS. Alterations of KCl- and ATP – induced increase in $[Ca^{2+}]_i$ in cardiomyocytes from vitamin B_6 deficient rats. Can. J. Physiol. Pharmacol. 1998; 67:1-6.
6. Grynkiewicz, G., Poenie, M., and Tsien, R.Y. A new generation of Ca^{2+} indicators with greatly improved fluorescence properties. J. Biol. Chem. 1985; 260: 3440-3450
7. Dhalla NS, Anand-Srivastava MB, Tuana BS. Solubilization of a calcium dependent adenosine triphosphatase from rat heart sarcolemma. J. Mol. Cell Cardiol. 1981; 13: 413-423.
8. Zhao DY, Dhalla NS, [^{35}S]ATP gamma S binding sites in the purified heart sarcolemma membrane. Am. J. Physiol., 1990; 258:C185-C188.
9. Litwack G. The glucorticoid receptor at the protein level. Cancer Res. 1988; 48:2636-2640.
10. Schmid A, Renaud J, Luxdunski M. Short term and long term effects of β adrenergic effectors and cyclic AMP on nitrendipine-sensitive voltage-dependent Ca^{2+} channels of skeletal muscle. J. Biol. Chem. 1985; 260:13041-13046.
11. Ferrante J, Triggle DJ. Drug and disease – induced regulation of voltage-dependent calcium channels. Parmacol. Rev. 1990; 42:29-44.
12. Lal KJ, Dakshinamurti K. Regulation of calcium influx into vascular smooth muscle by vitamin B_6. Clin. Exp. Hypertens. 1993; 15:489-500
13. Ellis JM and McCully KS. Prevention of myocardial infarction by vitamin B_6. Res. Commun. Chem. Pathol. Parmacol. 1995; 89:208-220.
14. Vermaak WH, Barnard HC, Potgieter GM, and Theron H. Vitamin B_6 and coronary artery disease. Epidemiological observations and case studies. Atherosclerosis, 1987; 63:235-238.

15. Ayback M, Sermet A, Ayyilidiz MO, and Karakilcik AZ. Effect of oral pyridoxine hydrochloride supplementation on arterial blood pressure in patients with essential hypertension. Drug Res. 1995;45:1271-1273.
16. Dubyak GR, El-Moatassium C. Signal transduction via P2-puinergic receptors for extracellular ATP and other nucleotides. Am. J. Physiol. 1993;265:C577-C606.
17. Trezise DJ, Bell NJ, Khakh BS, Michel AD, Humphrey PA. P2 purinoceptor antagonist properties of pyridoxal-5-phosphate. Eur. J. Pharmacol. 1994;259:295-300.

Biochemistry and Molecular Biology of Vitamin B_6
and PQQ-dependent Proteins
ed. by A. Iriarte, H.M. Kagan and M. Martinez-Carrion

GABA-aminotransferase, a target for antiepileptic drug therapy

Paola Storici[1], Guido Capitani[1,4], Daniela De Biase[2], Robert A. John[3], Johan N. Jansonius[1] and Tilman Schirmer[1]

[1]Structural Biology Division, Biozentrum, University of Basel, Klingelbergstrasse 70, CH-4056 Basel, CH.
[2]Dipartimento Scienze Biochimiche "Rossi Fanelli" Università La Sapienza P.le Aldo Moro 5, I-00185 Rome, IT.
[3]School of Molecular and Medical Biosciences, University of Wales, CF1 3US Cardiff, UK.
[4]*Current address:* Institut of Biochemistry, University of Zürich, Winterthurerstrasse 190, CH-8057 Zürich, CH.

Summary

The crystal structure of γ-aminobutyric acid aminotransferase (GABA-AT) from pig liver has been solved by molecular replacement at 3.0 Å resolution. In the brain, the enzyme catalyzes the degradation of the major inhibitory neurotransmitter GABA and is a recognized target for anticonvulsant drugs. The fold of the dimeric enzyme is similar to that of ornithine aminotransferase (OAT), dialkylglycine decarboxylase (DGD) and glutamate-1-semialdehyde aminomutase (GSAT). In the active site, the cofactor is covalently bound as an internal aldimine of pyridoxal phosphate and Lys329. The binding pocket of GABA-AT shows several structural homologies with that of OAT, with the exception of two side chain replacements which should be responsible for the distinct substrate specificities. Considering the high sequence identity with the human enzyme (96%) this structure represents a valuable target for the development of new anticonvulsant drugs.

Introduction

Gamma-aminobutyric acid (GABA) is the major inhibitory neurotransmitter in the mammalian central nervous system (Roberts & Kuriyama, 1968) and is closely linked to the metabolism of glutamate, glutamine and the tricarboxylic acid cycle. GABA is synthesized in neurons by glutamate decarboxylase, and is degraded by GABA-aminotransferase which catalyzes the reaction: GABA + α-ketoglutarate ↔ succinic semialdehyde + L-glutamate.

Several neurological and neuropsychiatric disorders, such epilepsy, Huntington's disease, anxiety and pain are known to be associated with low levels of GABA in the brain. Since, in the brain, inhibition of GABA-AT increases GABA concentration, the enzyme is an attractive target for antiepileptic drugs (Schwartz, 1991).

In this paper, we describe the structure of pig liver GABA-AT that was solved by molecular replacement, using OAT as a search model.

Materials and Methods

Crystals of pig liver GABA-AT were obtained as described by Markovic-Housley et al. (1990). They are of space group $P2_1$, a=69.9 Å, b=230.4 Å, c=72.8 Å, β=109.4°, and contain two dimers per asymmetric unit. Data were collected from a single crystal on an Enraf-Nonius FAST area detector diffractometer. The structure was solved by molecular replacement using a polyserine model of OAT dimer in which all loops and peripheral elements were removed. Phase improvement and refinement were performed using NCS constrains. The final model, refined to 3Å resolution, has R-factor and R-free of 18.6% and 21.7%, respectively (Storici et al., 1999).

Results and Discussion

The homodimeric enzyme is of compact shape and is composed of two intertwined subunits (Fig. 1; pdb code=1gtx). Each monomer contains one PLP molecule, and can be divided into three domains: the N-terminal domain (residues 11 to 71), the central, PLP-binding domain (residues 81 to 375), and the C-terminal domain (residues 376 to 472). The two cofactors are deeply buried in the core of the dimer and are located close to the subunit interface.

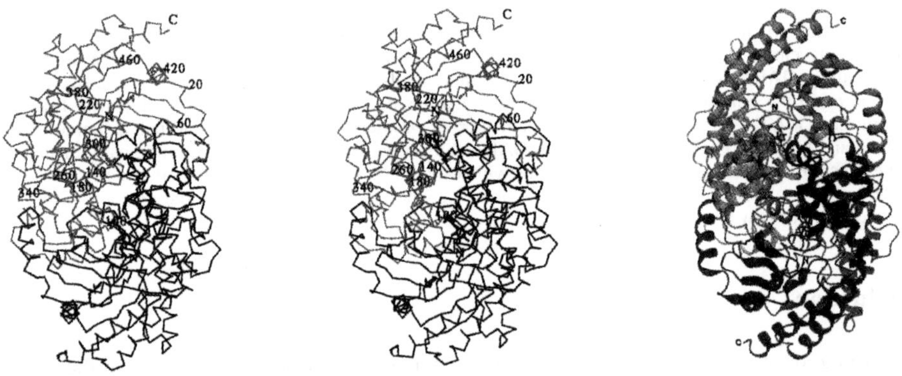

Fig.1 Spatial structure of GABA-AT. *Left:* Stereo Cα-trace of the dimer, viewed along the molecular 2-fold axis (every 40th residue is labeled). *Right:* Mono ribbon representation.

A unique feature of GABA-AT is the presence of a cluster of cysteines located at the dimer interface close to the two-fold symmetry axis. The cluster is composed of cysteines 135 and 138 and their symmetric related mates. The cysteines are located at the N-terminus of helix 5, close to the PLP phosphate group. The electron density seems consistent with two intra-chain disulphide bridges, although remaining density is found at the symmetry axis. Formation of an inter-chain disulphide is not possible because of geometrical restrains. A bridging metal ion cannot be accommodated without a substantial change of the cysteine conformations. Thus, the residual density may indicate the presence of partially oxidized cysteines, since no reducing agent was added to the crystallization solution. The physiological relevance of the cysteine cluster is unresolved, since formation of disulphide bridges is unusual for a mitochondrial protein. Further data are required to clarify this point.

Despite the very low sequence identity, the fold of GABA-AT is similar to that of OAT (Shen et al., 1998), DGD (Toney et al., 1995), and GSAT (Hennig et al., 1997) (see table).

Monomer Superposition	Sequence identity (%)	rms deviation (Å)	# of equivalent Cα
GABA-AT onto OAT	17.4	2.2	364 (79%)
GABA-AT onto DGD	18.0	2.1	396 (86%)
GABA-AT onto GSAT	12.3	2.4	376 (81%)
OAT onto DGD	28.4	1.6	362 (90%)

The major differences between the four structures are found on that part of the molecular surface where two sequence insertions are located. Ten residues after helix 5 are partially folded in an extra-helix, and there are about 20 residues, which extent helix 9 by three additional turns.

In the active site (Fig. 2), the cofactor is held in position by many specific interactions with main and side chains atoms of both subunits. The ε-amino group of Lys329 is covalently bound *via* a Schiff's base linkage to the aldehyde function of PLP. The phosphate group receives five H-bonds from Gly136, Ser137 and Thr353* (the asterisk indicates the residue contributed by the neighboring subunit). The helix macrodipole of helix 5 contributes to charge compensation. The pyridine ring is sandwiched between Phe189 and Val300, and its nitrogen forms a salt bridge with Asp298. The hydroxyl at C3, presumably ionized, receives one hydrogen bond from Gln301.

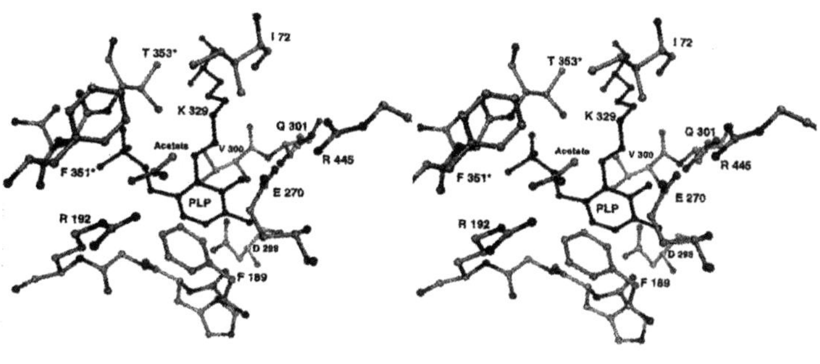

Fig.2 Stereo view of the active site of GABA-AT.

In close proximity to Arg192, remaining electron density could be modeled as an acetate molecule. Acetate was used as buffer in the crystallization set-up and is known to be a competitive inhibitor of GABA-AT (John & Fowler, 1976). This finding, together with the knowledge that, in

OAT, the homologous residue (Arg180) is responsible for binding the carboxylate group of the substrate (Shah et al., 1997; Storici et al., 1999), suggests that Arg192 fulfills the same role in GABA-AT. Based on this assumption a model of the external aldimine intermediate with GABA was built into the active site. Upon tilting of the cofactor ring, the substrate carboxylate can easily form a bifurcated salt bridge with the guanidino group of Arg192. This interaction allows the *pro-S* proton of the Cγ to be positioned appropriately for abstraction by the general base Lys329, in accordance with the known stereospecificity of the reaction (Bouclier et al., 1979).

Superposition reveals that the active site of GABA-AT is narrower and more hydrophobic compared to that of OAT. This is mainly due to the replacement of Tyr55 and Tyr85 of OAT by Phe351* and Ile72 in GABA-AT. These two residues are expected to be the major determinants of substrate recognition. Mutation studies would be useful to confirm this hypothesis.

Recently, we have co-crystallized GABA-AT with (S)-γ-vinyl-GABA; an enzyme-activated inhibitor that is already applied in the treatment of epilepsy (Davies, 1995). A 2.27 Å data set has been collected at the Elettra synchrotron (Trieste, IT) and processed (Rsym = 6.3, completeness 84.3%, I/σ 7.3). Refinement of the structure is underway. We expect that this model will clarify the inhibitory mechanism of vinyl-GABA, helping in optimizing its efficacy as antiepileptic drug.

Acknowledgements.
We are grateful to Dr. Markus Moser for collecting the data set of native GABA-AT and to Dr. Richard B. Silverman for kindly providing (S)-γ-vinyl-GABA.

References.

Bouclier M, Jung MJ, Lippert B. (1979) Stereochemistry of reactions catalysed by mammalian-brain L-glutamate 1-carboxy-lyase and 4-aminobutyrate: 2-oxoglutarate aminotransferase. *Eur. J. Biochem. 98* : 363-368.

Hennig, M., Grimm, B., Contestabile, R., John, R.A., & Jansonius, J.N. (1997) Crystal structure of glutamate-1-semialdehyde aminomutase: An α_2-dimeric vitamin B_6-dependent enzyme with asymmetry in structure and active site reactivity. *Proc. Natl. Acad. Sci. USA 94*, 4866-4871.

John, R.A., & Fowler, L.J. (1976) Kinetic and spectral properties of rabbit brain 4-aminobutyrate aminotransferase. *Biochem. J. 155*, 645-651.

Markovic-Housley, Z., Schirmer, T., Fol, B., Jansonius, J.N., DeBiase, D., & John, R.A. (1990) Crystallization and preliminary X-ray analysis of gamma-aminobutyric acid transaminase. *J. Mol. Biol. 214*, 821-832.

Roberts E & Kuriyama K. (1968) Biochemical-physiological correlations in studies of the gamma-aminobutyric acid system. *Brain Res. 8*: 1-35.

Schwartz, J.H. (1991) in *Principles of neural science* (Kandel, E.R., Schwartz, J.H., & Jessell, T.M., Eds.) pp 217-217, Prentice Hall International, Norwalk.

Shah, S.A., Shen, B.W. & Bruenger, A.T. (1997) Human ornithine aminotransferase complexed with L-canaline and gabaculine: structural basis for substrate recognition, *Structure 5*, 1067-1075.

Shen, B.W., Hennig, M., Hohenester, E., Jansonius, J.N., & Schirmer, T. (1998) Crystal structure of human recombinant ornithine aminotransferase. *J. Mol. Biol. 277*, 81-102.

Storici, P., Capitani, G., Mueller, R., Schirmer, T. and Jansonius, J.N (1999) Crystal structure of human ornithine aminotransferase complexed with the highly specific and potent inhibitor 5-fluoromethylornithine, *J.Mol.Biol. 285*, 297-309.

Storici, P., Capitani, G., De Biase, D., Moser, M., John, R.J., Jansonius, J.N. and Schirmer, T. (1999) Crystal Structure of GABA-aminotransferase, a target for antiepileptic drug therapy, *Biochemistry 38* 8628-8634.

Toney, M.D., Hohenester, E., Keller, J.W., & Jansonius, J.N. (1995) Structural and mechanistic analysis of two refined crystal structures of the pyridoxal phosphate-dependent enzyme dialkylglycine decarboxylase. *J. Mol. Biol. 245*, 151-179.

DECARBOXYLASES AND D-AMINO ACID TRANSAMINASE

Biochemistry and Molecular Biology of Vitamin B_6
and PQQ-dependent Proteins
ed. by A. Iriarte, H.M. Kagan and M. Martinez-Carrion

Mouse Ornithine Decarboxylase: Structural Comparisons to Other PLP-Dependent Enzymes

M. L. Hackert, A. D. Kern, M. A. Oliveira, J. J. Almrud, D. W. Carroll, and S. R. Ernst

Department of Chemistry and Biochemistry, University of Texas at Austin, Austin, TX 78712

Summary

Ornithine decarboxylases (ODCs) are PLP-dependent enzymes that initiate the first and rate-limiting step in the biosynthesis of polyamines. Sequence comparisons suggest four Groups of PLP-dependent decarboxylases, but there appear to be only two structural motifs. Decarboxylases of Groups I, II, and III possess a PLP-binding domain as seen in the structure of the bacterial ODC from *Lactobacillus* 30a. The eukaryotic ODCs belong to the Group IV decarboxylases, which includes biosynthetic ArgDC and diaminopimelate DC. mODC is active as a dimer formed through a head-to-tail interaction between monomers. Each monomer contains two domains: an α/β-barrel domain which binds the cofactor, and a second domain consisting mostly of β-structure. Analysis of the mODC active site provides insight into the stereochemical characteristics of PLP-dependent decarboxylation.

Introduction

Pyridoxal-5'-phosphate (PLP) is an enzyme cofactor utilized extensively in amino acid metabolism, performing a broad range of reactions [1]. PLP is incorporated into a variety of protein folds, as represented by aspartate aminotransferase (AspAT), tryptophan synthase, D-amino acid aminotransferase and alanine racemase (ALR) families, with the AspAT family being the most populated [2,3]. Decarboxylases (DCs) are arranged into four Groups based on sequence [4], but only two distinct PLP-binding folds have been reported [5-7]. Groups I, II, and III share the motif found in the bacterial ODC from *Lactobacillus 30a* (*L30a*ODC). The "PLP Binding" and "Specificity" domains resemble the structure of AspAT, characterized by a seven-stranded sheet and flanking helices that form the cofactor binding site. *L30a*ODC is observed as an active dodecamer composed of six dimers.

In contrast, Group IV DCs are represented by the structure (Fig. 1) of mouse ODC (mODC) [7] and fall into the ALR family. Despite the large evolutionary gap, mammalian ODC and bacterial ALR are strikingly similar in structure [7-9]. Both are obligate homodimers, the cofactors are similarly located, and each monomer has two domains. The *barrel* domain (residues 46-283 in mODC) forms an α/β-barrel containing the cofactor, while the remaining residues form the *sheet* domain, a mostly β-structure that includes two sheets (S1 and S2). Other Group IV DCs include eukaryotic ODCs, diaminopimelate DC (DAPDC) and biosynthetic arginine DC (bArgDC). All are predicted to possess an α/β barrel motif for binding the cofactor.

ODCs perform the first committed step in polyamine biosynthesis [2], required for normal cell growth. Constitutive ODC activity has been observed in cancer cells, and its uncontrolled expression confers a cancer phenotype in otherwise "normal" cells [10]. ODC inhibitors may be used to treat cancer and parasitic infections.

ODC Dimer Stabilities

Active mODC dimers are in rapid equilibrium with inactive monomers. Only 1,903 $Å^2$ of solvent-accessible surface area per mODC monomer is buried upon dimer formation, versus 6,166 $Å^2$ in *L30a*ODC. The symmetrical homodimer of mODC is formed by a "head to tail" interaction between the *barrel* of one monomer and the *sheet* domain of another. Two outlying salt bridges (K169/D364' and D134/K294') and a stack of aromatic residues (F397'/Y323'/Y331 and Y331'/Y323/F397) near the molecular 2-fold axis form the primary interactions that stabilize the dimer [primed residues are from the other monomer in the dimer]. The size of the interface,

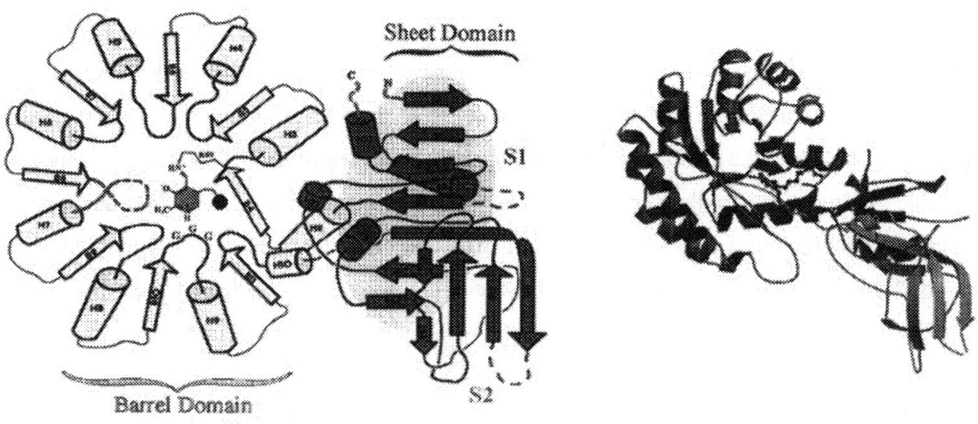

Figure 1: a) Schematic of the mODC monomer. b) Ribbon drawing of the mODC monomer.

and the small number of interactions stabilizing the dimer are consistent with the observed rapid equilibrium [11] between mammalian ODC monomers and dimers, which appears to be needed for ODC regulation by antizyme [12].

Active Site of mODC and Reaction Stereochemistry

The cofactor is found on the C-terminal end of the α/β-barrel, similarly to other cofactors utilizing this fold. The active site is formed primarily from residues donated by loops from the C-terminal end of the *barrel*, one loop from the *sheet* domain (connecting B14 and B15 of S2) and two additional loops (357'-364', 398'-403') from the *sheet* domain of the other monomer. One of the salt bridges (K169-D364') stabilizes the dimer and immobilizes the loop containing active site residues from the neighboring monomer (C360', D361' and D364').

The position and orientation of the cofactor in the active site is governed primarily through protein interactions with the 5'-phosphate and the pyridine nitrogen (Fig. 2). The phosphate binds in a pocket formed by two loops (235-241, 274-276) and the N-terminus of H10. One of these loops contains a GGG sequence (residues 235-237) that is conserved in all Group IV PLP-dependent DCs, while the other loop connects B11 to H10. Both loops contribute hydrogen bonds involving main chain nitrogens (G237, G276, R277). R277, at the N-terminus of H10, contributes a hydrogen bond to the phosphate via N_{ε}, while the N_{η} nitrogens are held in place by a salt bridge to D332. This salt bridge links the *barrel* and *sheet* domains. The 5' phosphate is further fixed by a hydrogen bond to Y389. Interestingly, a tyrosine is found with the same geometry in AspAT, the fold of which more closely resembles that of the bacterial *L30a*ODC.

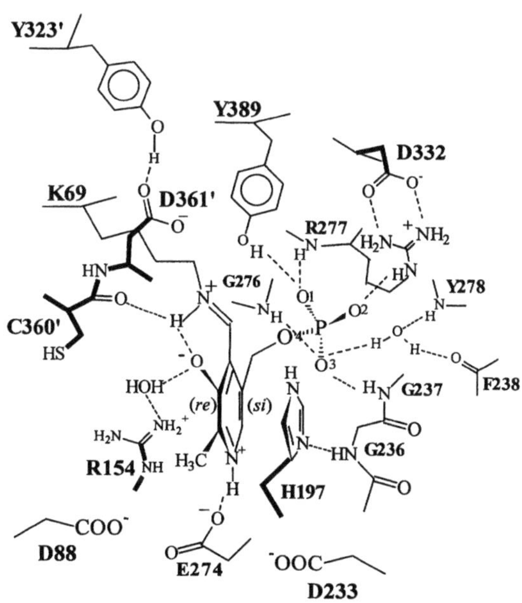

Figure 2: Schematic of the mODC active site.

The primary interaction of the pyridine nitrogen is with E274. It is part of a cluster of acidic residues (D88, E274, D233) which, with three bound water molecules, form a network of hydrogen bonds that likely increases the electron withdrawing properties of the cofactor. This view is consistent with data suggesting the pK_a of the pyridine nitrogen is 8.5 [13], and that the slowest step in the activity of *trypanosomal* ODC is product release [14]. In contrast to the acidic group found in most PLP-dependent enzymes, in ALR there is a basic group (R292). This likely diminishes the electron withdrawing capabilities of the cofactor relative to mODC.

mODC and ALR have distinct methods for immobilizing a histidine side chain next to the cofactor (H197 in mODC and H166 in ALR). The imidazole ring of H197 in mODC shares a hydrogen bond (via its $N_\varepsilon 2$ nitrogen) with the backbone nitrogen of G236, but its $N_\delta 1$ nitrogen is free, allowing the plane of the imidazole ring to be within 10 degrees of stacking parallel against the pyridine ring. In contrast, H166 of ALR requires a rotation of 35 degrees to make these rings parallel. This angle is enforced by having the cofactor pyridine nitrogen and $N_\delta 2$ of H166 both interact with the side chain of R219 as well as having a hydrogen bond between the imidazole $N_\delta 1$ and Y265'. The stereochemistry of PLP-dependent reactions involves coenzyme face selectivity with respect to the mode of cofactor binding (*re* or *si* face buried), the orientation of the labile sigma bond perpendicular to the pyridine ring, and the position of general acid/base groups on either or both faces of the cofactor. Transaminases use PLP for the interconversion of keto acids and amino acids involving a key 1,3-prototropic shift on the *si* face of PLP [15] by a lysine residue, which is also buried on the *si* face of the PLP cofactor [16] (Fig. 3).

In DCs, once the Michaelis-Menton complex is formed, transaldimination leads to the formation of a quinonoid intermediate with the release of CO_2 [2]. The protonation of the quinonoid intermediate at the C_α carbon may lead to retention or inversion of configuration. Retention is achieved with the proton donor on the same side of the pyridine ring as the leaving carboxylate group, whereas with inversion it is on the opposite side. It has been shown that mODC and *L30a*ODC both react with retention of configuration [17, 18].

There are three potential proton donors for the reaction in mODC: two on the *re* face of the cofactor (K69 and C360') and one on the *si* face (H197) (Fig. 2). Alanine mutations at positions K69 and C360 show a stronger k_{cat} effect for the K69A mutant than the C360A mutant [19]. This implies that the proton donor involved in the reprotonation of C_α is most likely K69, but C360'

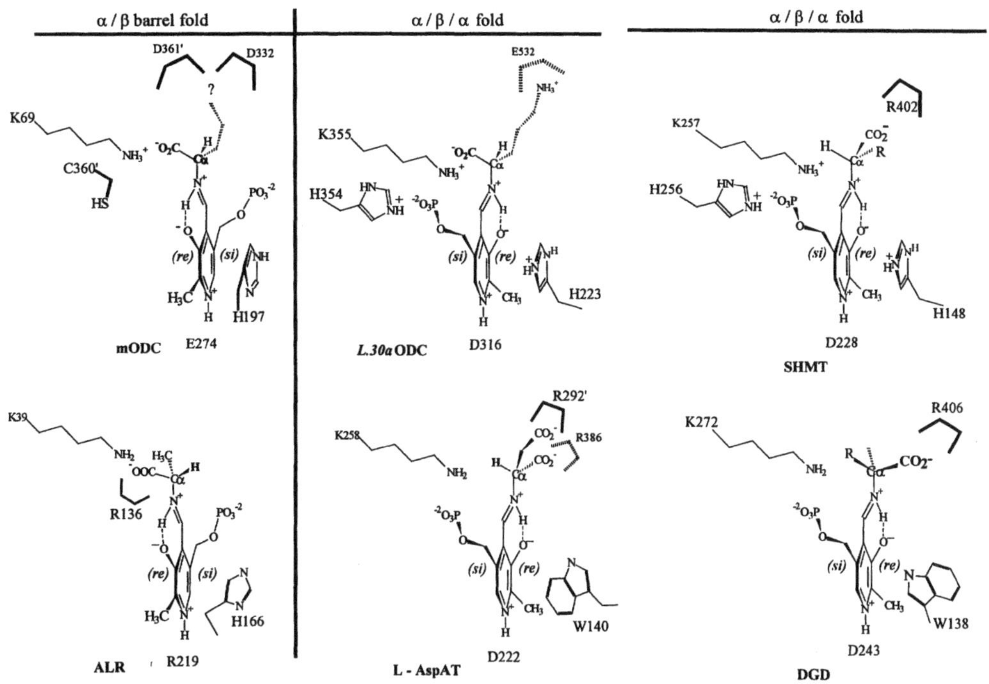

Figure 3: A comparison of cofactor positions in various folds. Note the placement of the *re* and *si* faces of the PLP. SHMT = Serine hydroxymethyl transferase, DGD = dialkylglycine decarboxylase

may serve that purpose in the K69A mutant. H197 is a potential proton donor, but is hydrogen bonded to the nitrogen of G236. This hydrogen bond connects H197 to a chain of four backbone hydrogen bonds capped by R107, making H197 an unlikely proton donor due to the higher pK_a of the arginine guanidinium with respect to the histidine imidazole. This anchors the imidazole ring of H197 in place, possibly stabilizing reaction intermediates. A similar histidine (H223) is observed in *L30a*ODC, where it too is found on the opposite side relative to the lysine forming a Schiff base with PLP (Fig. 3). The imidazole ring next to the cofactor most likely is needed for stabilizing PLP in its active site via ring stacking.

Acknowledgements
This work was supported in part by funding from NIH (GM30105) and the Foundation for Research.

References

1. Martell, A.E. (1982) Reaction pathways and mechanisms of pyridoxal catalysis. *Adv. Enzymol. Relat. Areas Mol. Biol. 53*: 163-199.
2. Jansonius, J.N. (1998) Structure, evolution and action of vitamin B6-dependent enzymes. *Current Opinion in Structural Biol. 8*: 759-769.
3. Grishin, N.V., Phillips, M.A. & Goldsmith, E.J. (1995) Modeling of the spatial structure of eukaryotic ornithine decarboxylase. *Prot. Science 4:* 1291-1304.
4. Sandmeier, E., Hale, T.I. & Christen, P. (1994) Multiple evolutionary origin of pyridoxal-5'-phosphate-dependent amino acid decarboxylases. *Eur. J. Biochem. 221*: 997-1002.
5. Momany, C., Ghosh, R. & Hackert, M.L. (1995) Structural motifs for pyridoxal-5'-phosphate binding in decarboxylases: an analysis based on the crystal structure of the *Lactobacillus 30a* ornithine decarboxylase. *Prot. Sci., 4:* 849-854.
6. Momany, C., Ernst, S., Ghosh, R., Chang, N.L. & Hackert, M.L. (1995) Crystallographic structure of a PLP-dependent ornithine decarboxylase from *Lactobacillus 30a* to 3.0 Å resolution. *J. Mol. Biol. 252*: 643-655.
7. Kern, A.D., Oliveira, M.A., Coffino, P. and Hackert, M.L. (1999) The Structure of Mammalian Ornithine Decarboxylase at 1.6Å Resolution: Stereochemical Implications of PLP-dependent Amino Acid Decarboxylases *Structure 7*: 567-581.
8. Almrud, J.J., Oliveira, M.A., Kern, A.D., Grishin, N.V., Phillips, M.A. & Hackert, M.L. (2000) Crystal structure of human ornithine decarboxylase at 2.1 Å resolution: structural insights to antizyme binding. *J. Mol. Biol., 295*:7-16.
9. Shaw, J. P., Petsko, G. A. & Ringe, D. (1997) Determination of the structure of alanine racemase from *Bacillus stearothermophilus* at 1.9 Å resolution. *Biochemistry 36*: 1329.
10. Auvinen, M., Paasinen, A., Andersson, L. C, & Holtta, E. (1992) Ornithine decarboxylase activity is critical for cell transformation. *Nature 360*: 355-358.
11. Tobias, K. E. Mamroud-Kidron, E. & Kahana, C. (1993) Gly387 of murine ornithine decarboxylase is essential for the formation of stable homodimers. *Eur. J. Biochem. 218*: 245-250.
12. Hayashi, S., Murakami, Y.& Matsufuji, S. (1996) Ornithine decarboxylase antizyme: a novel type of regulatory protein. *Trends Biochem. Sci., 21*: 27-30.
13. Osterman, A. L., Brooks, H. B., Rizo, J. & Phillips, M. A. (1997) Role of Arg-277 in the binding of pyridoxal-5'-phosphate to *Trypanosoma brucei* ornithine decarboxylase. *Biochemistry 36*: 4558-4567.
14. Brooks, H. B. & Phillips, M. A. (1997) Characterization of the reaction mechanism for *Trypanosoma brucei* ornithine decarboxylase by multiwavelength stopped-flow spectroscopy. *Biochemistry 36*: 15147-15155.
15. Hayashi, H., Mizuguchi, H. & Kagamiyama, H. (1998) The imine-pyridine torsion of the pyridoxal 5'-phosphate Schiff base of aspartate aminotransferase lowers its pK_a in the unliganded enzyme and is crucial for the successive increase in the pK_a during catalysis. *Biochemistry 37*: 15076-15085.
16. Gani, D. A (1991) Structural and mechanistic comparison of pyridoxal-5'-phosphate dependent decarboxylase and transaminase enzymes. *Phyl. Trans. R. Soc. Lond. B. Biol. Sci. 332*: 131-139.
17. Asada, Y., Tanizawa, K., Nakamura, K., Moriguchi, M. & Soda, K. (1984) Stereochemistry of ornithine decarboxylase reaction. *J Biochem. 95*: 277-282
18. Orr, G.R., Gould, S.J., Pegg, A.E., Seely, J.E., and Coward, J.K. (1984) Stereochemistry of the Decarboxylation of L-Ornithine with Ornithine Decarboxylase from Mouse Kidney. *Bioorganic Chemistry 12*: 252-258.
19. Coleman, C. S., Stanley, B. A. & Pegg, A. E. (1993) Effect of mutations at active site residues on the activity of ornithine decarboxylase and its inhibition by active site-directed irreversible inhibitors. *J. Biol. Chem. 268*: 24572-24579.

Biochemistry and Molecular Biology of Vitamin B_6
and PQQ-dependent Proteins
ed. by A. Iriarte, H.M. Kagan and M. Martinez-Carrion

Structural and mechanistic studies of *Trypanosoma brucei* ornithine decarboxylase.

Margaret A. Phillips

University of Texas Southwestern Medical Center, Dept. of Pharmacology, 5323 Harry Hines Blvd., Dallas, TX 75235, USA.

Summary

Ornithine decarboxylase (ODC) is a pyridoxal-5'-phosphate (PLP)-dependent enzyme that catalyzes the first committed step in the biosynthesis of polyamines. The polyamines are ubiquitous cell growth factors and the biosynthetic enzymes have been of interest as potential drug targets for the treatment of proliferative diseases. X-ray structural analysis of the eukaryotic ODCs shows that the enzyme is a homodimer with two identical active sites that are formed at the dimer interface. A number of active site residues have been identified that have similar mechanistic roles to the analogous residues in other PLP-dependent enzymes, whereas others provide the basis for the reaction and substrate specificity required by ODC.

Background

ODC catalyzes the PLP-dependent decarboxylation of L-Orn to produce the polyamine putrescine (Fig. 1). The polyamines, putrescine, spermidine and spermine are required for cell growth and differentiation. Inhibition or gene knockout of the biosynthetic enzymes causes cell growth arrest [1,2], while over-expression of ODC in mammalian cells leads to cell transformation, and in transgenic mouse models to increased tumor development [3]. Inhibitors of ODC have been used for the treatment of cancer and *Pneumocystis carinii* pneumonia, but to date the only significant clinical application of these inhibitors is for the treatment of African sleeping sickness, caused by the parasitic protozoa *Trypanosoma brucei gambiense*. α-difluoromethylornithine (DFMO; eflornithine), a suicide inhibitor of ODC was approved in 1990 for the treatment of both early and late stage human sleeping sickness, providing the first new chemotherapy against this disease in 40 years [4]. While toxicity of DFMO is relatively low, large doses (150 mg/kg i.v. every 6 hrs for 14 days) are required for efficacy. Additionally,

DFMO is ineffective against the second form of human sleeping sickness caused by *T. brucei* rhodesiense. ODC is therefore a proven drug target for the treatment of African sleeping sickness, however the development of new ODC inhibitors that overcome some of the shortcomings of DFMO would be highly desirable. Structure-based approaches for the design of new enzyme inhibitors have been used successfully to aid in the design of new drugs against a number of human diseases [5].

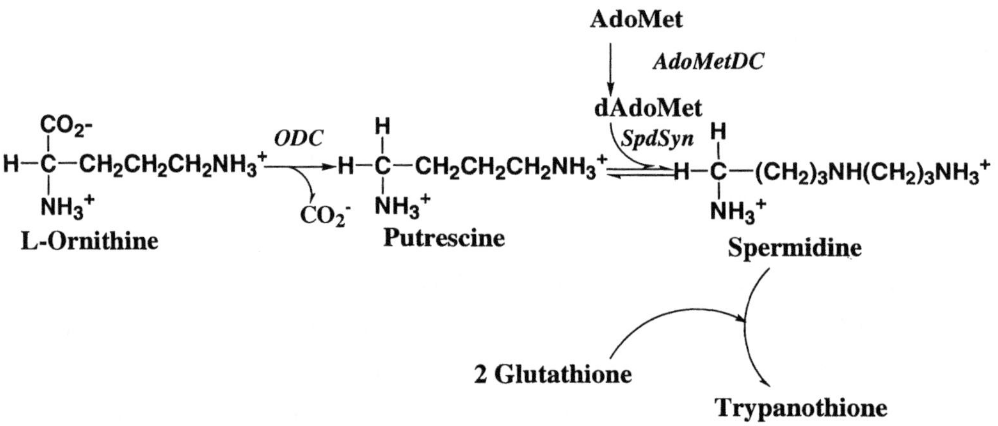

Fig. 1. The polyamine biosynthetic pathway in *T. brucei*. ODC, ornithine decarboxylase, AdoMetDC, S-adenosylmethionine decarboxylase, SpdSyn, spermidine synthase. Trypanothione is synthesized from glutathione and spermidine by glutathionylspermidine synthetase and trypanothione synthetase.

Structural analysis of the eukaryotic ODCs.

The X-ray structures of ODC from mouse [6], human [7] and *T. brucei* [8] have recently been solved. The eukaryotic ODCs are structural homologs of another PLP-dependent enzyme alanine racemase, while they are unrelated to bacterial ODC [9]. The N-terminal domain of ODC folds into a β/α barrel and PLP is bound to this domain close to strands N6-N8. The C-terminal domain forms a modified Greek key β-sheet. The enzyme is an obligate homodimer and the two identical active sites are formed at the subunit interface, with residues contributed from the β/α barrel domain from one subunit and from the C-terminal domain of the second subunit (Fig. 2). PLP forms several essential interactions with the β/α barrel domain: Lys-69 forms a Schiff base with PLP; the phosphate binding site is formed by Arg-277, Gly-237, Ser200 and Tyr389; the aromatic ring of PLP stacks against His-197; the pyridine nitrogen forms a hydrogen bond with Glu-274; and O_3 of PLP interacts with Arg-154 through a ordered water molecule.

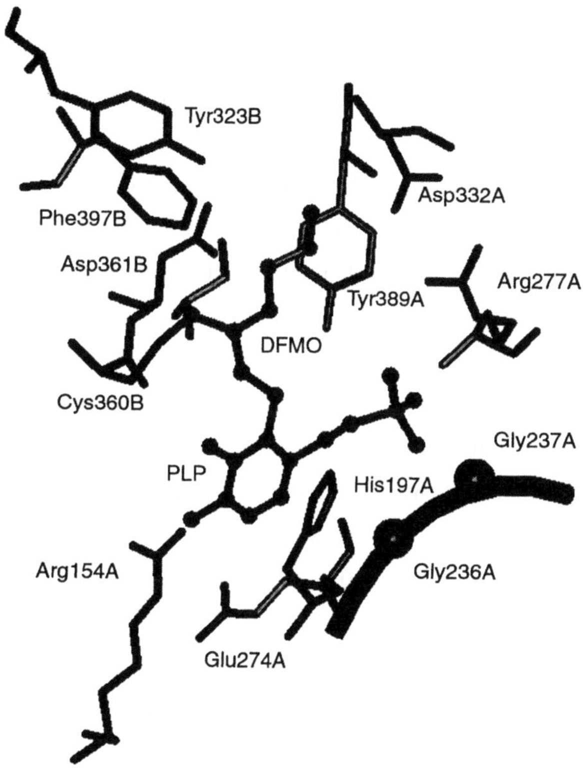

Fig. 2. Active site structure of *T. brucei* K69A ODC bound to DFMO. The structure was solved to 2.0 Å resolution [8]. The K69A mutant enzyme was used to avoid heterogeneity of labeling by DFMO. Residues contributed to the active site from the N-terminal domain of one subunit are labeled A and residues contributed from the C-terminal domain of the second subunit are labeled B. DFMO and PLP are displayed as ball and stick. The C_α of Gly236 and Gly237 are displayed as balls within a ribbon.

The substrate-binding site has been elucidated through the structure of the *T. brucei* K69A enzyme in complex with DFMO (Fig. 2). DFMO has been decarboxylated and the two fluoride molecules have been eliminated in the process of forming a covalent bond to Cys-360. The side chain of DFMO is in the extended conformation and the δ-amino group forms a hydrogen bond with Asp-332 from one subunit and with Asp-361 from the second subunit via an ordered water molecule. The aliphatic portion of the side chain is packed between the aromatic ring of Tyr-389 on one subunit and between Phe-397 and Asp-361 on the other subunit. The only significant structural difference between the DFMO-bound and the unliganded *T. brucei* ODC structures is the rotation of Cys-360 145° toward the active site upon ligand binding [8].

Mechanistic analysis of *T. brucei* ODC.

ODC catalyzed decarboxylation of L-Orn proceeds via the following steps: formation of a Schiff base species between L-Orn and PLP, decarboxylation to yield a quinoid intermediate (λ_{max} = 450 nm; [10]), protonation of C_α and product release (Schiff base hydrolysis). Schiff base formation with substrate occurs via a transimination reaction in which the internal aldimine with Lys-69 is displaced by the incoming substrate to form the external aldimine. Product release reverses this chemistry. The internal and external aldimine species have similar spectral properties (spectral bands of 340 and 420 nm are observed for each). Multi-wavelength stopped flow spectroscopy was used to demonstrate that the decarboxylation step occurs fast compared to the overall reaction (at 4°C, $k_{decarb} = 21\ s^{-1}$ and $k_{cat} = 0.5\ s^{-1}$), and that product release most likely limits the rate of the steady-state reaction [10].

Lys-69 is an essential catalytic residue. Pre-steady state analysis of the Lys-69 to Arg mutant of *T. brucei* ODC demonstrated that Lys-69 accelerates the rates of both Schiff base formation and Schiff base hydrolysis, minimally by 10^3-fold over the wild-type enzyme [11]. A similar role for the active site Lys residue in other PLP enzymes of differing structural classes has been described (e.g. aspartate aminotransferase [12] and tryptophan synthase [13]). In addition to its role in Schiff base interchange, the Lys residue in these enzymes is also the essential general base [14]. A general base is not required for in the decarboxylation reaction, yet surprisingly Lys-69 was also essential for the decarboxylation step, decreasing the rate of this step by minimally 10^4-fold. The external aldimine between putrescine and K69R ODC is 2 kcal/mol more stable than for the wild-type enzyme [11]. These data suggest that attainment of the transition-state during decarboxylation is more difficult because of the greater stability of the prior enzyme-substrate intermediate. Additionally, the X-ray structure of *T. brucei* ODC suggests that Lys-69 forms part of the CO_2 binding site [8] and may play a role in orienting the CO_2 for efficient decarboxylation.

The roles of several other active site residues have been studied in detail. Glu-274 forms a hydrogen bond with the pyridine nitrogen of PLP. Mutation of Glu-274 to Ala decreases k_{cat} by 50-fold in the presence of PLP as the cofactor. Substitution of PLP with N-methylPLP, which fixes a positive charge on the pyridine nitrogen, restores the k_{cat} to near wild-type levels in the mutant enzyme [15]. These data demonstrate that the role of Glu-274 is to stabilize the positive charge on the pyridine nitrogen of PLP, thereby enhancing the ability of the ring to stabilize the carbanion generated upon decarboxylation. This result is analogous to that observed for aspartate aminotransferase [16] and indeed, a negatively charged residue at this position is common to most PLP enzymes. An exception is alanine racemase where an Arg residue occupies this position [17]. Finally, Arg-277 interacts with the 5'phosphate on PLP and was shown to be essential for high affinity PLP binding [18], while Asp-361 interacts with the δ-nitrogen of DFMO and was demonstrated to be essential for substrate binding [15].

Structural differences between the mammalian and *T. brucei* enzymes.

Mouse and *T. brucei* ODC share 61% sequence identity, however comparison of the X-ray structures reveals that the active sites are invariant [8]. Residues in the dimer interface share 70% sequence identity, but fully functional cross-species heterodimers form between mouse and *T. brucei* ODC, suggesting that the residues in the interface that are essential for the energetics of the interaction are conserved [19]. These results suggest that it will be difficult to design species selective inhibitors to the eukaryotic ODCs. However, DFMO is not a selective inhibitor of *T. brucei* ODC and yet, selective toxicity of the drug is achieved. Several metabolic differences between the host and parasite cells have been identified that likely account for the selective toxicity [1]. These observations suggest that new inhibitors that are designed against ODC will also be able to achieve selective drug toxicity towards *T. brucei,* despite the similarities between the host and parasite enzymes. The availability of the X-ray structures for both human and *T. brucei* ODC will allow structure-based approaches to be utilized towards the discovery of new ODC inhibitors.

References.

1. Phillips, M.A. *Ornithine decaboxylase*, in *The encyclopedia of molecular biology*, T. Creighton, Editor. 1999, John Wiley & Sons: New York. p. 1726-1730.
2. Marton, L.J. and Pegg, A.E. (1995) Polyamines as targets for therapeutic intervention. *Annu. Rev. Pharmacol. Toxicol.*, *35*: 55-91.

3. O'Brien, T.G., Megosh, L.C., Gilliard, G., Soler, A.P. (1997) Ornithine decarboxylase overexpression is a sufficient condition for tumor promotion in mouse skin. *Cancer Res.*, *57*: 2630-2637.

4. Wang, C.C. (1995) Molecular mechanisms and therapeutic approaches to the treatment of African Trypanosomiasis. *Annu. Rev. Pharmacol. Toxicol.*, *35*: 93-127.

5. Amzel, L.M. (1998) Structure-based drug design. *Current opinion in Biotech.*, *9*: 366-369.

6. Kern, A.D., Oliveira, M.A., Coffino, P., and Hackert, M. (1999) Structure of mammalian ornithine decarboxylase at 1.6 Å resolution: Stereochemical implications of PLP-dependent amino acid decarboxylase. *Structure*, *7*: 567-581.

7. Almrud, J.J., Oliveira, M.A., Grishin, N.V., Phillips, M.A., and Hackert, M.L. (1999) Crystal structure of human ornithine decarboxylase at 2.1 Å resolution: structural perspectives of antizyme binding. *J. Mol. Biol.*, *295*: 7-16.

8. Grishin, N.V., Osterman, A.L., Brooks, H.B., Phillips, M.A., and Goldsmith, E.J. (1999) The X-ray structure of ornithine decarboxylase from *Trypanosoma brucei*: the native structure and the structure in complex with α-difluoromethylornithine. *Biochemistry*, *38*: 15174-15184.

9. Grishin, N.V., Phillips, M.A., and Goldsmith, E.J. (1995) Modeling of the spatial structure of eukaryotic ornithine decarboxylases. *Protein Science*, *4*: 1291-1304.

10. Brooks, H.B. and Phillips, M.A. (1997) Characterization of the reaction mechanism of *Trypanosoma brucei* ornithine decarboxylase by multiwavelength stopped-flow spectroscopy. *Biochemistry*, *36*: 15147-15155.

11. Osterman, A.L., Brooks, H.B., Jackson, L., Abbott, J.J., and Phillips, M.A. (1999) Lys-69 plays a key role in catalysis by *T. brucei* ornithine decarboxylase through acceleration of the substrate binding, decarboxylation and product release steps. *Biochemistry*, *38*: 11814-11826.

12. Toney, M.D. and Kirsch, J.F. (1993) Lys-258 in asp aminotransferase: enforcer of the Circe effect for amino acid substrates and general-base catalyst for the 1,3-prototropic shift. *Biochemistry*, *32*: 1471-1479.

13. Lu, Z., Nagata, S., McPhie, P., and Miles, E.W. (1993) Lysine 87 in the beta subunit of tryptophan synthase that forms an internal aldimine with pyridoxal phosphate serves critical roles in transimination, catalysis and product release. *J. Biol. Chem.*, *268*: 8727-34.

14. Toney, M.D. and Kirsch, J.F. (1989) Direct Bronsted analysis of the restoration of activity to a mutant enzyme by exogenous amines. *Science*, *243*: 1485-1488.

15. Osterman, A., Kinch, L.N., Grishin, N.V., and Phillips, M.A. (1995) Acidic residues important for substrate binding and cofactor reactivity in eukaryotic ornithine decarboxylase identified by alanine scanning mutagenesis. *J. Biol. Chem.*, *270*: 11797-11802.

16. Onuffer, J.J. and Kirsch, J.F. (1994) Characterization of the apparent negative co-operativity induced in *Escherichia coli* aspartate aminotransferase by the replacement of Asp222 with alanine. Evidence for an extremely slow conformational change. *Prot. Eng.*, *7*: 413-424.

17. Shaw, J.P., Petsko, G.A., and Ringe, D. (1997) Determination of the structure of alanine racemase from Bacillus stearothermophilus at 1.9-Å resolution. *Biochemistry*, *36*: 1329-1342.

18. Osterman, A.L., Brooks, H.B., Rizo, J., and Phillips, M.A. (1997) The role of Arg-277 in the binding of pyridoxal 5'-phosphate to *Trypanosoma brucei* ornithine decarboxlyase. *Biochemistry*, *36*: 4558-4567.

19. Osterman, A.L., Grishin, N.V., Kinch, L.N., and Phillips, M.A. (1994) Formation of functional cross-species heterodimers of ornithine decarboxylase. *Biochemistry*, *33*: 13662-13667.

Biochemistry and Molecular Biology of Vitamin B_6
and PQQ-dependent Proteins
ed. by A. Iriarte, H.M. Kagan and M. Martinez-Carrion

Mechanistic Analysis of Dialkylglycine Decarboxylase

Michael D. Toney*, Xianzhi Zhou, and Shaoxian Sun
Department of Chemistry, University of California-Davis
One Shields Ave., Davis, CA 95616, U.S.A.

Summary
Dialkylglycine decarboxylase catalyzes the oxidative decarboxylation of small 2,2-dialkylglycines in the first half-reaction of its ping-pong kinetic mechanism and the transamination of pyruvate to L-alanine in the second. We have analyzed the reactions of several alternate decarboxylation substrates to define the specificity of binding subsites in the active site, and provide evidence for the existence of stereoelectronic control of reactivity. We have also examined the 2-aminoisobutyrate, L-alanine, and pyruvate half-reactions by stopped-flow spectrophotometry. This and other data provide free energy profiles for the half-reactions, which show that the decarboxylation step largely limits the rate of the overall reaction.

Introduction

Dialkylglycine decarboxylase (DGD) is an interesting pyridoxal phosphate (PLP) dependent enzyme due to its natural ability to catalyze both decarboxylation and transamination reactions in its normal catalytic cycle (1,2). This suggests that one might be able to use it to obtain direct evidence for the role of stereoelectronic effects in the control of reaction specificity in PLP enzymes. Dunathan (3) originally proposed that stereoelectronic effects are important, with the bond that is held perpendicular to the plane of the coenzyme ring being more reactive due to orbital overlap between the nascent *p* orbital at Cα and the aldimime/pyridine ring conjugated π system.

The sequence and structural homology of DGD with subgroup II aminotransferases (4) suggests that DGD acquired decarboxylation abilities on top of pre-existing transaminase abilities. Mechanistically, it is interesting to determine how good DGD has become at decarboxylation compared to its native transamination ability. We have collected stopped-flow data on the decarboxylation of 2-aminoisobutyrate and on the transamination of both L-alanine and pyruvate. These data and ^{13}C kinetic isotope effects show that the C-C bond breaking step in the decarboxylation half-reaction largely limits the normal catalytic cycle.

Materials and Methods

DGD was expressed and purified as previously described (5). Sun et al. (5) described the analysis of the reactions of alternate decarboxylation substrates. The stopped-flow analysis of the DGD half-reactions was performed under conditions similar to those employed by Sun et al. (5) and used the same instrument. The X-ray structures of inhibitors bound to DGD have been reported by Malashkevich et al. (6). The measurement of the ^{13}C kinetic isotope effects on k_{cat} was as reported by Zhou and Toney (7), except using [1-^{13}C]-AIB.

Results and Discussion

The origin of the stereoelectronic control of reactivity in PLP enzymes is illustrated in Figure 1. The carboxylate is labilized here since the *p* orbital formed during its scission will be aligned

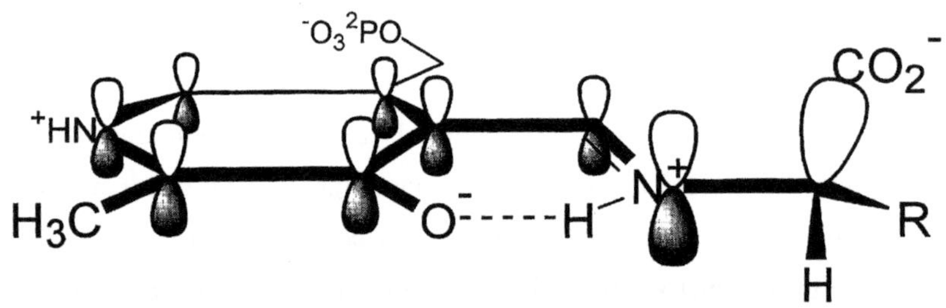

Figure1. Substrate aldimine with PLP illustrating the origin of stereoelectronic effects.

with those of the conjugated π system of the aldimine/pyridine ring. There is little direct kinetic evidence for this mechanism of reaction specificity control in the literature.

We have analyzed the reactions of alternate decarboxylation substrates and, through these data, validated the functional model of the DGD active site shown in Figure 2 (5). The A subsite is where bond making and breaking in both the decarboxylation and transamination half-reactions occurs. The B subsite accepts either alkyl groups in the decarboxylation reaction or a carboxylate group in the transamination half-reaction. The C subsite is specific for a methyl group.

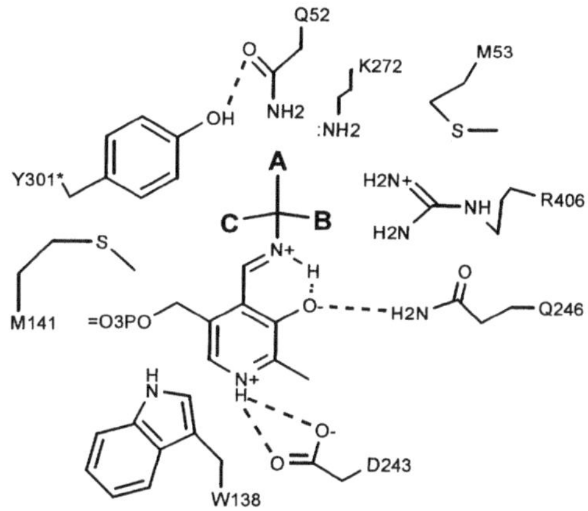

Figure 2. Binding subsite model for the DGD active site. It was proposed, based on the active site structure and Dunathan's hypothesis, that the A subsite is where all bond making and breaking occurs, while the B subsite accepts either a carboxylate or an alkyl group and the C subsite accepts only small alkyl groups.

Strong evidence in support of the A subsite being the sole reactive locus was obtained from the decarboxylation of a series of L-amino acids of increasing side chain size. Figure 3 schematically depicts the experiment and Figure 4 presents the data.

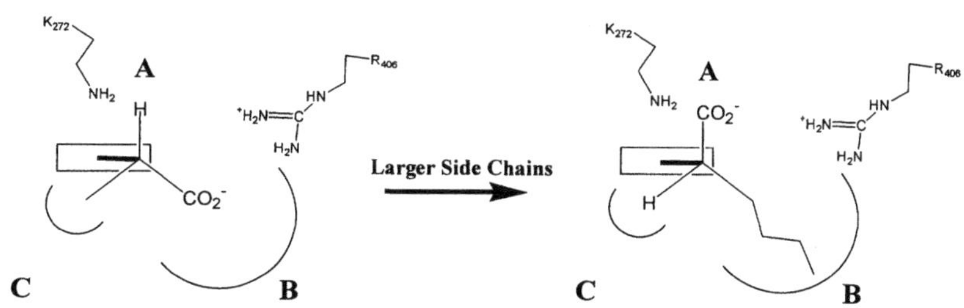

Figure 3. Experiment designed to probe for nonproductive binding, with respect to decarboxylation, of amino acids in the DGD active site.

The increase in k_{cat} is due to the nonproductive binding mode that exists with the smaller substrates, in which the carboxylate is bound in the B subsite, the productive site for transamination. This increase in k_{cat} could occur only if the A subsite is the sole site for bond breaking, since the other reactions examined (e.g. transamination) all show decreased rate constants in response to increasing side chain size.

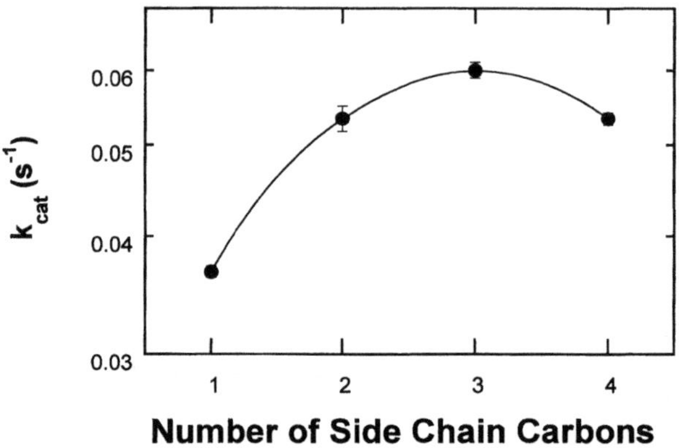

Figure 4. Increase in k_{cat} for decarboxylation of L-amino acids with side chain size.

Pre-steady-state analysis of the DGD half-reactions shows that, as expected from previous results (8), the enzyme reacts in two kinetic processes. These were shown to be due to the slow interconversion of two enzyme forms that differ in reactivity (8). We have concentrated our analysis on the fast reacting form since it contributes the majority of activity in steady-state turnover. Figure 5 and 6 present rate constant and spectral data, respectively, obtained from an analysis of AIB decarboxylation.

We find that the external aldimines of both AIB and L-alanine are formed very rapidly, in the dead-time of the instrument. Absorbance at ~500 nm, characteristic of a quinonoid intermediate, is observed only in the L-alanine/pyruvate transamination half-reaction. The ^{13}C kinetic isotope effect (KIE) measured on k_{cat} for AIB is 1.043. A value of 1.045 is calculated based on the kinetic parameters of the two half-reactions and assuming an intrinsic KIE of 1.06. Thus, the decarboxylation reaction largely, but not fully limits the overall catalytic cycle; the transamination of pyruvate significantly contributes to rate limitation as seen by the reduction of the observed KIE from the intrinsic.

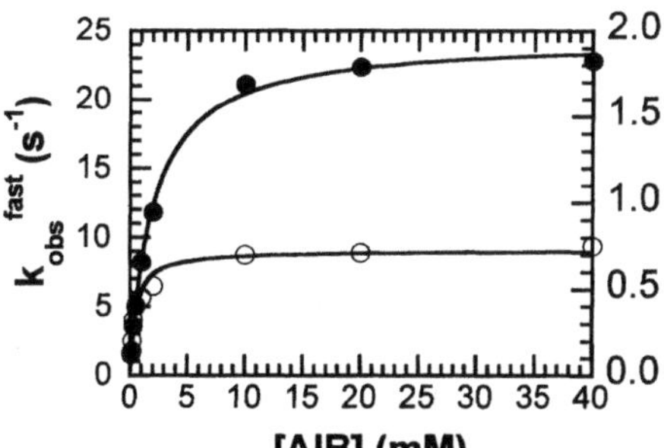

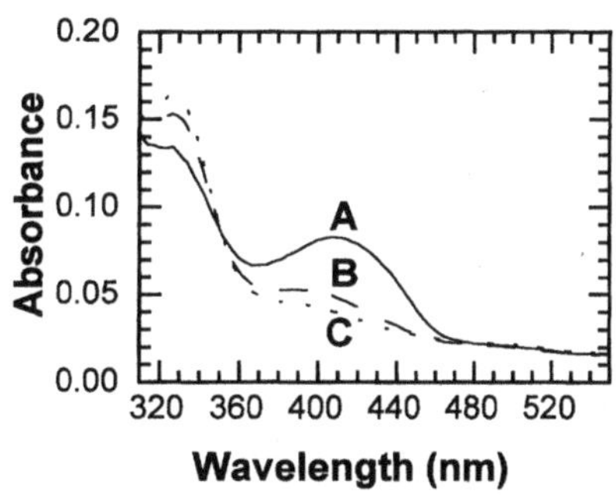

Figures 5 (left) and 6 (right). The dependence of the observed pseudo-first order rate constants on AIB concentration for the fast and slow reacting enzyme forms is shown in Figure 5, while Figure 6 shows the spectra of the three species in the two exponential process. These spectra were obtained from global analysis of spectral data sets collected with a diode-array detector using the program SPECFIT.

The pre-steady-state data as well as steady-state data and KIE's allow the construction of the free energy profiles shown in Figures 7 and 8.

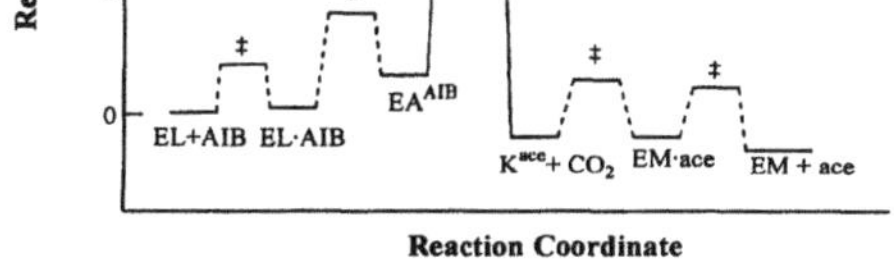

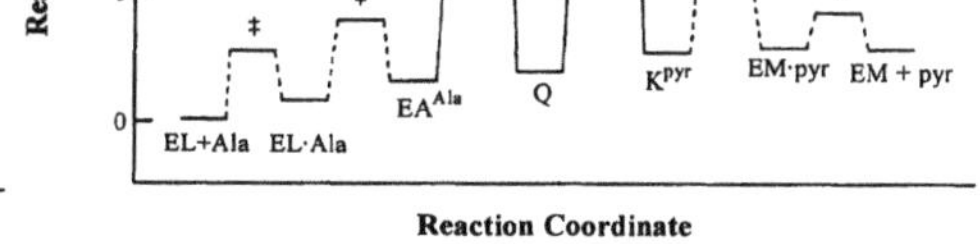

Figures 7 (left) and 8 (right). Free energy profiles for the AIB decarboxylation and the L-alanine/pyruvate transamination half-reactions. The C-C breaking step limits the

These profiles show that the C-C breaking step is fully rate limiting for the decarboxylation of AIB, as might be expected given that DGD has transaminase origins. The transamination half-reaction is limited by the rate of protonation of the quinonoid intermediate in the L-alanine to pyruvate direction while deprotonation of the ketimine limits the rate of the reverse transamination.

References

1. Bailey, G. B., and Dempsey, W. B. (1967) *Biochemistry 6*, 1526-1533.
2. Honma, M., and Shimomura, T. (1974) *Agr. Biol. Chem. 38*, 953-958.
3. Dunathan, H. C. (1966) *Proc. Natl. Acad. Sci. U. S. A. 55*, 712-716.
4. Keller, J. W., Baurick, K. B., Rutt, G. C., O'Malley, M. V., Sonafrank, N. L., Reynolds, R. A., Ebbesson, L. O. E., Vajdos, F. F. (1990), *J. Biol. Chem. 265*, 5531-5539.
5. Sun, S., Bagdassarian, C., and Toney, M.D. (1998) *Biochemistry 37*, 3876-3885.
6. Malashkevich V.N., Strop P., Keller J.W., Jansonius J.N., Toney M.D. (1999) *J. Mol. Biol. 294*, 193-200
7. Zhou, X. and Toney, M.D. (1998) *J. Am. Chem. Soc.120*, 13282-13283.
8. Zhou, X., Kay, S., and Toney, M.D. (1998)*Biochemistry 37*, 5761-5769.

Biochemistry and Molecular Biology of Vitamin B_6
and PQQ-dependent Proteins
ed. by A. Iriarte, H.M. Kagan and M. Martinez-Carrion

Studies on an Active Site residue, E177, That Affects Binding of the Coenzyme in D-Amino Acid Transaminase, and Mechanistic Studies on a Suicide Substrate

Peter W. van Ophem#, Bryan W. Lepore*, Kazuhisa Kishimoto#, Dagmar Ringe* and James M. Manning#.

#Northeastern University, Biology Department, Boston, MA 02115, USA
*Brandeis University, Rosenstiel Basic Medical Sciences Center, Waltham, MA 02154, USA

Summary

While the E177S mutation in D-amino acid transaminase has a reduced ability to transaminate D-alanine, the efficiency of β-elimination of β-chloro-D-alanine remains almost intact. Interestingly, the latter reaction showed the appearance of an intermediate absorbing around 460 nm with this attenuated enzyme. The protein CD spectrum of the apo-enzyme of E177S resembled that of wild-type enzyme, but that of E177K showed a negative elipticity around 280 nm.

Introduction

D-amino acid transaminase converts ∝-keto acids into their corresponding D-amino acids, which are important components of the bacterial cell wall. Thus, it is an attractive candidate for the development of anti-microbial agents. The enzyme contains pyridoxal 5'-phosphate (PLP) as cofactor and its crystal structure has been solved (1). The structure showed that the pyridinium nitrogen of the cofactor is stabilized by a Glu-177 whose importance was clearly demonstrated when this negatively charged residue was changed into a positively charged lysine (2). This enzyme converts the suicide substrate, β-chloro-D-alanine, to pyruvate through β-elimination. Here, we present a study on the effect of replacing Glu-177 by serine on the overall performance of the enzyme and on the β-elimination of β-chloro-D-alanine, to provide information on the types of intermediates that are involved in this process.

Experimental procedures

The mutagenesis, expression, purification and activity measurements have been described (2,3). Time-dependent conversion of L-alanine into D-alanine ("racemase" activity) was performed at room temperature as described (2). Incubation with β-chloro-D-alanine was performed in the following manner. Enzyme (approx. 0.5 mg/ml) was incubated with various concentrations of β-chloro-D-alanine in 50 mM potassium phosphate, pH 7.0. The reaction was followed by recording spectral changes with time. In addition, samples were taken and analyzed for pyruvate and for remaining activity (2,3). Titration of SH-groups was done using 5,5'-dithio-bis(2-nitrobenzoic acid) (DTNB, Ellman's Reagent) as described (3). Cofactor analysis was performed as described (4).

The structure of the E177S mutant was determined using general methods described elsewhere (1,2). The enzyme was concentrated to 30 mg/ml in 50 mM potassium phosphate buffer, pH 7.3, containing 0.13 M KCl, 0.2 mM EDTA, 50 ,lM PLP and 0.01 % , β-mercaptoethanol. The protein was crystallized by the hanging drop method with 26% polyethyleneglycol 3350, 0.3 M sodium acetate and 0.1 M Tris/HCl, pH 8.5. Two microliters of protein solution were mixed with 2ul of the above crystallization solution and suspended at room temperature over 0.5 ml of crystallization solution. Within two days, yellow crystals appeared. X-ray data were collected with a scan width of 1° per frame and an exposure time of 20 minutes per frame on a R-Axis IIC image plate system mounted on a Rigaku RU-200 X-ray generator running at 45 kV and 140 mA. Frames were integrated and scaled using the program DENZO (5). (Table 1).

Table 1. Data for Solution of the Crystal Structure

Crystal Data	
protein	E177S
space-group	P212121
a	77.9 A
b	91.9 A
c	89.4 A
unit cell volume	620,893 A^3
Data Collection	
temperature	4°C
number of crystals used	1
reflections, observed/unique	303,396/96,070
Rmerge (%, on I)	12.3
resolution	27 – 1.8 A
completeness, overall/1.86-1.80 A (%)	85.4/24.4
Rsym (%, on I), overall/1.86-1.80 A	12.3/97.5
Refinement	
resolution	27 – 2.0 A
I/σ(I) cutoff	1.0
R-factor (maximum likelihood target function, %)	18.8
R-free	22.7
protein atoms	4,464
cofactor atoms	30
water molecules	204
B-factor model	individual, restrained
restraints (rms observed)	
bond length (A)	0.006
bond angles (deg)	1.278
improper angles (deg)	0.769
dihedral angles (deg)	22.59

The structure was solved by direct application of phases from the L201A mutant orthorhombic structure of D-aAT (PDB Code 2DAB)(10) using the CCP4 package (6). The program CNS (7) was then used for all remaining refinement. Rigid body refinement was followed by several cycles of positional refinement, grouped temperature factor refinement and placement of water molecules. Water molecules were placed with the water pick script in CNS. The final model has an R-factor of 19% to 2.0A resolution with good geometry (Table 1). Maps were calculated using all data to 1.8A resolution.

Results and Discussion

The residue at position 177 of the wild-type D-amino acid aminotransferase is conserved in kind in all transaminases (2). However, in other types of pyridoxal-phosphate dependent enzymes, the chemical nature of this residue changes. We have speculated that these differences may be correlated with the specific chemistry that is catalyzed by the cofactor (2). For instance, the presence of a negative charge in this position seems to be correlated with releasing keto-group containing compounds. In comparison, the equivalent residue is a serine in tryptophan synthase, where the reaction catalyzed is β-elimination (8). Consequently, we have replaced residue 177 in DATase by a serine in order to test this premise in an enzyme that originally evolved to be a transaminase.

We have studied the kinetic properties of this mutant by several methods. Replacing Glu-177 by a serine has a significant effect on the ability to efficiently convert the natural substrates of D-amino acid transaminase (Table 2). Thus, when tested under standard assay conditions in the presence of 25 mM α-ketoglutarate (2) the specific activity of E177S (0.15 U/mg) is largely reduced compared to wild-type (200 U/mg). The K'_m of α-ketoglutarate of E177S is approx. 10-fold lower than that of wild-type, while at concentrations above 5 mM the activity decreased, probably due to substrate inhibition. Conversion of L-alanine to D-alanine ("racemase" activity) was at a similar level found for other active-site mutants (2).

Table 2: Kinetic properties of E177S, E177K and wild-type enzymes.

Enzyme	Sp. act. with D-Ala as substrate (units/mg)	K'_m α-ketoglutarate (mM)	Racemase act. with L-Ala as substrate (units/mg)	β-chloro-D-alanine conversion (units/mg)
E177S	0.18	–	0.050	1.60
E177K	0.15	5.0	0.054	0.12
Wild-type	200	1.0	0.004	2.50

The activity of the enzyme can also be measured by following the spectral changes induced by the addition of substrate. The spectrum of E177S shows a maximum around 415 nm similar to that of freshly purified wild-type or E177K enzymes (2), suggesting that the cofactor is in the PLP-form. Addition of 0.1 M D-alanine resulted in conversion of the cofactor into the PMP-form absorbing at 330nm. Unlike wild-type, where this conversion is instantaneous (2), with E177S this is a slow process taking several hours for partial conversion. This was also observed with E177K.

The effect of the serine therefore can be compared to that of lysine at position 177 where a significant decrease in activity was also observed (2). The presence of Lys-177 changes some of the interactions of the residues surrounding the active site due to the nature of this amino acid. Therefore, it can be argued that some or most of the changes in activity observed are caused by the different interactions. Specifically, in the E177K structure, there is no interaction between the residue and the pyridinium nitrogen of the cofactor. No other interaction has taken the place of this residue. Replacing Glu-177 with Ser changes the type of interactions observed with the pyridinium nitrogen (see below), but the impact on performance of the enzyme is similar to that of E177K. Clearly, the importance of the stabilization of the pyridinium nitrogen of the cofactor by a negatively charged residue, Glu-177, is important for stabilization of intermediates leading to transamination. .

The wild type enzyme shows a small amount of β-elimination when reacting with D-amino acid substrates capable of this reaction. We have studied this reaction with the two active site mutants, E177S, in order to determine the effect of the mutation on this reaction. The wild-type enzyme catalyzes a significant production of pyruvate from β-chloro-D-alanine with no obvious intermediate being formed. The β-elimination reaction catalyzed by the E177S mutant seems to be quite different. Although the mutant enzyme also produces pyruvate from the same substrate, it does so with significant accumulation of an intermediate that has a spectral signature similar to that of the presumed quininoid intermediate of the transaminase reaction (9).

Spectral studies of wild-type enzyme or the E177S mutant enzyme with 1 mM β-chloro-D-alanine at room temperature showed that rapid changes occur within minutes, thereby preventing an accurate spectral analysis for intermediates. Lowering the temperature slowed the reaction, while increasing the β-chloro-D-alanine to 10 mM also allowed intermediates to persist. This showed that in the case of wild-type (Fig. 1a) an intermediate with an absorption maximum around 330 was present for 10 - 15 min before the 415 nm peak quickly re-appeared.

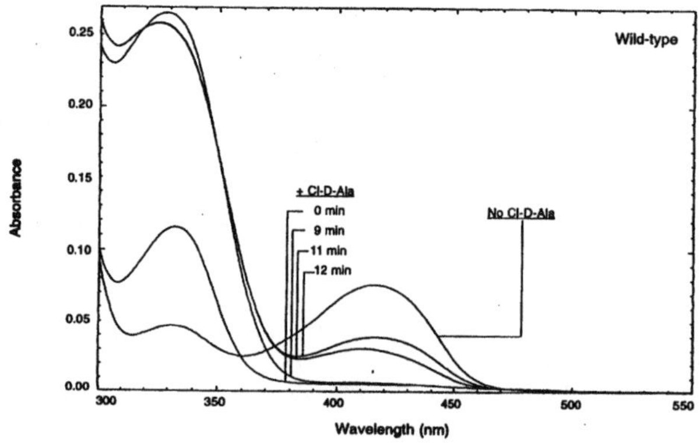

Fig. 1a: Time-dependent spectral changes of the wild-type enzyme in the presence of β-chloro-D-alanine. Enzyme (0.5 mg/ml) was mixed with 10 mM β-chloro-D-alanine in 50 mM potassium phosphate buffer, pH 7.0, and spectra were recorded at the indicated times.

This result was also found for E177K although at a much slower rate (not shown), taking hours for completion. For the E177S mutant, however, an intermediate with a maximum around 460 nm was detected (Fig. 1b), which, in time, disappeared with a concomitant increase around 415 nm.

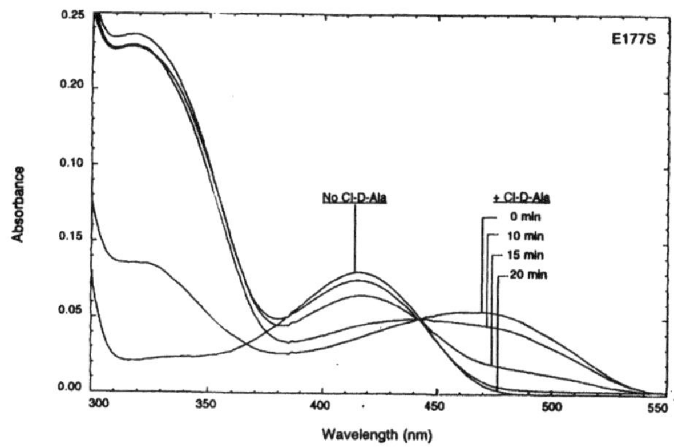

Fig. 1b: Time-dependent spectral changes of the E177S mutant enzyme in the presence of β-chloro-D-alanine. Enzyme (0.5 mg/ml) was mixed with 10 mM β-chloro-D-alanine in 50 mM potassium phosphate buffer, pH 7.0, and spectra were recorded at the indicated times.

Kinetic studies of pyruvate production through β-elimination of β-chloro-D-alanine showed that for all three enzymes - wild type, E177K and E177S - pyruvate was produced rapidly and prior to the re-appearance of the 415 nm peak (Fig. 2).

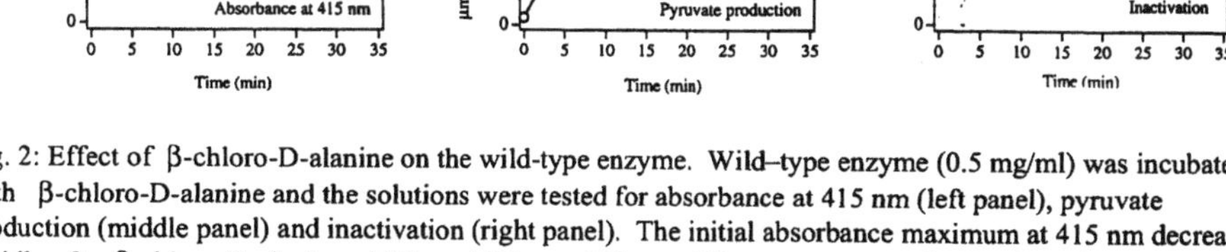

Fig. 2: Effect of β-chloro-D-alanine on the wild-type enzyme. Wild-type enzyme (0.5 mg/ml) was incubated with β-chloro-D-alanine and the solutions were tested for absorbance at 415 nm (left panel), pyruvate production (middle panel) and inactivation (right panel). The initial absorbance maximum at 415 nm decreases rapidly after β-chloro-D-alanine addition, shifting to 330 nm. When pyruvate production from β-chloro-D-alanine is finished, the absorbance at 415 nm begins to increase. At the same time, inactivation ceases because all of the β-chloro-D-alanine has been consumed.

Inhibition of transaminase activity in wild-type enzyme by β-chloro-D-alanine occurred while β-chloro-D-alanine was being converted into pyruvate and prior to the reappearance of the 415 nm peak. Interestingly, while the transaminase activity of wild-type was inhibited about 50 % in the presence of β-chloro-D-alanine, the E177S mutant was fully active under the same conditions..

The effect of the mutations on the reactivity of the enzyme may be due to changes in accessability of the active site. This effect had been noted for the E177K mutant, for which SH titrations showed that the sulfhydryl groups are more accessible than in the wild type enzyme, based on the rate at which they are titrated by DTNB (2). The rate at which accessible cysteine side chains were titrated by this reagent in the E177S mutant was greater than in the wild type enzyme, but not as fast as in the E177K mutant enzyme (Fig. 3), suggesting that the accessibility for DTNB in E177S is between that in E177K and wild-type. Examination of the structure of the mutant enzyme (see below) suggests that the titratable residue whose accessibility has increased may be Cys 142.

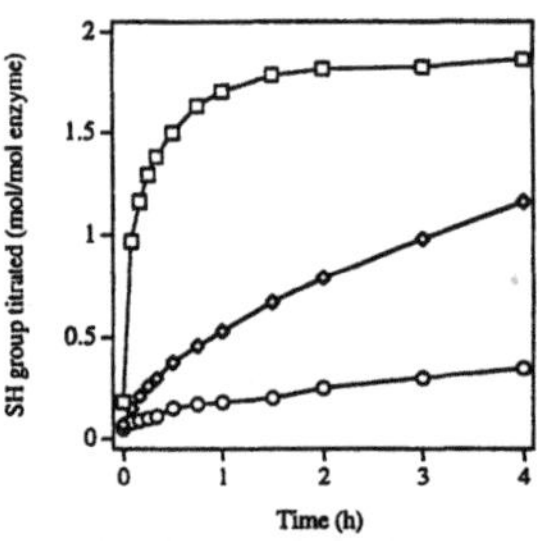

Fig. 3: SH-titration of E177S, E177K and wild type enzymes with DTNB. E177S enzyme (0.4 mg/ml) was incubated at room temperature in 0.1 M Tris/HCl. 2 mM EDTA, pH 7.4, in the presence of 1mM DTNB (◊). The wild type (o) and E177K values ( ) are from reference 2. Given are the values per dimer.

Ultimately, these biochemical properties must be explained by structural comparisons of the mutant and the wild type D-amino acid aminotransferases. The overall structure of the E177S mutant enzyme is the same as that of the wild-type enzyme, with significant differences only in the active site (Fig. 4), where small conformational changes have occurred. Noteworthy are the tilt of the imidazole ring of His100 and the conformation of the Arg138 side chain. The most important conformational change is the tilt of the cofactor ring backwards and away from the solvent accessible side of the active site. This tilt decreases the angle that the cofactor makes with the internal aldimine. Such a conformational change might be expected to decrease the reactivity of the cofactor with incoming substrate by reducing the strain of this linkage and increasing the extent of conjugation between the pyridinium ring of the cofactor and the aldimine.

In the wild-type enzyme, the glutamate at position 177 is held in place by a specific interaction with Arg 138, insuring that the salt-bridge between the carboxylate and the protonated pyridinium nitrogen of the cofactor is optimally positioned (1). The serine side chain is shorter than the glutamate side chain so that a direct interaction with the pyridinium nitrogen does not occur. Instead, a water molecule appears where the carboxylate of glutamate would have been. This water molecule is held tightly in place, stabilized by interactions with the hydroxyl group of Ser177, the backbone nitrogen of residue 201, and the amide nitrogen of Asn182. This water molecule interacts directly with the pyridinium nitrogen of the cofactor. Since the pKa of such a water molecule is expected to be significantly different than that of a carboxylate, it is expected that the pyridinium nitrogen of the cofactor is unprotonated in the mutant, reducing its ability to stabilize the requisite intermediates during catalysis leading to transamination. In contrast, this form of the enzyme may be more capable of stabilizing intermediates, such as the quinonoid intermediate, that are not normally observed in the reaction catalyzed by the wild type enzyme.

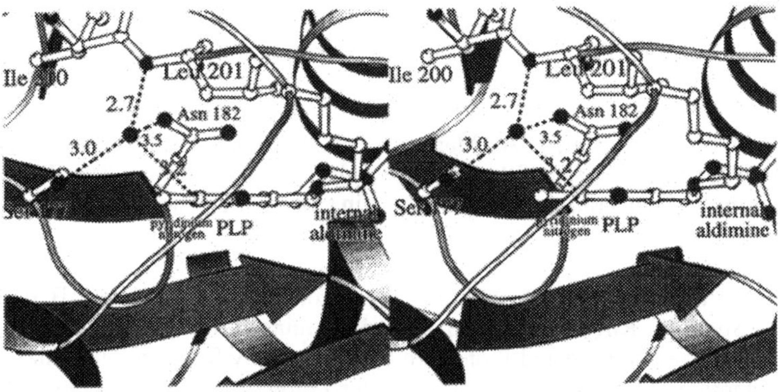

Fig. 4: Active site region taken from the crystal structure of the E177S mutant of D-amino acid aminotransferase. Shown in stereo are the interactions between the new water molecule observed in the mutant in the position of the residue that interacts with the pyridinium nitrogen of the cofactor in the wild type enzyme. The position of the new water molecule is shown with distances indicated in Angstroms.

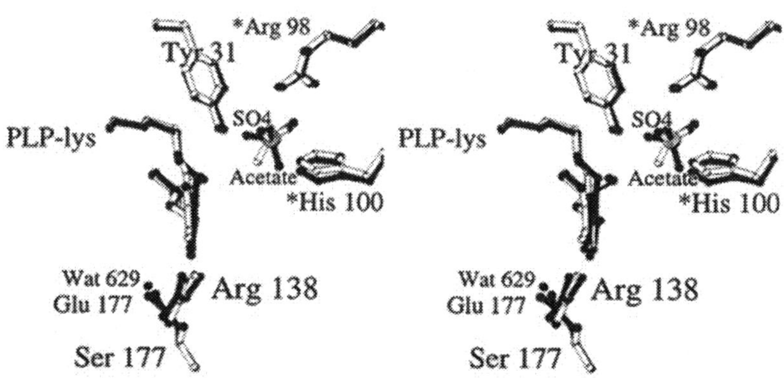

Fig. 5: Overlay of the active sites of wild-type (black) and E177S (white) D-amino acid aminotransferase. Residues from the second subunit of the enzyme are indicated with an asterisk. An acetate ion has been modeled into the substrate carboxylate binding site of the enzyme. In the structure of the wild type enzyme a sulfate ion is observed at this position.

Acknowledgements
This work was supported in part by NSF Grants MCB9317373 (D.R.) and MCB9727613 (J.M.M.). Thanks to Andre White for help with figures.

References

1. Sugio, S., Petsko, G.A., Manning, J.M., Soda, K. and Ringe, D. (1995) Crystal Structure of a D-Amino Acid Aminotransferase: How the Protein Controls Stereoselectivity. *Biochemistry 34*: 9661 - 9669.

2. Van Ophem, P.W., Peisach, D., Erickson, S.D., Soda, K., Ringe, D. and Manning, J.M. (1999) Effects of the E177K mutation in D-amino acid transaminase. Studies on an essential coenzyme anchoring group that contributes to stereochemical fidelity. *Biochemistry 38*:1323 - 1331.

3. Van Ophem, P.W., Pospischil, M.A., Manning, J.M., Petsko, G.A., Peisach, D., Ringe, D. and Soda, K. (1995). Comparison of the Stability and Catalytic Ability of Two Recombinant Mutants of D-Amino Acid Transaminase Involved in Coenzyme Binding. *Prot. Sci. 4*:2578 - 2586.

4. Van Ophem, P.W., Erickson, S.D., Martinez del Pozo, A., Haller, I., Chait, B.T., Yoshimura, T., Soda, K., Ringe, D., Petsko, G. and Manning, J.M. (1998) Substrate Inhibition of D-Amino Acid Transaminase and Protection by Salts and by Reduced Nicotinamide Adenine Dinucleotide. Isolation and Initial Characterization of a Pyridoxo Intermediate Related to Inactivation. *Biochemistry 37*:2879 - 2888.

5. Otwinowski, Z. and Minor, W. (1997) Processing of X-ray Diffraction Data Collected in Oscillation Mode. *Methods Enzymol. 276*:307-326.

6. Collaborative Computational Project 4 (1994) *Acta Crystallogr. D50*:760-763.

7. Brunger, A.T., Adams, P.D., Clore, G.M., DeLano, W.L., Gros, P., Grosse-Kunstleve, R.W., Jiang, J-S., Kuszewski, J., Nilges, M., Pannu, N.S., Read, R.J., Rice, L.M., Simonson, T. and Warren, G.L. (1998) Crystallography & NMR System: A new Software Suite for Macromolecular Structure Determination. *Acta Cryst. D54*:905-921.

8. Miles, E.W. (1991) Structural Basis for Catalysis by Tryptophan Synthase. *Adv. Enzymol. 64*:83-172.

9. Kallen, R.G., Korpela, T., Martell, A.E., Matsushima, Y., Metzler, C.M., Metzler, D.E., Morozov, Y.V., Ralston, I.M., Savin, F.A., Torchinsky, Y.M. and Ueno, H. (1985) in *Transaminases* (Christen, P. and Metzler, D.E., eds.), pp 38-108 Wiley Interscience, NY.

10. Sugio, S., Kashima, A., Kishimoto, K., Peisach, D., Petsko, G.A., Ringe, D., Yoshimura, T. and Esaki, N. (1998) Crystal structures of the L201A mutant of D-amino acid aminotransferase at 2.0A resolution: implication of the structural role of Leu201 in transamination. *Prot. Eng. 11*:613-619.

BIOTECHNOLOGY AND APPLICATIONS OF EMERGING TECHNOLOGIES

Biochemistry and Molecular Biology of Vitamin B_6
and PQQ-dependent Proteins
ed. by A. Iriarte, H.M. Kagan and M. Martinez-Carrion

Functional properties of immobilized pyridoxal 5'-phosphate-dependent enzymes probed by absorption microspectrophotometry

Andrea Mozzarelli*[§], Barbara Campanini[§], Stefano Bettati*[§] and Alessio Peracchi[§]
[§]Institute of Biochemical Sciences and [§]INFM, University of Parma, 43100 Parma, Italy.

Summary

Functional properties of the pyridoxal 5'-phosphate-dependent enzymes tryptophan synthase and *O*-acetylserine sulfhydrylase immobilized by crystallization or encapsulation in wet porous silica gels were investigated by absorption microspectrophotometry. The enzyme behavior exhibits a few striking differences in the crystals and silica gels compared to solution, likely due the constraints of either lattice forces or silica matrix on the conformational equilibria that accompany catalytic events. The results define the experimental conditions to accumulate metastable catalytic species in the crystalline state, to develop bioreactors with immobilized enzymes and to slow down protein folding processes.

Introduction

Immobilization of biomolecules is a powerful strategy for investigating structure, dynamics and function and for developing biodevices (Mosbach, 1988, and references therein). Immobilization of a protein via crystallization is the critical step in the determination of the three dimensional structure, and, recently, in the preparation of effective vaccines (St. Clair et al., 1999). Immobilization via encapsulation in porous silica gels provides an easy way to confine biologically active proteins in a defined environment, to select individual tertiary and quaternary states, to stabilize native vs unfolded states, and to prepare biosensors (Lan et al., 1999; Das et al., 1998; Mozzarelli and Bettati, 2000, and references therein).

Several PLP-enzymes have been crystallized and the catalytic competence in the crystalline state has been assessed either by determining enzyme activity on microcrystalline suspensions or ligand binding and transient reactions by single crystal polarized absorption microspectrophotometry (Mozzarelli and Rossi, 1996, and references therein). To our knowledge PLP-enzymes have never been encapsulated in silica gels or linked to other inert matrices. In this work, we have immobilized in the crystal and in silica gels the tryptophan synthase $\alpha_2\beta_2$ complex (TS; Miles, 1991) and *O*-acetylserine sulfhydrylase (OASS; Cook and Wedding, 1976) in order i) to assess the catalytic competence of these immobilized enzymes, ii) to define the experimental conditions for the accumulation of the quinonoid intermediate of TS and the aminoacrylate Schiff base of OASS, key catalytic species for which structural determination is still lacking, and iii) to evaluate the effect of denaturants on the stability of encapsulated OASS.

Materials and Methods

The tryptophan synthase $\alpha_2\beta_2$ complex from *Salmonella typhimurium* was purified as previously reported (Yang et al., 1996). The A-isozyme of *O*-acetylserine sulfhydrylase, obtained from a plasmid-containing *Salmonella typhimurium* overproducing strain (LT-2) (Tai et al. 1993), was a kind gift of Dr. Paul F. Cook (Department of Chemistry and Biochemistry, University of Oklahoma, Norman, OK, USA). All reagents were of the best available quality and were used without further purification. Crystals of TS and OASS were grown from PEG solutions as previously described (Mozzarelli et al., 1989; 1998). Encapsulation of TS and OASS in silica gels was carried out as described (Bettati and Mozzarelli, 1997 and references therein). A solution containing tetramethyl orthosilicate, water and hydrochloric acid was thoroughly mixed in a glass vial and sonicated for 20 minutes, at 4°C. An equal volume of a solution containing either 50 mM potassium phosphate, pH 8.0 for TS or 10 mM potassium phosphate, pH 5.9 for OASS, is added to the sol. A mixture is made using equal volumes of the sol and a solution containing either 29 mg/ml TS, 50 mM Bicine, pH 8, or 26 mg/ml OASS, 10 mM Hepes, pH 8, within cylindrical 20 µl wells. When the gel is formed, it is suspended in a solution of either 50 mM Bicine, pH 7.8, for encapsulated TS or 100 mM Hepes, pH 7, for encapsulated OASS. The silica wet gels were stored at 4°C for at least 12 hours before use.

Microspecrophotometric measurements were carried out with a Zeiss MPM03 instrument, equipped with a thermostated stage. Spectra on enzyme single crystals were recorded using linearly polarized light, as previously described (Mozzarelli et al., 1989; 1998). Spectra on enzyme silica gels were recorded using unpolarized light since silica gels are optically isotropic. Measurements were carried out at 20°C.

To form the quinonoid species of TS, the reaction mixture contained 25 mM Bis-Tris propane, 1 mM EDTA, 50 mM L-serine and indoline, in the absence and presence of sodium ions, pH 8.0, and, in crystal experiments, 20 % w/v PEG 8000. The concentration of indoline was estimated on the basis of an extinction coefficient of 2,600 M^{-1} cm^{-1} at 289 (Roy et al., 1988).

To form the α-aminoacrylate species of OASS, the reaction mixture contained 100 mM Hepes, *O*-acetylserine, pH 7.2, and, in crystal experiments, 34% w/v PEG 4000, 0.076 M Tris-HCl, 0.152 Li_2SO_4.

Results and Discussion

Formation of the quinonoid species of tryptophan synthase in the crystal and in silica gels. Quinonoid species are key intermediates of the catalytic reactions of many PLP-enzymes. Whereas the structure of other PLP-intermediates has been determined, the structure of a quinonoid has not yet been reported. In solution, the reaction of tryptophan synthase with the indole analog indoline leads to the formation of a metastable quinonoid species absorbing at 476 nm (Roy et al., 1988). The presence of Na^+ significantly decreases the dissociation constant of indoline and increases the accumulation of the quinonoid (Mozzarelli et al., 2000). Similarly to solution, the reaction with indoline leads to the formation of a stable quinonoid species in the crystal (Figure 1a). However, the influence of sodium ions is strikingly different. In fact, the amount of quinonoid is 4-5 fold higher in the absence than in the presence of sodium ions (Figure 1b). These results define the experimental conditions that lead to the accumulation of the quinonoid in the crystal opening the way to its X-ray crystallographic analysis. They also indicate the relevance of carrying out functional studies of enzymes in the same physical state where structural data are collected.

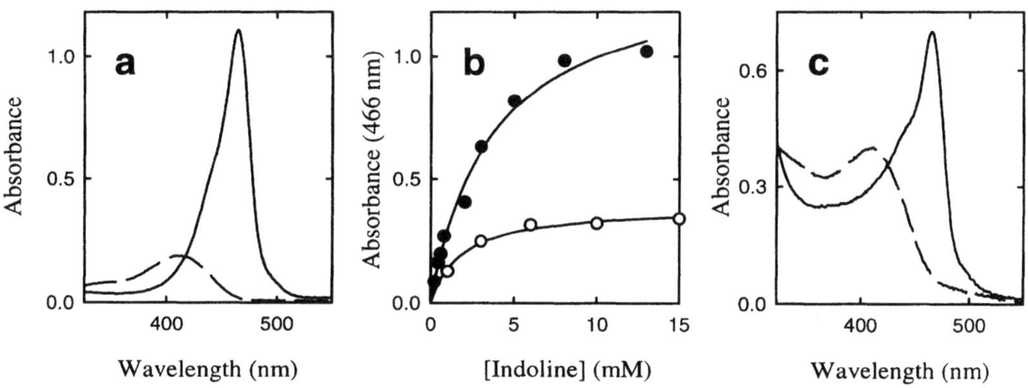

Figure 1. Quinonoid species is formed in the reaction of indoline with L-serine-TS $\alpha_2\beta_2$ complex in the crystal (panel a, b) and in silica gel (panel c). (a) Polarized absorption spectra of a TS crystal suspended in a solution containing 20% w/v PEG8000, 1mM EDTA, 50 mM Bicine-Na, pH 7.8, in the absence (dash line) and presence of 10 mM indoline, 50 mM L-serine (solid line). (b) Binding of indoline to TS crystals suspended in a solution containing 50 mM L-serine, 25 mM Bis-Tris, pH 8, in the absence (closed circles) and presence of 250 mM sodium chloride (open circles). (c) Absorption spectra of TS silica gels suspended in a solution containing 25 mM Bis-Tris propane, pH 7.8 (dash line) and 25 mM indoline, 50 mM L-serine, in the absence of monovalent cations (solid line).

Tryptophan synthase encapsulated in silica gels forms the indoline quinonoid (Figure 1c). Since tryptophan synthase has been used in bioreactors for the production of L-tryptophan (Soda et al., 1987; Makiguchi et al., 1987), this finding indicates that encapsulated tryptophan synthase can be used in enzyme bioreactors, as encapsulation provides a convenient system to separate the enzyme from the products and, usually, increases protein stability. Similarly to crystalline TS, the amount of quinonoid is higher in the absence than in the presence of sodium ions (data not shown).

Formation of the α-aminoacrylate of O-acetylserine sulfhydrylase in the crystal and in silica gels. The reaction of *O*-acetylserine (OAS) with OASS leads to the formation of the α-aminoacrylate species absorbing at 470 nm (Cook et al., 1992). The same absorption band was observed when OASS crystals were suspended in a solution containing OAS (Figure 2a). However, the affinity of the enzyme for OAS is about 500 fold lower than in solution (Mozzarelli et al., 1998). These findings are likely explained by lattice constraints imposed to the open to closed conformational transition associated with catalytic events (Schnackerz et al., 1995).

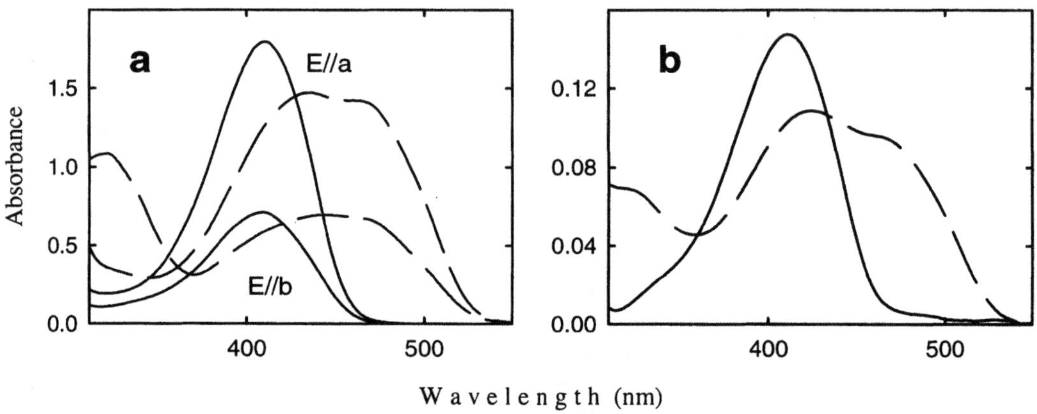

Figure 2. α-Aminoacrylate Schiff base is formed in the reaction of *O*-acetylserine (OAS) with OASS in the crystal (a) and in the gel (b). (a) Polarized absorption spectra of crystals of OASS suspended in a solution containing 34% w/v PEG4000, 0.076 M Tris, 0.152 M Li_2SO_4, pH 6.9, in the absence (solid line) and presence of 18 mM OAS (dash line). (b) Absorption spectra of OASS encapsulated in silica gel suspended in a solution containing 100 Hepes, pH 7, in the absence (solid line) and presence of 20 mM OAS (dash line)

Since OASS, as TS, can be used in bioreactors (Soda et al., 1987) to produce L-cysteine, as well as, eventually, the selenium derivative, the reactivity of the enzyme encapsulated in silica gels was investigated. It was found that the enzyme reacts with OAS to form the α-aminoacrylate (Figure 2b). The affinity of OAS is significantly lower than in solution and more similar to that observed on the crystalline enzyme. This might be due to a decrease of the conformational flexibility of the enzyme within the gel matrix.

Unfolding of OASS in solution and in silica gels induced by guanidinium hydrochloride (GdnHCl). PLP-enzymes belong to different functional and structural families (Grishin et al., 1995, and references therein). We are addressing the question whether enzymes belonging to the same structural family exhibit the same folding mechanism and which role is played by the coenzyme. TS and OASS belong to the β-family and share a sequence identity of about 20%. Folding-unfolding studies of the β-subunit of TS have been carried out (Chaffotte et al., 1991 and references therein) whereas no information is available for folding of OASS. The unfolding of OASS in solution has been studied using absorption (Figure 3a) and fluorescence spectroscopy (data not shown). The reaction is biphasic (Figure 3a, inset). The first step is the formation of a PLP-denatured enzyme complex. The second step is the slow dissociation of the coenzyme, as evidenced by the blue shift

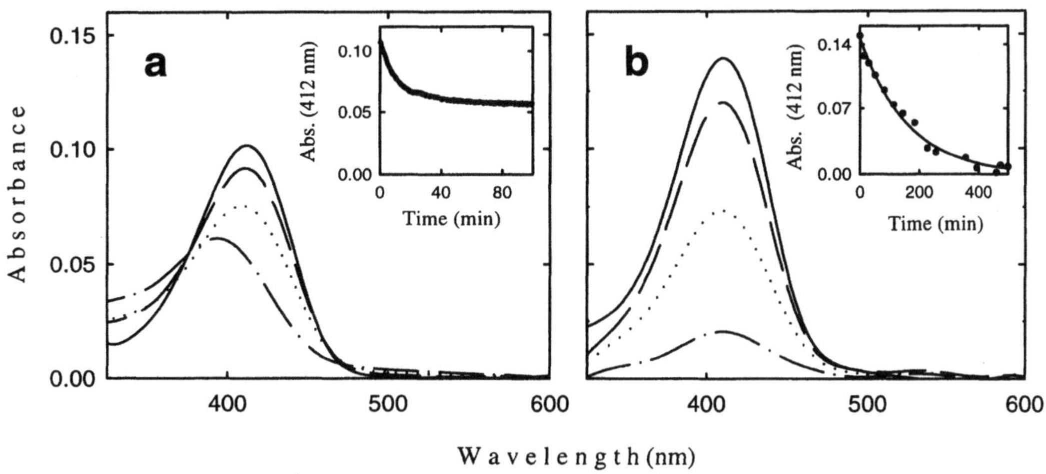

Figure 3. Unfolding of OASS in solution (a) and encapsulated in silica gels (b) by exposure to guanidinium hydrochloride (GdnHCl). (a) Absorption spectra of OASS, 100 mM Hepes, pH 7.0, in the absence (solid line) and presence of 3.5 M GdnHCl, after 3 (long dash), 10 (dotted line) and 90 min (dash dot dash line). Inset: the time course of the reaction is fitted to a biphasic decay with lifetimes of 5 (amplitude 67%) and 23 min. (b) Absorption spectra of OASS encapsulated in silica gels, 100 mM Hepes, pH 7.0, in the absence (solid line) and presence of 3.5 M GdnHCl, after 15 (long dash), 114 (dotted line) and 355 min (dash dot dash line). The absorption of the protein-free gel has been subtracted. Inset: the time course of the reaction is fitted to a monophasic decay with lifetime of 108 min.

of the band (Figure 3a). When the same unfolding process is carried out on OASS encapsulated in silica gels, the spectral changes are somewhat different (Figure 3b). In fact, there is only a decrease of the band absorbing at 412 nm without any spectral shift, suggesting that the slow unfolding of the enzyme is followed by a relatively fast release of PLP, diffusing out of the gel in the suspending medium. Remarkably, the rate of unfolding is about 20 fold lower than in solution (Figure 3b, inset). This finding indicates that encapsulation restrains OASS conformational changes, as observed for other proteins (Mozzarelli and Bettati, 2000, and references therein; Das et al., 1998), allowing to slow down processes and, eventually, to isolate folding intermediates.

Acknowledgments

This work was supported by funds from C.N.R. (A.M. 98.01117.CT14) and Target Project on Biotechnology.

References

Bettati, S. and Mozzarelli, A. (1997) T state hemoglobin binds oxygen noncooperatively with allosteric effects of protons, inositol hexaphosphate and chloride. J. Biol. Chem. 272, 32050-32055

Chaffotte, A., Guillou, Y., Delepierre, M., Hinz, H.-J. and Goldberg, M. E. (1991) The isolated C-terminal (F_2) fragment of the *Escherichia coli* tryptophan synthase β_2-subunit folds into a stable, organized nonnative conformation. Biochemistry 30, 8067-8074

Cook, P. F. and Wedding, R. T. (1976) A reaction mechanism from steady-state kinetic studies for *O*-acetylserine sulfhydrylase from *Salmonella typhimurium* LT-2. J. Biol. Chem. 251, 2023-2029

Das, T. K., Khan, I., Rousseau, D. L. and Friedman, J. M. (1998) Preservation of the native structure in myoglobin at low pH by sol-gel encapsulation. J. Am. Chem. Soc. 120, 19268-10269

Grishin, N. V., Phillips, M. A. and Goldsmith, E. J. (1995) Modeling of the spatial structure of eukaryotic ornithine decarboxylases. Protein Sci. 4, 1291-1304

Lan, E. H., Dave, B. C., Fukuto, J. M., Dunn, B., Zink, J. I. and Valentine, J. S. (1999) Synthesis of sol-gel encapsulated heme proteins with chemical sensing properties. J. Mater. Chem. 9, 45-53

Makiguchi, N., Fukuhara, N., Shimada, M. Asai, Y., Nakamura, T and Soda, K. (1987) Industrial production of L-tryptophan from indole and DL-serine with two PLP-dependent enzymes. In "Biochemistry of vitamin B_6" (eds. Korpela, T. and Christen, P.) Birkhauser, Basel, 457-460

Miles, E. W. (1991) Structural basis for catalysis by tryptophan synthase. Adv. Enzymol. 64, 93-172

Mosbach, K. (ed.) (1988) Immobilized enzymes and cells. Meth. Enzymol. 137

Mozzarelli, A. Peracchi, A., Rossi, G. L., Ahmed, S. A. and Miles, E. W. (1989) Microspectrophotometric studies on single crystals of the tryptophan synthase $\alpha_2\beta_2$ complex demonstrate formation of enzyme-substrate intermediates. J. Biol. Chem. 264, 15774-15780

Mozzarelli, A. and Rossi, G. L. (1996) Protein function in the crystal. Annu. Rev. Biophys. Biomol. Struct. 25, 343-365

Mozzarelli, A., Bettati, S., Pucci, A. M., Burkhard, P. and Cook, P. F. (1998) Catalytic competence of *O*-acetylserine sulfhydrylase in the crystal probed by polarized absorption microspectrophotometry. J. Mol. Biol. 283, 135-146

Mozzarelli, A., Peracchi, A., Rovegno, B., Dalè. G., Rossi, G. L. and Dunn, M. F. (2000) Effect of pH and monovalent cations on the formation of quinonoid intermediates of the tryptophan synthase $\alpha_2\beta_2$ complex in solution and in the crystal. J. Biol. Chem. In press

Mozzarelli, A. and Bettati, S. (2000) Functional properties of immobilized proteins. In "Advanced functional molecules and polymers" Gordon and Breach Science Publishers, Tokio, in press

Roy, M., S. Keblawi, S. and Dunn, M. F. (1988) Stereoelectronic control of bond formation in *Escherichia coli* tryptophan synthase: substrate specificity and enzymatic synthesis of the novel aminoacid dihydroisotryptophan. Biochemistry 27, 6698-6704

Schnackerz, K. D., Tai, C.-H., Simmons, J. W., Jacobson, T. M., Rao, J. G. S. and Cook, P. F. (1995) Identification and spectral characterization of the external aldimine of the *O*-acetylserine sulfhydrylase reaction. Biochemistry 34, 12152-12160

Soda, K., Tanizawa, K., Esaki, N. and Tanaka, H. (1987) Enantiomer synthesis of aminoacids with pyridoxal enzymes. In "Biochemistry of vitamin B_6" (eds. Korpela, T. and Christen, P.) Birkhauser, Basel, 435-444

St. Clair, N., Shenoy, B., Jacob, L. D. and Margolin, A. L. (1999) Cross-linked protein crystals for vaccine delivery. Proc. Natl. Acad. Sci. USA 96, 9469-9474

Tai, C.-H., Nalabolu, S. R., Jacobson, T. M., Hinter, D. E. and Cook, P. F. (1993) Kinetic mechanism of the A and B isoenzymes of *O*-acetylserine sulfhydrylase from *Salmonella typhimurium* LT-2 using the natural and alternative reactants. Biochemistry 32, 6433-6442

Yang, L., Ahmed, S. A., and Miles, E. W. (1996) PCR mutagenesis and overexpression of tryptophan synthase from *Salmonella typhimurium*: on the roles of β_2 subunit Lys-382. Protein Expr. Purif . 8, 126-36

Biochemistry and Molecular Biology of Vitamin B_6
and PQQ-dependent Proteins
ed. by A. Iriarte, H.M. Kagan and M. Martinez-Carrion

Characterization of Recombinant Porcine Pyridoxal Kinase using Surface Plasmon Resonance Biosensor Technique

Fong C.C.[+], Lai W.P[++]., Yang M.[+], Leung Y.C. [++]and Wong M.S.[++]
[+] Department of Biology and Chemistry, City University of Hong Kong and [++]the Open Laboratory of Asymmetric Synthesis, Department of Applied Biology and Chemical Technology, The Hong Kong Polytechnic University, Kowloon, Hong Kong, P.R.C.

Summary: In the present study, surface plasmon resonance biosensor was employed to characterize the binding interaction between wild type porcine Pyridoxal Kinase (PK) and an immobilized substrate, pyridoxamine. The binding of PK to the immobilized pyridoxamine was followed in real time and the kinetic parameters were derived from the sensorgram using the BIAevaluation™ software. At pH 7.4, the association rate constant and dissociation rate constant of the wild-type PK for immobilized pyridoxamine are $7.25 \times 10^3 \pm 0.35 \times 10^3\ M^{-1}s^{-1}$ and $5.01 \times 10^{-3} \pm 0.93 \times 10^{-3} s^{-1}$, respectively. The equilibrium affinity constant of the wild-type PK is $1.45 \times 10^6\ M^{-1}$. The effect of buffer pH on the binding kinetic parameters was also determined.

Introduction: PK is an important enzyme involved in the metabolic activation of vitamin B6. It has been detected in virtually all important mammalian tissues and is made up of two identical subunits of 40 kDa each. It catalyzes the formation of PLP from PL and ATP in the presence of Zn^{2+}. The recent cloning of the porcine PK cDNA allows us to study recombinant PK protein at the molecular level (1). Binding of PK to its substrate is the prerequisite step for the catalytic phosphorylation of the substrate. In the present study, we aim to employ surface plasmon resonance (SPR) biosensor to characterize the binding kinetics between pyridoxal kinase and its substrate, pyridoxamine. The long term significance of this study is to characterize enzyme-substrate binding interactions using pyridoxal kinase as a model, as well as to develop the SPR biosensor as a tool for understanding the mechanism in enzymatic activation of vitamin and the structure-function relationship of pyridoxal kinase.

Materials and Methods:

Immobilization of pyridoxamine on sensor surface: Commercial CM5 sensor chip was reconditioned to generate bare gold surface for subsequent modification according to the method by Kalb et al (2). Briefly, the carboxyl-dextran methylated (CMD) chip were removed from its plastic holder and were cleaned with 70% H_2SO_4 and 30% H_2O_2 mixture by heating to about 90°C for 5 min. They were then rinsed with deionized water, washed with methanol and dried for 1h at 150°C. Immediately before use, the gold chips were further cleaned in argon plasma cleaner for 10 min (Model PDC 3 X G). 11-Mercaptoundecanoic acid (MUA) was introduced to elongate the linker as described by Duschl et al (3). Pyridoxamine and MUA was first linked by EDC/NHS activation pathway in solution. Thin layer chromatography (TLC) was used for the separation of MUA and its derived product, and the product containing pyridoxamine was further separated from the mixture using silica gel chromatography (eluent: 2:1 ethyl acetate and petroleum ether mixture). The formation of monolayer on gold surface was tested using cyclic voltametry in the presence of 10mM $K_3Fe(CN)_6$ buffer. The gold working electrode and a saturated calomel reference electrode (SCE) were used. The result showed that the gold surface was covered with pyridoxamine-linked monolayer (data not shown).

Real-time binding experiments using surface plasmon resonance biosensor: The binding experiments were carried out using a SPR biosensor (BIAcore) at 25°C with a constant flow rate of 5 μl/min of 5 mM potassium phosphate buffer (pH 7.4). Twenty microliter of different concentration of pyridoxal kinase (500nM to 6μM in PB buffer) was injected over immobilized pyridoxamine surface, followed by 10-min wash by PB buffer. For each experiment, another SAM presenting MUA was used as control surface in order to correct for the non-specific binding and bulk refractive index change. The sensorgram were recorded, and the specific binding profiles of PK to immobilized pyridoxamine were obtained after subtracting the control surface.

Data Analysis: The sensorgram data were analyzed using BIAevaluation™ Software version 2.2.4. Several binding models were used to fit the pyridoxyl kinase and pyridoxamine interaction data. The simple and the most appropriate model for the binding interaction was used to derive the kinetic parameters. In this case, the one-site model seems to be sufficient in fitting the data, so the binding kinetic constants i.e. k_a (association rate constant) and k_d (dissociation rate constant) were also derived using the equation 1&2. The equilibrium affinity constant (K_A) is calculated based on the ratio of k_a/k_d.

for association (1)

$$R = R_{eq}\{1 - \exp[-(k_a[A] + k_d)(t - t_0)]\} + R_I$$

for dissociation (2)

$$R = R_0 \exp[-k_d(t - t_0)]$$

Results and Discussion:

Specific binding curve is obtained by subtracting the sensorgram reading of the control surface presenting MUA from the reading of the surface presenting pyridoxamine as shown in Figure 1.

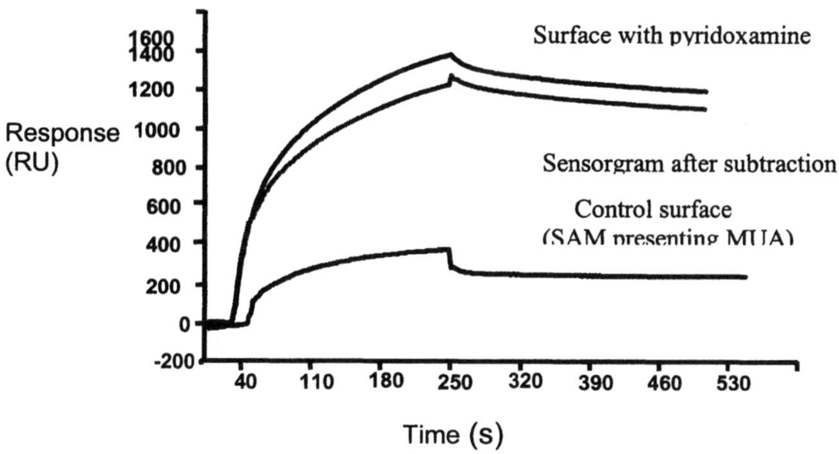

Figure 1: The sensorgram of the binding interaction of PK to immobilized pyridoxamine. The biosensor chip was prepared by creating self-assembled monolayer presenting pyridoxamine that formatted by MUA linking through solution organic synthesis.

The association and dissociation of various concentrations of pyridoxal kinase with immobilized pyridoxamine were measured and recorded as sensorgram by the biosensor. After obtaining all the sensorgrams, the data were analyzed using BIAevaluation™ Software version 2.2.4. All fitting were satisfactory as judged by the chi square value (0.3) and randomness of the residual. In our case, one-site kinetic model is sufficient in describing the binding (Figure 2). The dissociation curves and association curves for each concentration of PK were fitted with Eq.1 and Eq. 2. The obtained kinetic parameters are summarized in Table 1. The effect of pH on inding between PK and pyridoxamine was studied. For the range of 6.5 to 8.0, the binding kinetics varied. (Table 1)

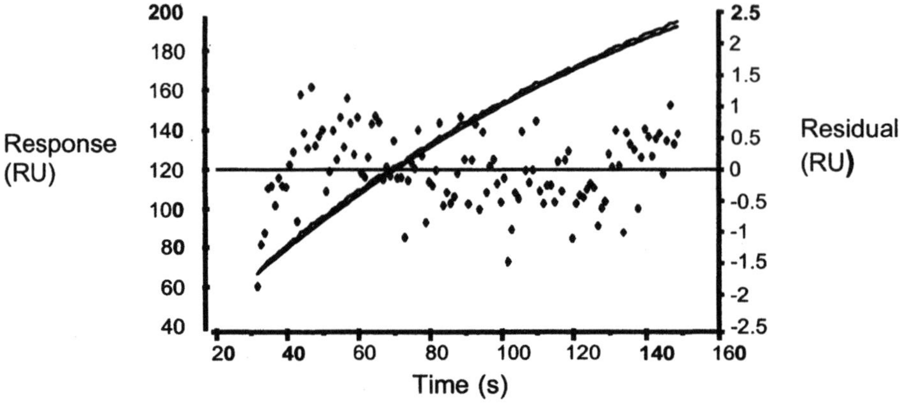

Figure 2: The fitting curve and residual plot of fitting the binding interaction between pyridoxyl kinase and immobilized pyridoxamine using 1:1 simple model

Table 1: Kinetic parameters of pyridoxal kinase and pyridoxamine using SPR biosensor

PK-pyridoxamine interaction	Association rate constant (k_a) (x 10^3 $M^{-1}s^{-1}$)	Dissociation rate constant (k_d) (x 10^{-3} s^{-1})	Equilibrium affinity constant (K_A) (x 10^5 M^{-1})
pH 7.4	7.25 ± 0.35	5.01 ± 0.93	1.45 ± 0.28
pH 6.5	5.23 ± 0.23	20.3 ± 3.25	0.26
pH 7.0	5.55 ± 0.61	6.17 ± 0.02	0. 90
pH 8.0	0.85 ± 0.13	4.79 ± 0.09	0.18

Acknowledgments The authors would like to thank Professor J. E. Churchich for his advice and guidance throughout the planning and execution of the experiments. The study was supported by the Area of Strategic Development Grant A008.

References

1. Gao ZG, Lau CK, Lo SCL, Choi SY, Churchich JE & F Kwok. (1998) Porcine pyridoxal kinase c-DNA cloning, expression, and confirmation of its primary sequence. Int. J. Biochem. & Cell Biol., 30: 1379-1388.
2. Kalb E, Frey S and LK Tamm. (1992) Formation of supported planar bilayers by fusion of vesicles to supported phospholipid monolayers. Biochim Biophys Acta, 1103 (2): 307-316.
3. Duschl C, Sevin-Landais A-F, and H Vogel. (1996) Surface engineering: optimization of antigen presentation in self-assembled monolayers. Biophy. J. 70: 1985-1995.

Biochemistry and Molecular Biology of Vitamin B_6
and PQQ-dependent Proteins
ed. by A. Iriarte, H.M. Kagan and M. Martinez-Carrion

Three-Dimensional Model Of The ATP-Binding Domain Of Pyridoxal Kinase

Lau CK, Leung AKM, Chau FT, Kwok F and Lo SCL.
Department of Applied Biology and Chemical Technology, The Hong Kong Polytechnic University, Hung Hom, Kowloon, Hong Kong SAR, China.

Summary

With a combination of multiple sequence alignment, secondary structure prediction and fold recognition techniques, a molecular model of the ATP-binding domain of pyridoxal kinase was constructed. The overall fold of the model consists of a central β-sheet of five parallel strands sandwiched between six helices. It shows high resemblance with the typical Rossmann fold in nucleotide binding. The quality of this model was assessed with the program PROCHECK using conformation criteria. PROCHECK locates over 90% of the residues of the current model in the "most favored" and "additional allowed" regions in Ramachandran plot.

Introduction

Pyridoxal kinase (PK) isolated from sheep brain was found to consist of two subunits of identical size each with 36 kDa [1]. Although the primary structure of PK from different species have been identified independently using amino acid sequencing of the protein [2] and nucleotide deduction from the DNA sequence [3,4], only limited information with respect to its tertiary structure could be retrieved from the primary structure. Meanwhile, X-ray diffraction study of PK crystal is yet to be successful. Therefore, in order to obtain further insight into the architecture of protein with respect to the substrate binding sites, a molecular model of the ATP-binding domain of PK (ATPBD-PK) is constructed.

Materials and Methods

Folding of the ATPBD-PK

With the information identifying globular domains by peptide sequencing, the program THREADER 2 [5] was adopted to search for a possible fold for the ATPBD-PK. Amino acid sequences (V138-L322) was threaded with the standard library of 256 folds provided by the software. Results were sorted according to the Z score of the weighted sum of pairwise and solvation energy.

Multiple sequence alignment and secondary structure calculation

NAD-dependent formate dehydrogenase (NDFD) was chosen for further processing (see results below). Families of NDFD and PK sequences retrieved from SWISS-PROT were aligned separately with the program CLUSTAL W aided by manual adjustment. Insertion or deletion was only carried out in the loop region of the template structure. Both alignments were submitted to the PHD server for secondary structure prediction.

Model building for the ATP-binding domain

Based on the combined sequence alignment of the two families, a model was built using the HOMOLOGY module of INSIGHT II (MSI, U.S.A.). Coordinates of NDFD were assigned to the ATP-binding domain sequence. Loop search program in HOMOLOGY module was employed in search for peptide structure of a given length which connects structure elements with the geometry that matches ends of the assigned secondary cores in the loop databases. The completed model was refined by automated rotamer optimization followed by energy minimization (steepest descent and conjugate gradient algorithms to a final derivative of 0.1 kcal $mol^{-1}Å^{-1}$). The force field in DISCOVER module for energy minimization was CVFF [4]. The relatively reliability of the model was evaluated by the structure verification program, PROCHECK [6].

Results and Discussion

After the processing by the THREADER 2 program, two proteins (D-galactose/D-glucose, Z=4.82; NAD-dependent formate dehydrogenase (NDFD), Z=4.02) were found. Based on the substrate similarity between NAD+ and ATP and better sequence alignment between NDFD and ATPBD-PK (15% amino acid identity), structure of chain A of NDFD was selected as a template for the construction of the ATPBD-PK model. The ATPBD-PK sequence was threaded onto the structure of the catalytic domain of NDFD. The proposed folds were verified by separate multiple sequence alignments and secondary structure prediction of NDFD and ATPBD-PK. Ten out of the 12 secondary structure elements (helices and strands) contained in NDFD were correctly predicted and aligned in ATPBD-PK. In the present alignment, a total of seven gaps all located in the loop region were required to be introduced into the sequence. The model was constructed by assigning coordinates of the template structure to the query sequence according to the sequence alignment using the HOMOLOGY module. After construction of the model, minimization of steric hindrance in macromolecule was carried out. The overall fold of the model consists of a central β-sheet of five parallel strands flanked by several helices. The architecture of the model showed high resemblance to the classical Rossmann fold of nucleotide-binding [7]. Topology plot of the model is shown in Figure 1.

Figure 1. ATP-binding domain of PK showing the three-layered (αβα) sandwich: box=helices, arrows=strands. Helices in front of the sheet are drawn higher and helices behind the sheet are shown lower. The dotted line represents the region where modeling was not performed.

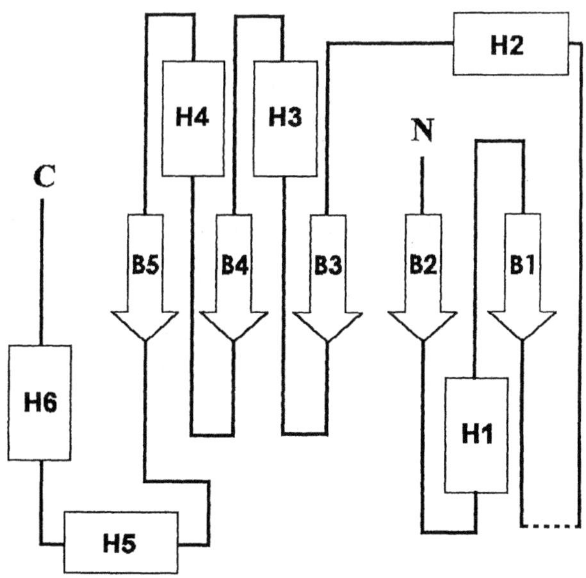

In the model constructed, the proposed degenerate P-loop motif, GTGDL [5] was found to be in a close proximity to the lysine residue (K232) which was suggested to be involved in the binding of phosphate group of ATP [8]. This segment of peptide (G217-K232) formed a helix-loop-strand supersecondary structure in the model. Residues closed to this motif were T113-N135, S172-N174, P195-D196 and H233-N235. Quality of the model was assessed with the program PROCHECK using conformation criteria. PROCHECK located >90% of the residues of the current model in the "most favored" and "additional allowed" regions in the Ramachandran plot. In the absence of an ideal homologous template structure for model building, the fold recognition program THREADER 2 was used successfully to generate a molecular model of the ATPBD-PK.

Acknowledgment

This work is supported by a grant from The Hong Kong Polytechnic University (A/C:AP-135).

References

1. Kwok F, Kerry JA & Churchich JE. (1986) Sheep brain pyridoxal kinase: fluorescence spectroscopy of the dimeric enzyme. Biochim. Biophys. Acta 874: 167-173.
2. Gao ZG, Lau CK, Lo SCL, Choi SY, Churchich JE & F Kwok. (1998) Porcine pyridoxal kinase: c-DNA cloning, expression, monoclonal antibody production and confirmation of its primary sequence. Int. J. Biochem. & Cell Biol.
3. Hanna MC, Turner AJ & Kirkness EF. (1997) Human pyridoxal kinase: cDNA cloning, expression, and modulation by ligands of the benzodiazepine receptor. J. Biol. Chem. 272: 10758-10780.
4. Yang Y, Tsui HC, Man TK & Winkler ME. (1998) Identification and function of the pdxY gene, which encodes a novel pyridoxal kinase involved in the salvage pathway of pyridoxal 5'-phosphate biosynthesis in *E.Coli* K-12. J. Bacteriol. 180: 1814-1821.
5. Jones DT, Taylor WR & Thornton JM. (1992) A new approach to protein fold recognition. Nature 358: 86-89.
6. Laskowski RA, MacArthur MW, Moss DS & Thornton JM. (1993) PROCHECK: A program to check the stereochemical quality of protein structure. J. Appl. Cryst. 26: 283-291.
7. Rossmann MG, Moras D & Olsen KW. (1974) Chemical and biological evolution of a nucleotide-binding protein. Nature 250: 194-199.
8. Dominici P, Scholz G, Kwok F & Churchich JE. (1988) Affinity labeling of pyridoxal kinase with adenosine polyphosphoyridoxal. J. Biol. Chem. 263: 14712-14716.

Biochemistry and Molecular Biology of Vitamin B_6
and PQQ-dependent Proteins
ed. by A. Iriarte, H.M. Kagan and M. Martinez-Carrion

Structural Fluctuations of Pyridoxal Kinase: Effect of Viscogen Cosolvents †

Y. C. Leung and J. E. Churchich

Department of Applied Biology and Chemical Technology, The Open Laboratory of Asymmetric Synthesis, The Hong Kong Polytechnic University, Hung Hom, Kowloon, Hong Kong, China

Summary

The k_{cat} values of pyridoxal kinase are attenuated when the viscosity of the solution is varied between 0.89 and 13.75 cp at room temperature. This effect may be due to fluctuations of the tertiary structure of the enzyme in response to changes in the viscosity of the medium. A plot of log k_{cat} against log viscosity according to Kramer equation yields $\varepsilon = 0.60$ and $E_A = 7.8$ kcal/mol. Structural fluctuations of the protein in response to the viscosity of the medium were detected by emission anisotropy measurements. A sharp change in emission anisotropy occurs upon viscosity changes from 2.7 to 0.89 cp. Two rotational correlation times $\phi_1 = 17$ ns and $\phi_2 = 3.2$ ns fit the Perrin plot. The shortest rotational correlation time reflects the rotational mobility of tryptophanyl groups of the protein. It is postulated that enzymatic activity is correlated to a dynamic coupling between the protein matrix and the surrounding medium.

Introduction

Pyridoxal kinase is a flexible protein which changes its conformation upon addition of substrates and substrate analogues (1). The protein undergoes a process of reversible unfolding in the presence of moderate concentrations of guanidinium chloride (mid-point 0.6 M Gnd HCl). The magnitude of the free energy associated with the folding transition ($\Delta G_0 = 1.6$ kcal/mol) suggests that the protein is easily destabilized by denaturing agents (2). Although the effect of viscogens on the conformational dynamics has not been investigated, it was observed that the rate of cleavage of the protein by the proteolytic enzyme chymotrypsin is influenced by the viscosity of the medium (3). In view of these observations, it is pertinent to ask whether viscogen agents would influence the rotational dynamics of side chain groups of the kinase. Is there any correlation between structural fluctuations and catalytic activity?

†This paper is dedicated to the memory of Professor Jorge Churchich.

Materials and Methods

Pyridoxal kinase was purified as reported (1). Enzyme kinetic parameters were measured in 0.1 M potassium phosphate buffer at pH 6.3 at 25 °C by following the change in absorbance at 388 nm, at which pyridoxal-5-P has an absorption maximum with an absorption coefficient of 4900 cm^{-1} M^{-1}.

Fluorescence emission spectra were recorded by a Perkin-Elmer LS-50B spectrofluorometer using a 1-cm light-path quartz cuvette. Emission anisotropy was measured in the same instrument. Illumination was provided by a Xenon-Hg lamp (150 W) with wavelengths selected by a quartz prism monochromator. The band-width for excitation was 5 nm. Emission anisotropy values were measured with a precision of +/- 0.002.

Results and discussion

The catalytic activity of pyridoxal kinase in the presence of varied concentration of sucrose was investigated at pH 6.3 in 0.1 M potassium phosphate. It can be seen from the results included in Fig. 1 that the k_{cat} values are attenuated when the viscosity of the solution is varied between 0.89 and 13.75 cp at a fixed temperature of 25 °C. Regardless of the concentration of sucrose used to increase the viscosity of the solution, the enzyme obeys Michaelis-Menten kinetics with K_m values for the substrates ATP and pyridoxal practically invariant with the viscosity of the medium (Table 1).

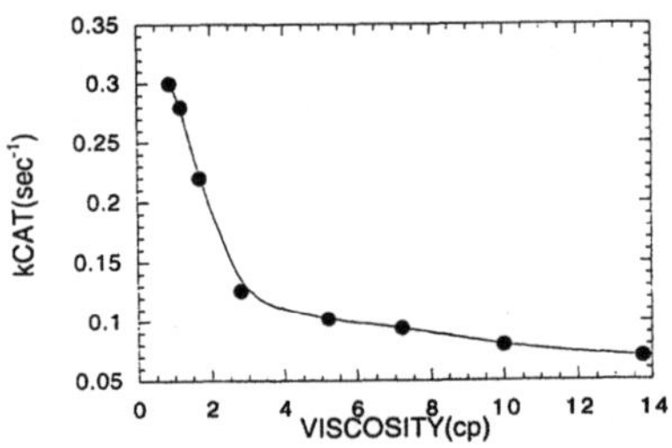

Fig. 1. Effect of viscosity of sucrose solutions in 0.1 M potassium phosphate buffer on the k_{cat} of pyridoxal kinase. Experiments conducted at pH 6.3 and 25 °C. The assay mixture contained 0.1 mM ATP, 0.05 mM Zn chloride and 0.1 mM pyridoxal. The rate of formation of PLP at 388 nm was monitored for 2 min.

Table 1. Effect of viscosity on the K_m values for ATP and pyridoxal. Assays were conducted at pH 6.3 in 0.1 M potassium phosphate at 25 °C.

Viscosity (cp)	$K_m$pyridoxal (μM)	K_mATP (μM)
0.893	20	70
1.704	16	72
2.78	17	69

The presence of large quantities of viscogenic solute required to modify the viscosity of the solution might affect the rates catalysed by the enzyme by shifting the pKa of some ionizable groups of the protein involved in catalysis. To test this possibility, the pH profiles of the kinase in the absence and presence of sucrose were determined. As shown by the results included in Fig. 2, the pH profile of pyridoxal kinase is not altered by addition of 20% sucrose at 25 °C. This observation implies that catalytic activity measurements cannot detect subtle changes in the pKa values of ionizable groups induced by addition of sucrose.

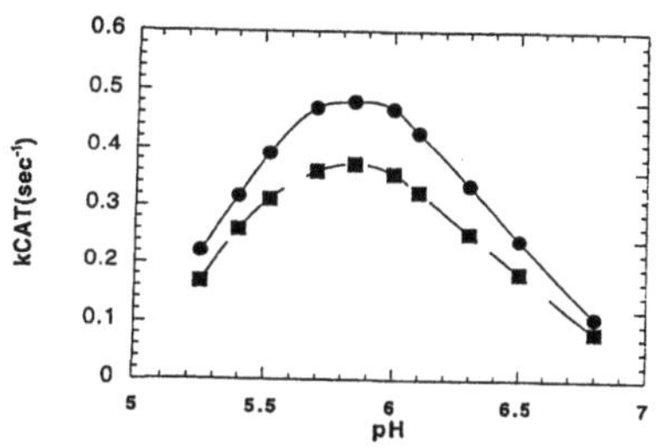

Fig. 2. Effect of pH on the catalytic parameters of pyridoxal kinase in the absence (•) and presence (♦) of 20% sucrose. Experiments conducted in 0.1 M potassium phosphate buffer at the indicated pH values. The temperature of the system was 25 °C.

The effect of the physicochemical properties of the additives on the catalytic activity of the kinase was investigated at pH 6.3. According to results published in the literature (4), sucrose and glucose differ in their molar volumes and hydration numbers. The molar volume and hydration number of sucrose are practically two-fold greater than the corresponding values of glucose. However, at a viscosity of 2.80 cp, the kinase shows the same k_{cat} value, suggesting that the chemical nature of the viscogen does not influence the catalytic behavior of the enzyme (Table 2). The effect of glycerol, a viscogen solute commonly used in viscosity studies, could not be tested because pyridoxal kinase undergoes a slow process of irreversible denaturation in the presence of glycerol concentrations higher than 10% at room temperature.

Table 2. Effect of sucrose, glucose and fructose on the k_{cat} values of pyridoxal kinase assayed at pH 6.3.

Sugar	Viscosity (cp)	Apparent molar volume (ml/mol)	Degree of hydration	k_{cat} (s^{-1})
Sucrose	2.80	210.2	7.03	0.13
Glucose	2.82	112.8	3.45	0.12
Fructose	2.81	108.8	3.24	0.12

The absence of any observed effect on the K_m values for ATP and pyridoxal, suggests that the reaction catalysed by the kinase is not diffusion controlled by reactants transport (Table 1). Therefore, the effect of sucrose on the k_{cat} values must be sought in fluctuations of the tertiary structure of pyridoxal kinase in response to changes in the viscosity of the medium. If fluctuations of the protein matrix is essential for the activation process, then the dynamic state of the macromolecular structure of the enzyme-substrate complex is governed by its interaction with the solvent molecules through the action of two competing processes; i.e., random collisions and frictional forces (5, 6, 7). For reactions involving the formation of activated complex, the rate constant of the reaction (k) is related to the viscosity (η) of the medium by the equation derived by Kramers (8) (Equation 1).

$$k = A \,.\, \eta^{-\varepsilon} \,.\, e^{-E_A/RT} \qquad (1)$$

where E_A is the activation energy; A, a constant; ε, the viscosity coupling constant and η, the viscosity of the medium. $A \,.\, \eta^{-\varepsilon}$ characterizes the dynamic interaction of the protein with the solvent, and E_A is the potential barrier.

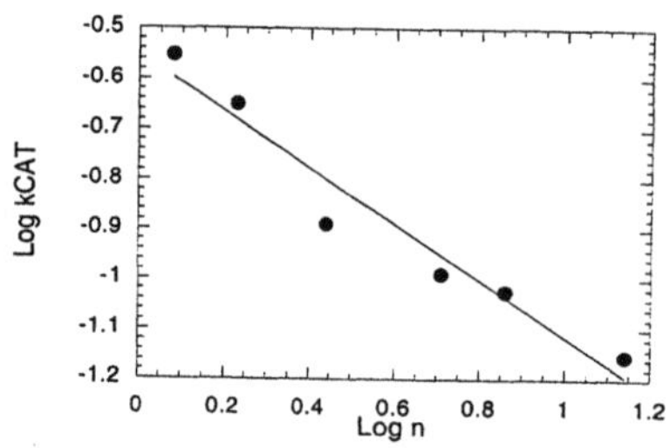

Fig. 3. Plot of log k_{cat} against log viscosity using Kramer equation. The plot yields $\varepsilon = 0.6$ and an energy of activation of 7.8 kcal/mol.

A plot of log k_{cat} against Log η is linear for pyridoxal kinase with $\varepsilon = 0.60$ and $E_A = 7.8$ kcal/mol (Fig. 3). In these calculations a value $A = 0.83 \times 10^6$ $cm^{-1}s^{-2}g$ (6) was assumed. According to this analysis, dynamic coupling between the protein matrix and the surrounding medium induces structural fluctuations inside pyridoxal kinase. In an effort to detect structural fluctuations of the protein in response to the viscosity of the medium, the polarization excitation spectra of pyridoxal kinase in the presence of increasing concentrations of sucrose were examined over the spectral range 270-310 nm. Fig. 4 shows a typical excitation polarization spectrum of pyridoxal kinase in 50% sucrose at pH 6.4 in 0.1 M potassium phosphate. Under illumination at 297 nm, which excites tryptophanyl groups, the emission anisotropy reaches a value of 0.26. The fluorescence maximum in 50% sucrose is identical to that of the enzyme in the absence of viscogen solutes.

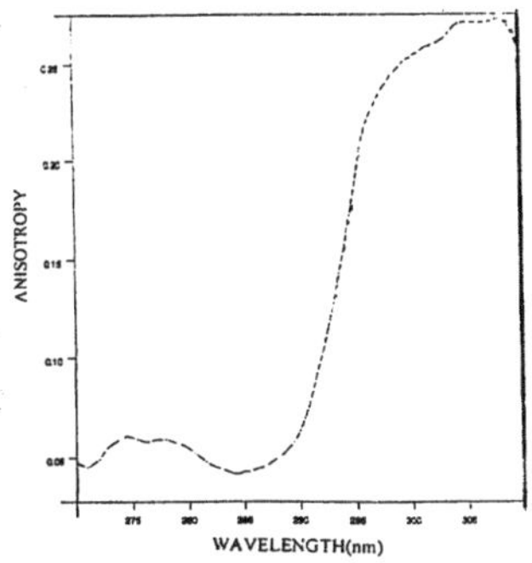

Fig. 4. Polarization excitation spectrum of pyridoxal kinase in 50% sucrose at pH 6.3 in 0.1 M potassium phosphate buffer. A maximum of the emission anisotropy of 0.275 is attained in the region coinciding with the absorption of tryptophanyl residues.

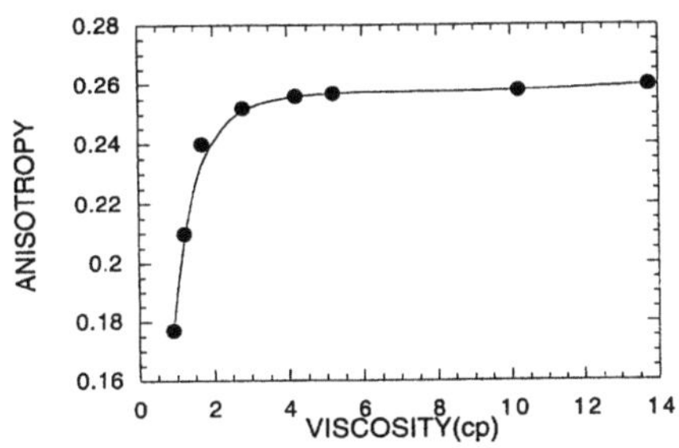

Fig. 5. Plot of the anisotropy as a function of the viscosity of sucrose solutions at 25 °C. The excitation wavelength was 297 nm, and the emission wavelength was set up at 350 nm. Excitation and emission slits 5 nm.

A sharp change in emission anisotropy values is detected when the viscosity changes from 2.7 to 0.89 cp. Fig. 5 shows how the emission anisotropy values varies as a function of the viscosity of the medium. The rotational mobility of the tryptophanyl residues can be estimated from fluorescence anisotropy measurements obtained at various viscosity values. A Perrin plot of 1/A against T/η is shown in Fig. 6 where it can be seen that the curve intersects the ordinate at A = 0.27, which is lower than the emission anisotropy observed for the model compound N-acetyltryptophan (A_0 = 0.32) in glycerol. When the Perrin plot was analysed using a function containing two rotational correlation times, it was found that ϕ_1 = 17.3 ns and ϕ_2 = 3.2 ns, which characterize the rotational dynamics of the protein. The longest correlation time could be assigned to the overall rotational motion of the macromolecule since its magnitude is similar to that calculated for the equivalent spherical model of 35 kDa using equation (2):

$$\phi_0 = [M \,.\, (\bar{\nu} + \delta)\, \eta] / RT \qquad (2)$$

A monomer of pyridoxal kinase yields a $\phi_0 = 14$ ns, whereas a dimer yields $\phi_0 = 25$ ns.

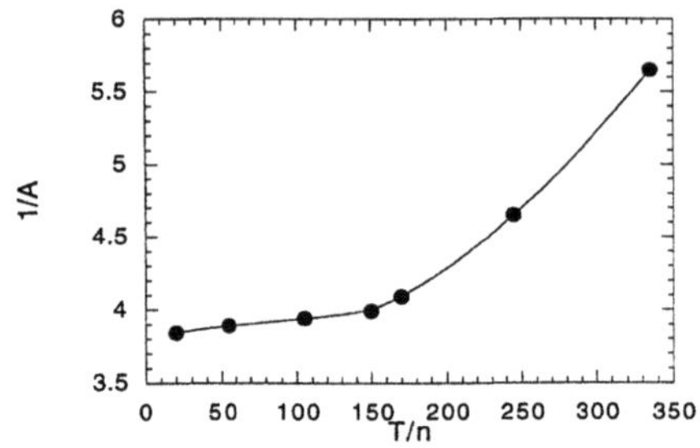

Fig. 6. Perrin plot of the reciprocal of the emission anisotropy as a function of the temperature/ viscosity. The viscosity was increased by addition of sucrose and the temperature was kept at 25 °C. Two rotational relaxation times fit the plot. A longer rotational correlation time of 17.3 ns corresponds to the rotation of the whole macromolecule in solution, whereas a rotational correlation time of 3.2 ns corresponds to the rotation of short polypeptides bearing tryptophanyl groups.

The shortest rotational correlation time reflects the rotational mobility of tryptophanyl groups of the protein. Thus, tryptophanyl residues participate in high frequency intramolecular mobility during the brief lifetime of the excited state [$\tau = 3$ ns]. The sharp increase in emission anisotropy from 0.18 to 0.25 ns occurs over a narrow range of viscosity values (Fig. 5); whereas the catalytic activity of pyridoxal kinase is reduced by 60% over the same range of viscosity values (Fig. 1). Hence, enzymatic activity seems to be correlated to a dynamic coupling between the protein matrix and the surrounding medium. High concentrations of viscogen, inhibit side chain motion leading to more compact globular structure.

This work was supported by the Area of Strategic Development Grant (A016).

References

1. Churchich, J. E. and Wu, C. (1981). *J. Biol. Chem.* 256, 780-784.
2. Pineda, T. and Churchich, J. E. (1993). *J. Biol. Chem.* 268, 20218-20222.
3. Churchich, J. E. (1990). *J. Prot. Chem.* 9, 613-621.
4. Lamy, L., Pertmann, M. O., Mathlouthi, M. and Larreta, G. V. (1990). *Biophys. Chem.* 36, 71-76.
5. Gavish, B. and Weber, M. M. (1979). *Biochemistry* 18, 1269-1275.
6. Ng, K and Rosenberg, A. (1991). *Biophys. Chem.* 39, 57-68.
7. Doster, W. (1983). *Biophys. Chem.* 17, 97-106.
8. Kramers, H. A. (1940). *Physica (Utrecht)* 7, 284-291.

Biochemistry and Molecular Biology of Vitamin B_6
and PQQ-dependent Proteins
ed. by A. Iriarte, H.M. Kagan and M. Martinez-Carrion

The Aspartate Aminotransferase Folding Intermediates Recognized by GroEL are Partially Folded Monomers that Bind Pyridoxal Phosphate.

M.J. McNeill, S. Scherrer, F. Doñate, C. Torella, A. Iriarte and M. Martinez-Carrion

Division of Molecular Biology and Biochemistry, School of Biological Sciences, University of Missouri-Kansas City, Kansas City, MO 64110.

Summary
To gain insight into the mechanism of chaperonin-assisted folding of mitochondrial aspartate aminotransferase (mAAT), we have analyzed the interaction of chemically unfolded mAAT with GroEL using a variety of spectroscopic techniques. Since native mAAT is a homodimeric pyridoxal phosphate (PLP)-dependent protein, we also examined the oligomeric state and the ability of the GroEL-associated protein to bind PLP. The folding intermediates that bind to GroEL are monomeric but have substantial structure and are already able to bind PLP with low affinity. Trapping of these transient intermediates by GroEL may prevent their aggregation and increase the efficiency of the folding reaction.

Introduction

Mitochondrial aspartate aminotransferase (mAAT) is a pyridoxal 5'-phosphate (PLP)-dependent enzyme composed of two identical 45-kDa subunits. Although mAAT can refold spontaneously after chemical denaturation only at low temperatures, it folds rapidly following its translocation into isolated mitochondria at 30°C, probably with the assistance of chaperones present in the matrix such as the hsp60/hsp10 chaperonins. The *E. coli* chaperonin system GroEL/GroES greatly extends the temperature range at which mAAT successfully refolds *in vitro* (1). GroEL, the best characterized chaperonin, is a tetradecamer of identical 57-kDa subunits arranged in two heptameric rings with a central cavity lined with hydrophobic residues where non-native substrate polypeptides bind (2). Many structurally unrelated proteins have been shown to bind to GroEL in addition to mAAT.

Chaperonins facilitate ATP-dependent protein folding by reversibly binding and preventing aggregation of partially structured folding intermediates. One of the many open questions regarding the mechanism by which GroEL assists protein folding relates to the conformation of the polypeptide substrate stably bound to the chaperonin, and whether there are conformational changes in the substrate associated with binding by GroEL. Results on several spectroscopic properties of the partially structured species together with their coenzyme-binding ability and association state of the subunits of mAAT bound to GroEL are presented.

Materials and Methods

Rat liver mAAT was purified as previously described (3). Apoenzyme was prepared by incubating the holoenzyme (pyridoxal form) with a 20-fold molar excess of cysteine sulfinate as previously reported (4). Acid-induced reversible unfolding of mAAT at pH 2.0 and formation of the GroEL complex was as previously described (5,1).

mAAT was labeled at Cys166 by incubation with a 10-fold molar excess of the spin label, S-(1-oxy-2,2,5,5-tetramethylpyrroline-3-methyl)methanethiosulfate (MTSSL) in the presence of 20 mM α-methyl aspartate at 4°C for 60 min. The unreacted spin label was removed by ultracentrifugation and the EPR spectra were collected at room temperature on a Varian E9 spectrometer. Fluorescence spectra were recorded at 10°C on a SLM 8000C Aminco spectrofluorometer settled on photon counting mode.

Results and Discussion

The high mobility of the nitroxide side chain attached to Cys166 in the mAAT-GroEL complex is equivalent to that observed in mAAT denatured at pH 2.0 suggesting that the local structure around this residue is as flexible and devoid of tertiary interactions as in the unfolded state. In the native protein, Cys166 is located in a cleft at the interface between the two domains (6) and the EPR spectrum of the spin-label attached to its side chain is characteristic of nearly fully immobilized nitroxides (Fig. 1).

The fluorescence maximum of the seven tryptophan residues of mAAT shifts from 336 to 352 nm upon unfolding whereas the fluorescence intensity decreases relative to that of the apo-form. (Table I), consistent with a transfer of tryptophans to a more polar environment. The emission

λ_{max} and fluorescence intensity of the GroEL-bound protein are identical to those of the native apoenzyme suggesting that tryptophan side chains are present in an environment of similar polarity to the native state.

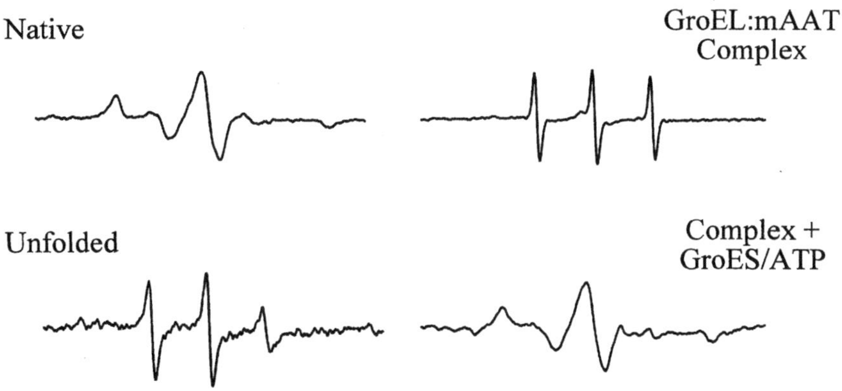

Figure 1. EPR spectra of spin-labeled mAAT. Microwave power of 2 mW was used and a peak-to-peak modulation amplitude of ca. 1 G. Spectral width is 100 G.

The accessibility of the spin labeled-Cys166 or the tryptophan side chains to relaxing agents or quenchers, respectively, is intermediate between those of the native and unfolded proteins (Table I). These data are compatible with a more unfolded conformation of the protein bound to GroEL than in the native conformation. Yet, addition of PLP results in a decrease of the intrinsic fluorescence of the bound protein suggesting that the mAAT folding intermediates is able to bind coenzyme. It remains to be determined whether binding involves a primitive version of the native active site or occurs at an unrelated location in the protein.

To determine the state of oligomerization of mAAT bound to GroEL, we followed the formation of hybrids between two forms of mAAT of different size: mature mAAT and precursor form which contains a N-terminal extension of 29 residues and a tag of 6 histidine residues at its C-terminal end (pmAAT-His). Stable complexes of these two mAAT forms with

GroEL were formed separately by diluting the guanidine-denatured proteins (0.67 μM final concentration) into refolding buffer containing 1 μM GroEL at 10 °C. Release and refolding of the bound protein from GroEL could be induced by addition of the co-chaperonin GroES and MgATP. Because of the differences in size between pmAAT-His and mAAT monomers, formation of mAAT:pmAAT-His hybrid dimers could be followed by trapping of the His-containing dimers with Ni-NTA resin and analysis of their subunit composition by SDS-PAGE. The results showed that when the preformed complexes were mixed prior to addition of the releasing factors, the fraction of hybrid dimer in the mixture was as expected for a random dimerization of mAAT and pmAAT-His monomers (data not shown). By contrast, only homodimers were observed when prior to mixing, the two types of complexes were separately incubated with GroES/MgATP to allow refolding and dimerization of each mAAT form. These results indicate that the mAAT form sequestered by GroEL is predominately monomeric.

Table I. Intrinsic Fluorescence (λ_{ex}=296 nm) and EPR Power Saturation Parameters for mAAT in Different Folding States.

	Native holoenzyme	Native apoenzyme	Unfolded	GroEL-bound
Fluorescence[a]				
λ_{max} (nm)	336	338	352	338
Intensity at λ_{max}	1	2.7	1.6	2.5
K_{sv} (M^{-1})				
Iodide	0.4±0.21	----	3.9±0.57	2.4±0.20
Acrylamide	1.8±0.01	----	7.8±0.08	5.8±0.05
EPR [b]				
$\Delta P_{1/2}$ (mW)	4.8		10.0	6.1

[a] The fluorescence quenching data were analyzed according to the Stern-Volmer equation (7). F_0/F values were estimated at the corresponding emission maxima after correcting for dilution, fluorescence of a blank without protein and for the inner filter effect due to acrylamide. mAAT concentration, 30 μg/ml.

[b] Power saturation data was collected at 20 +/- 1 °C. The microwave power was varied from 0.01 mW to 40 mW, while the peak to peak amplitude of the $m_I = 0$ line ($A_{mI=0}$) was measured at regular power intervals. $P_{1/2}$, the power required to reduce the resonance amplitude to half its unsaturated value, was calculated according the equation by Altenbach *et al.*(8). Values for $\Delta P_{1/2}$ were calculated from the difference in $P_{1/2}$ values in the presence and absence of relaxing agent (NiEDDA, 10 mM).

In summary, the mAAT folding intermediates stably bound to GroEL are substantially more structured than the unfolded state but lack the compact nature of the native conformation. The differences observed between the complex with GroEL and the native mAAT support earlier proposals that GroEL recognizes predominately molten globule-like folding states of proteins that could otherwise engage in undesirable interactions leading to aggregation.

References

1. Sigler, P.B., Xu, Z., Rye, H.S., Burston, S.G., Fenton, W.A. and Horwich, A.L. (1998) Structure and Function in GroEL-Mediated Protein Folding. *Annu. Rev. Biochem. 67*: 581-608
2. Mattingly, Jr., J.R., Iriarte, A. and Martinez-Carrion, M. (1995) Homologous Proteins with Different Affinities for GroEL. The Refolding of the Aspartate Aminotransferase Isozymes at Varying Temperatures. *J. Biol. Chem. 270*: 1138-1148
3. Altieri, F., Mattingly, Jr., J.R., Rodriguez-Berrocal, F.J., Youssef, J., Iriarte, A., Wu, T., and Martinez-Carrion, M. (1989) Isolation and Properties of a Liver Mitochondrial Precursor Protein to Aspartate Aminotransferase Expressed in *Escherichia coli*. *J. Biol. Chem. 264*: 4782-4786
4. Reyes, A.M., Iriarte, A., and Martinez-Carrion, M. (1993) Refolding of the Precursor and Mature Forms of Mitochondrial Aspartate Aminotransferase after Guanidine Hydrochloride Denaturation. *J. Biol. Chem. 268*: 22281-22291
5. Artigues, A., Iriarte, A. and Martinez-Carrion, M. (1994) Acid-Induced Reversible Unfolding of Mitochondrial Aspartate Aminotransferase. *J. Biol. Chem 269*: 21990-21999
6. McOhalen, C.A., Vincent, M.G. and Jansonius, J.N. (1992) X-ray Structure Refinement and Comparison of Three Forms of Mitochondrial Aspartate Aminotransferase. *J. Mol. Biol. 225*: 495-517
7. Lakowicz, J.R. (1984) Quenching of Fluorescence. In *"Principles of Fluorescence Spectroscopy"*, pp. 257-301 Plenum Press, New York.
8. Altenbach, C., Greenhalgh, D.A., Khorana, H.G., and Hubbell, W.L. (1994) A Collision Gradient Method to Determine the Immersion Depth of Nitroxides in Lipid Bilayers: Application to Spin-Labeled Mutants of Bacteriorhodopsin. *Proc. Natl. Acad. Sci. U.S.A. 91*: 1667-1671

AUTHOR INDEX

MIX
Papier aus verantwortungsvollen Quellen
Paper from responsible sources
FSC® C105338

If you have any concerns about our products,
you can contact us on
ProductSafety@springernature.com

In case Publisher is established outside the EU,
the EU authorized representative is:
Springer Nature Customer Service Center GmbH
Europaplatz 3, 69115 Heidelberg, Germany

Printed by Libri Plureos GmbH
in Hamburg, Germany